INTEGRATED NEMATODE MANAGEMENT IN HORTICULTURAL CROPS

P. PARVATHA REDDY
Former Director
Indian Institute of Horticultural Research,
Bangalore - 560 089 (Kar.)

SCIENTIFIC PUBLISHERS (INDIA)
P.O. BOX 91 JODHPUR

Published by:
Pawan Kumar
Scientific Publishers (India)
5-A, New Pali Road, P.O. Box 91
Jodhpur - 342 001 (Raj.)
Tel.: +92-291-2433323, 2624154
E-mail: info@scientificpub.com
www.scientificpub.com

Print: 2021

ISBN: 978-81-7233-523-6

Lasertype Set: Rajesh Ojha
Printed in India

Affectionately dedicated to
my co-workers for their
unstinted support

M.S. Rao
M. Nagesh
R.M. Khan
D.B. Singh

PREFACE

Nematodes continue to threaten horticultural crops throughout the world, particularly in tropical and sub-tropical regions. For century's man's essential crop plants have been plagued by these microscopic organisms that feed on roots, buds, stems, crowns, leaves and developing seed. Estimated overall average annual yield loss of the world's major horticultural crops due to damage by plant parasitic nematodes is 13.54%. For the seven life sustaining horticultural crops (banana, cassava, coconut, field bean, potato, sugar beet and sweet potato) that stand between man and starvation, an estimated annual yield loss of 12.77% is reported. The 16 economically important horticultural crops (cocoa, citrus, coffee, cowpea, eggplant, grapevine, guava, melons, okra, ornamentals, papaya, pepper, pineapple, tea, tomato and yam) that represent a group important for food or export value were reported to have an estimated annual yield loss of 14.31%. Monetary losses due to nematodes on 10 horticultural crops, six of which are life sustaining were estimated at US $ 19.37 billion annually based on 1984 production figures and prices.

The nematodes form a major constraint in the production of quality horticultural products. The farmer in his anxiety to contain the nematode pest may resort to indiscriminate use of nematicides posing hazard to the environment. Since the horticultural produce especially fruits and vegetables are consumed afresh, consumers expect residue-free produce both for internal and export markets. This brings into focus the importance of managing the nematode pests with minimal hazard to the ecosystem and ensuring absolute safety to human beings. Therefore, intensive research efforts for exploring alternate eco-friendly nematode management strategies have become imperative. In this context, developing **'Integrated Nematode Management (INM)'** strategy is the challenge before the nematologists. INM is a systems approach that, in the context of the associated environment and the population dynamics of the nematode species, utilizes all suitable techniques and methods in as compatible manner as possible and maintains the nematode populations below those causing economically unacceptable damage or loss. The criteria for judging the effectiveness of INM are productivity, stability, sustainability and equity. Sound knowledge of nematodes, cropping systems and the environment are important for the implementation and success of INM.

The information on INM, especially crop-wise, is very much scattered and there is no book which deals with both principles and practices of INM entirely on horticultural crops. Hence, the present book is an attempt which comprehensively deals with both principles and practices. This book is divided into two parts. The first part deals with the principles of INM covering aspects such as introduction, role of nematodes in horticulture, interactions with other micro-organisms and nematode management options such as regulatory, physical, cultural, chemical, biological and integrated methods including host resistance. The second part deals with practices for integrated nematode management in horticultural crops such as fruit, vegetable, ornamental, medicinal, aromatic, plantation, spice and tuber crops. An entire chapter is devoted for sources of availability of critical inputs used for INM which would be very useful to the farmers.

This book is a practical guide to practicing farmers of horticultural crops. Further, it is a useful reference to policy makers, research and extension workers and students. The material can also be used for teaching undergraduate and post-graduate courses. Suggestions to improve the contents of the book are most welcome (E-mail: reddy_parvatha@yahoo.com). The publisher, Scientific Publishers (India), Jodhpur deserves commendation for their professional contribution.

Bangalore
November 13, 2007

P. Parvatha Reddy

CONTENTS

Part – I: Principles of Integrated Nematode Management

PART – I

PRINCIPLES OF INTEGRATED NEMATODE MANAGEMENT

Chapter 1

INTRODUCTION

Nematology is an important branch of biological science, which deals with the study of a group of animals known as nematodes. The name nematode is derived from two Greek words (*nema*–thread, *eidos*–resembling) which indicates the external appearance of the organisms (thread-like forms) in this group.

Nematodes constitute one of the most important groups of organisms which inhabit the soil about the roots of plants and which frequently play a vital role in their growth and production. Rarely any crop is free from their attacks, yet we usually are unaware of their presence because of their microscopic size and protected position within the soil. These slender, active, worm-like creatures are so numerous that Cobb (1914) aptly remarked "...... if all the matter in the universe except the nematodes were swept away, our world would still be dimly recognizable .... We would find its mountains, hills, valleys, rivers, lakes, and oceans represented by a film of nematodes."

1.1 Historical

Berkeley (1855) determined a root-knot nematode, *Meloidogyne* sp. to be the cause of galls on the roots of greenhouse grown cucumber plants in England. Kuhn (1857) found *Ditylenchus dipsaci* infecting the heads of teasel, *Dipsacus fullonum*. Schacht (1859) reported the sugar beet cyst nematode, *Heterodera schachtii,* from Germany. Goeldi (1887) published a description of *Meloidogyne exigua*, which produced galls on the roots of coffee in Brazil. Ritzema-Bos (1891) discovered the foliar nematode, *Aphelenchoides fragariae* on strawberry. Liebscher (1892) reported pea cyst nematode, *Heterodera goettingiana* on peas. Kuhn (1913-14) observed cyst-forming nematode, *Globodera rostochiensis* attacking potatoes in Germany. The burrowing nematode, *Radopholus similis* was discovered in 1953 as the causal organism of 'spreading decline' of citrus in Florida, USA (Suit and DuCharme, 1953) and 'yellows' disease of pepper in Bangka islands, Indonesia. In 1955, the lesion nematode, *Pratylenchus penetrans* was discovered as the causal organism of 'peach replant problem' in Canada. Hewitt *et al.* (1958) first reported the transmission of 'grapevine fan leaf virus' by the dagger nematode, *Xiphinema index* in USA.

The study of plant parasitic nematodes on horticultural crops in India dates back to 1901 when Barber reported the root-knot nematode on tea from southern part of the country. Butler (1906) reported the root-knot nematode on black pepper from Kerala. Ayyar (1926, 1933, 1934) reported the root-knot nematode on vegetable crops from south India. Thirumala Rao (1956) reported root-knot nematodes on citrus from Andhra Pradesh. Jones (1961) detected the presence of the potato cyst nematode, *Globodera rostochiensis* in Nilgiris, a pest of great economic importance. It was this discovery, which triggered the nematological research in India. Siddiqi (1961) reported the citrus nematode, *Tylenchulus semipenetrans* for the first time from India. Another important parasite, the burrowing nematode, *Radopholus similis* was reported on banana from Kerala by Nair *et al.* (1966). Weischer (1967) noticed *Radopholus similis* infestation in root-wilt affected coconut plantations in Kerala. These are some of the early records of plant parasitic nematodes on horticultural crops, which accelerated the growth of Nematology in India.

1.2 Nematodes in General

Nematodes are triploblastic (body formed from 3 germ layers, i.e., the ectoderm - outer body wall, the endoderm - inner body wall and the mesoderm - body cavity), bilaterally symmetrical, unsegmented, pseudocoelomate (body cavity not lined by true epithelium) and vermiform animals. They are microscopic in size with an average body length of 1 mm. They have nervous, excretory and reproductive systems, but lack respiratory and circulatory systems.

Nematodes are a sharply differentiated group of invertebrates usually ranked as a Phylum Nematoda of the Animal Kingdom.

1.3 Life Cycle

The life cycle of most plant parasitic nematodes is simple and direct. Females lay eggs that hatch into larvae and these larvae are much like adults in appearance and structure. The nematodes usually undergo the first molt within the egg shell. During their growth and development, larvae undergo a series of four molts and the periods of growth between molts are called larval stages. The fourth molt terminates the larval stage, after which the individual is an adult.

1.4 Feeding Habits

About 95% of the plant parasitic nematodes feed on the roots and other structures that grow below ground, while some (about 5%) feed on above ground plant parts such as buds, stems and leaves. All plant parasitic nematodes have at the anterior end a stylet or spear, which is a sharp pointed structure that may be used to puncture plant cells and suck out the protoplasmic contents. Nematodes feed on individual cells.

Plant parasitic nematodes can be divided into 3 broad categories according to their feeding habits:

(i) Ectoparasitic Nematodes: (a) Migratory ectoparasites (Species of *Belonolaimus, Trichodorus, Paratylenchus, Criconemoides)* (b) Sedentary ectoparasites (*Hemicycliophora arenaria, Cacopaurus pestis*)

(ii) Endoparasitic Nematodes: a) Migratory endoparasites (Species of *Praylenchus, Radopholus, Hirschmanniella*) (b) Sedentary endoparasites (Species of *Meloidogyne, Heterodera, Globodera)*

(iii) Semiendoparasitic Nematodes: (a) Migratory semiendoparasites (Species of *Helicotylenchus, Hoplolaimus, Tylenchorhynchus)* (b) Sedentary semiendoparasites (Species of *Tylenchulus, Rotylenchulus, Nacobbus*)

1.5. Spread

Nematodes on their own move only a meter or so in a year and can spread by anything which push or carry particles of soil or plant material. Farm equipments, irrigation, flood or drainage water; domestic or wild animals, birds and insects; nursery transplants, seeds and bulbs; and dust storms can spread nematodes from one place to another. The active stages of nematodes are susceptible to desiccation; the resistant or resting stages are the most important in long distance spread.

1.6. Survival

Nematodes survive during unfavourable conditions such as cold, dry and hot periods and during absence of hosts in a quiescent or inactive state that is often associated with lowered metabolic rate. The length of quiescent period depends on the amount of food reserves in the nematodes and the environmental conditions. Nematodes survive longer at lower than at higher temperatures. Quiescence may serve to extend comparatively short life cycle of 20 to 40 days to periods varying from a year for many nematodes to 20 to 32 years as in case of the stem and bulb nematode, *Ditylenchus dipsaci* and the seed gall nematode, *Anguina tritici*, respectively.

1.7. Symptoms

A considerable injury caused by nematodes is due to the plant tissue reaction (symptoms) to the secretion injected into them while the parasites are feeding. The above – ground symptoms are mostly non specific such as poor growth, stunting, patchiness and discoloured foliage in a crop.

1.7.1. Below-Ground Symptoms

Some of the common symptoms produced by below-ground feeders are as follows:

(i) Root Galls: *Meloidogyne* spp. on several crops, *Nacobbus serendipiticus* on tomato, *N batatiformis* on sugarbeet, *Hemicycliophora arenaria* on citrus, *Xiphinema diversicaudatum* on rose, *Longidorus* sp. on marigold.

(ii) Reduced Root System: (a) Stubby root: *Trichodorus chrisitei* on corn. (b) Coarse root: *Belonolaimus longicaudatus* on corn.

(iii) Root Lesions: *Pratylenchus* spp. on several crops; *Radopholus similis* on banana, citrus, black pepper, coconut, arecanut; *Helicotylenchus multicinctus* on banana.

(iv) Rots of Fleshy Parts: *Ditylenchus destructor* on yam, *Scutellonema bradys* on yam.

(v) Excessive Root Branching: *Meloidogyne hapla* on onion.

1.7.2. Above-Ground Symptoms

(i) Dead or Devitalized Bud: *Aphelenchoides fragariae* on strawberry.

(ii) Necrosis and Discolouration: *Rhadinaphelenchus cocophilus* on coconut.

(iii) Leaf Lesions: *Aphelenchoides ritzemabosi* on chrysanthemum.

(iv) Leaf Galls: *Anguina balsomophila* on balsam.

The recognition of symptoms is important in diagnosis of nematode diseases. However, complete diagnosis can not be based on symptoms alone. In all cases, the causal organism should be identified.

1.8. Quotable Quotes

"If all the matter in the universe except the nematodes were swept away, our world would still be dimly recognizable .... we would find its mountains, hills, valleys, rivers, lakes and oceans represented by a film of nematodes".

— N.A. Cobb (1914)

"Some say roundworms, others thread or eelworms
Sexes are separate, producing eggs and sperms
They reproduce sexually, have habits strange
A millimeter to a few meters, is their body range
Esophagus with nerve ring, lumen triradiate
Body cavity false, hence pseudocoelomate
Circulatory organs lacking, yet triploblastic
Respiration through cuticle, which is elastic
Amphids and phasmids are their eyes and ears
Plant feeders possess – either stylets or spears".

— R.S. Kanwar

"Whatever approaches we take to achieve the goal of controlling diseases associated with plant nematodes, knowledge of the nematodes themselves as well as the plants which they parasitize, will increase the chances of success...... In other words, there must always be room for the specialist who studies different aspects of the biology of plant nematodes".

— H.R. Wallace

"The extraction of food from one organism, the host, by another organism, the parasite, inevitably entails some disturbance of the host's integrity".

— H.R. Wallace

"Plants are rarely, if indeed ever, subject to association with only one potential p :thogen".

— N.T. Powell

"Each year these minute organisms (nematodes) exact an ever-increasing toll from almost every cultivated acre in the world: a bag of rice in Burma, a pound of tea in Ceylon, a ton of sugar beets in Germany, a bag of potatoes in England, a bale of cotton in Georgia, a bushel of corn in Iowa, a box of apple in New York, a sack of wheat in Kansas, or a crate of oranges in California...... which the farmers can ill afford to lose and which the undernourished millions on earth would welcome if made available to them ".

— G. Thorne (1961)

"The most severe nematode problems occur where good host crops are grown too frequently for too long on the same land".

— B.G. Chitwood

"If at all there was one single factor that triggered nematological research in India, I would say it was this discovery (relating to the report of potato cyst nematode from Ooty, Tamil Nadu by Jones, 1961)".

— Abrar M. Khan (1974)

"A convergence of technical, environmental and social forces is moving agriculture towards more non-pesticide pest management alternatives like biological control, host plant resistance and cultural management".

— Michael Fitzner, National IPM Program Leader, USDA Extension Service

1.9. Nematode Diseases of Horticultural Crops

Radopholus similis – **'Yellows'** of black pepper in Bangka Island, Indonesia.

– **'Slow Wilt'** of black pepper in Kerala.

– **'Black Head'** or **'Toppling'** or **'Rhizome Rot'** of banana.

Pratylenchus coffeae – **'Slump Disease'** of Citrus.

Radopholus citrophilus – **'Spreading Decline'** of citrus in Florida State, USA.

Tylenchulus semipenetrans – **'Slow Decline'** of citrus.

Rhadinaphelenchus cocophilus – **'Red Ring'** of coconut.

Macroposthonia xenoplax – **'Peach Tree Short Life'** disease of peach.

Hemicriconemoide spp. – **"Crinkle Leaf"** disorder of coffee.

Ditylenchus dipsaci – **'Bloat'** in onion.

Ditylenchus destructor – **'Dry Rot'** of potato.

Aphelenchoides besseyi – **'Summer Crimp'** or **'Summer Dwarf'** of chrysanthemum.

Aphelenchoides fragariae – **'Cauliflower Disease'** or **'Spring Dwarf'** or **'Crimp'** or **'Strawberry Bunch'** or **'Red Plant'** of strawberry.

Chapter 2

THE ROLE OF PLANT PARASITIC NEMATODES IN HORTICULTURE

Plant parasitic nematodes present some of the most difficult pest problems encountered in our horticultural economy. According to Thorne (1961) "Each year these minute organisms exact an ever-increasing toll from almost every cultivated acre in the world; a bag of potatoes in England, a bale of cotton in Georgia, a bushel of corn in Iowa, a box of apples in New York, a sack of wheat in Kansas, or a crate of oranges in California".

The destructive plant-parasitic nematodes are one of the major limiting factors in the production of horticultural crops throughout the world. For centuries, man has been plagued by these microscopic organisms feeding on the roots of crop plants essential to his survival and well being. Roots damaged by the nematodes are not efficient in the utilization of available moisture and nutrients in the soil resulting in reduced functional metabolism. Visible symptoms of nematode attack often include reduced growth of individual plants, varying degrees of chlorosis, wilting of the foliage and sometimes death of plants. Furthermore, roots weakened and damaged by nematodes are easy prey to many types of fungi and bacteria which invade the roots and accelerate root decay. These deleterious effects on plant growth result in reduced yields and poor quality of crops. Nematode management is therefore, important for high yields and quality that are required by the high cost of modern crop production.

2.1. Economic Importance

2.1.1. International Scenario

On a worldwide basis, the 10 most important genera of plant parasitic nematodes were reported to be as follows:

i. Meloidogyne *vi. Tylenchulus*

ii. Pratylenchus *vii. Xiphinema*

iii. Heterodera | *viii. Radopholus*
iv. Ditylenchus | *ix. Rotylenchulus*
v. Globodera | *x. Helicotylenchus*

Estimated overall average annual yield loss of the world's major horticultural crops due to damage by plant parasitic nematodes is 13.54% (Table 2.1). For the 7 horticultural crops (left-hand column) that stand between man and starvation (life sustaining horticultural crops), an estimated annual yield loss of 12.77% is reported.

The 16 economically important horticultural crops (right-hand column) that represent a miscellaneous group important for food or export value were reported to have an estimated annual yield loss of 14.31%.

Monetary losses due to nematodes on 10 horticultural crops, 6 of which are life sustaining, were estimated at US $ 19.37 billion annually based on 1984 production figures and prices (Table 2.2).

These figures are staggering and the real figure, when all horticultural crops are considered, probably exceeds US $ 50 billion annually. The losses are 5.8% greater in developing countries than in developed countries.

Table 2.1. Estimated annual yield losses due to damage by plant parasitic nematodes – world basis

Life sustaining horticultural crops	Loss (%)	Economically important horticultural crops	Loss (%)
Banana	19.7	Cocoa	10.5
Cassava	8.4	Citrus	14.2
Coconut	17.1	Coffee	15.0
Field bean	10.9	Cowpea	15.1
Potato	12.2	Eggplant	16.9
Sugar beet	10.9	Grapes	12.5
Sweet potato	10.2	Guava	10.8
		Melons	13.8
		Okra	20.4
		Ornamentals	11.1
		Papaya	15.1
		Pepper	12.2
		Pineapple	14.9
		Tea	8.2
		Tomato	20.6
		Yam	17.7
Average	**12.77 %**	**Average**	**14.31 %**
Overall average – 13.54%			

Table 2.2. Estimated annual losses due to nematodes for selected world horticultural crops

Crop	No. of estimates/ crop	FAO production estimates ('000 MT)	Estimated price / MT (US $)	Estimated yield loss due to nemas (%)	Estimated monetary loss due to nematodes (US $)
Banana	78	2097	431	19.7	178049979
Cassava	25	129020	90	8.4	975391200
Citrus	102	56100	505	14.2	4022931000
Cocoa	13	1660	2584	10.5	450391200
Coffee	36	5210	3175	15.0	2481262500
Field bean	70	19508	544	10.9	1156746300
Potato	141	312209	152	12.2	5789403696
Sugar beet	51	293478	37	10.9	1183596774
Sweet potato	67	117337	219	10.2	2621073906
Tea	16	2218	2807	8.2	510562300
				Total	**19369408865**

2.1.2. Indian Scenario

The ten major nematode pests in India have been identified and the priority list is as follows:

i. *Meloidogyne incognita*

ii. *M. javanica*

iii. *Rotylenchulus reniformis*

iv. *Heterodera avenae*

v. *H. cajani*

vi. *Tylenchulus semipenetrans*

vii. Radopholus similis

viii. *Pratylenchus zeae*

ix. *M. graminicola*

x. *Hirschmanniella* spp.

The best guess rankings of nematode problems of important crops in India made on the basis of survey by highly experienced nematologists of the country are presented in Table 2.3.

Table 2.3. Best guess rankings of nematode problems of important horticultural crops in India*

State	Vegetable crops	Temperate fruit crops	Tropical fruit crops	Plantation crops
Andhra Pradesh	3	---	5	---
Assam	---	---	3	---
Bihar	3	---	7	---
Delhi	3	---	7	---
Gujarat	1	---	5	---

Haryana	1	---	5	---
Himachal Pradesh	1	3	---	---
Jammu & Kashmir	5	7	---	---
Karnataka	1	---	3	---
Kerala	5	---	3	3
Madhya Pradesh	3	---	---	---
Maharashtra	3	---	5	---
Manipur	---	---	7	---
Punjab	1	---	3	---
Rajasthan	3	---	---	---
Tamil Nadu	3	---	5	---
Uttar Pradesh	1	---	---	---
West Bengal	3	---	---	---

* Ranking on 1-9 scale: 1-Most important, 3-Very important, 5- Important, 7- Moderately important, 9-Least important.

The avoidable yield losses due to plant parasitic nematodes in horticultural crops have been presented in Table 2.4.

Table 2.4. Avoidable yield losses in horticultural crops due to plant parasitic nematodes.

Crop	Nematode(s)	Yield loss (%)	Reference(s)
Banana	*Radopholus similis*	38.00	Rajagopalan & Naganathan, 1977a
		32.00	Parvatha Reddy *et al.*, 1996
		41.00	Nair, 1979
	Meloidogyne incognita	30.90	Jonathan & Rajendran, 2000
Sweet orange	*Tylenchulus semipenetrens*	69.00	Baghel & Bhatti, 1983a
Lemon	*T. semipenetrans*	29.00	Mukhopadhyaya & Suryanarayana, 1969
Sweet lime	*T. semipenetrans*	19.00	Mukhopadhyaya & Dalal, 1971
Grapevine	*M. incognita*	55.00	Rajagopalan & Naganathan, 1977b
	M. javanica	53.00	Baghel & Bhatti, 1983b
Papaya	*Rotylenchulus reniformis*	28.00	Rajendran & Naganathan, 1981
Pomegranate	*Meloidogyne* sp.	24.64-27.45	Singh *et al.*, 2003
Peach	*Macroposthonia xenoplax*	33.00	Anon, 1990a
Plum	*M. xenoplax*	10.00	Anon, 1990a
Potato	*M. incognita*	42.50	Prasad, 1989
	Globodera rostochiensis	99.50	Prasad, 1989

Tomato	*M. incognita*	30.57- 46.92	Bhatti & Jain, 1977; Reddy, 1985; Darekar & Mahse, 1988
	M. javanica	77.50	Anon, 1993
	R. reniformis	42.25 – 49.02	Subramaniyam *et al.*, 1990
Brinjal	*M. incognita*	27.30- 48.55	Bhatti & Jain, 1977 Parvatha Reddy & Singh, 1981 Darekar & Mahse, 1988
Chilli	*Meloidogyne* sp.	24.54 – 28.00	Singh *et al.*, 2003
Okra	*M. incognita*	90.90 28.08	Bhatti & Jain, 1977 Parvatha Reddy & Singh, 1981
	M. javanica	20.20-41.20	Jain *et al.*, 1986
French bean	*M. incognita*	19.38 - 43.48	Das, 1994; Patel *et al.*, 2004 Parvatha Reddy & Singh, 1981
	M. javanica	30-40	Sharma *et al.*, 2002
Cowpea	*M. incognita*	28.60	Parvatha Reddy & Singh, 1981
	R. reniformis	13.20-32.00	Palanisamy & Sivakumar, 1981 Hasan & Jain, 1998
Peas	*M. incognita*	20.0- 50.61	Parvatha Reddy, 1985b; Upadhyay & Dwivedi, 1987 Sharma, 1989
	R. reniformis	15.8	Dalal & Vats, 1998
Carrot	*M. incognita*	56.64	Devi, 1993
Bitter gourd	*M. incognita*	36.72	Darekar & Mahse, 1988
Pointed gourd	*M. incognita*	30-40	Verma, 2001
Water melon	*M. incognita*	18-33	Hasan & Jain, 1998
Mushroom	*Ahelenchoides sacchari*	40.6-100.0	Singh *et al.*, 2003
	A. composticola	35-60	Luqman Khan, 2001
Crossandra	*M. incognita*	21.64	Khan & Parvatha Reddy, 1994
Tuberose	*M. incognita*	13.78	Khan & Parvatha Reddy, 1994
Patchouli	*M. incognita*	47.00	Prasad & Reddy, 1984
Davana	*M. incognita*	50	Haseeb & Pandey, 1989
Cymbopogon	*M. incognita*	20	Pandey, 1994
Carnation	*M. incognita*	27	Nagesh & Parvatha Reddy, 2000
Gerbera	*M. incognita*	31	Nagesh & Parvatha Reddy, 2000
Coleus forskohlii	*M. incognita*	70.2	Senthamarai *et al.*, 2006c
Chethiko-duveli (*Plumbago rosea*)	*M. incognita*	29.00-43.96	Santhosh Kumar & Sheela, 2004

Kacholam (*Kaempferia galanga*)	*M. incognita*	64	Sheela & Rajani, 1998
Colocasia	*M. incognita*	24	Anon, 1990b
Menthol mint	*M. incognita*	30	Pandey, 2003
Cardamom	*M. incognita*	32-47	Ali, 1986
Betelvine	*M. incognita*	21.1 - 38	Saikia, 1992
			Jonathan *et al.*, 1990
Ginger	*M. incognita*	29.60-33.35	Ramana *et al.*, 1999
		74.10	Koshy, 2002
		46.40	Charles & Kuriyan, 1980
	R. similis	39-73	Sundararaju *et al.*, 1979
Turmeric	*M. incognita*	18.6-25.0	Poornima & Vadivelu, 1998
	R. similis	46-76	Sosamma *et al.*, 1979
Coriander	*M. incognita*	52	Midha & Trivedi, 1991
Cumin	*M. incognita*	43	Midha & Trivedi, 1991
Fennel	*M. incognita*	42	Midha & Trivedi, 1991
Black pepper	*R. similis* & *M. incognita*	38.5-64.6	Mohandas & Ramana, 1991
	R. similis	59	Mohandas & Ramana, 1991
	M. incognita	46	Mohandas & Ramana, 1991
Coconut	*R. similis*	30	Koshy and Geetha, 1992

2.2. Major Nematode Pests

2.2.1. Root-Knot Nematodes, Meloidogyne spp.

(i) Fruit Crops: The root-knot nematode, *M. incognita* is a serious problem on grapevine, papaya, pineapple and passion fruit. *M. indica* causes considerable damage to citrus in Andhra Pradesh, Gujarat and Delhi. *M. javanica* is a major problem on citrus in Andhra Pradesh (Mani, 1986).

(ii) Vegetable Crops: The root-knot nematodes are by far the most important pests of vegetable crops. At least four species viz., *M. incognita, M. javanica, M. hapla* and *M. arenaria* have been reported and all form conspicuous root galls. *M. incognita* and *M. javanica* are most widespread in distribution and have a wide host range among vegetables, whereas *M. hapla* attacks potato and *M. arenaria* infects chillies. Four races of *M. incognita* have been reported from India (Khan *et al.*, 1988).

Potato: In case of potato, galls are formed both on roots and tubers. Tuber surface becomes uneven and warty because of numerous blisters like galls. When an infested tuber is transversely cut, the white glistening swollen females of the size of pin head can be easily seen embedded in the potato tissue. Such tubers are invariably small, unfit for marketing, rot quickly and cause considerable yield losses.

Onion: Vadivelu and Rajendran (1986) reported *M. arenaria* on onion from Palladam district of Tamil Nadu. The infected plants showed patchiness and unthrifty growth with 'neck twisting' symptoms. The root system was very much reduced and bulbs were sleek and elongated.

(iii) Ornamental Crops

Crossandra: Root-knot nematodes are found in all crossandra (*Crossandra undulaefolia*) growing areas like Tamil Nadu, Karnataka and Andhra Pradesh. Infected plants are stunted with dried peripheral branches bearing smaller chlorotic leaves almost turning to white in later stages (Rajendran *et al.*, 1976).

Tuberose: *M. incognita, M. javanica* and *M. arenaria* have been reported as the major limiting factors in successful tuberose (*Polyanthes tuberosa*) cultivation in Tamil Nadu (Sundarababu and Vadivelu, 1988b), while *M. incognita* and *M. javanica* were potential pests in Karnataka (Khan and Parvatha Reddy, 1992). Affected plants exhibit stunting, yellowing and drying up of leaves and rotting of bulbs (Jayaraman *et al.*, 1975).

Gladiolus: Parvatha Reddy *et al.* (1979) reported *M. incognita* causing heavy damage to gladiolus. Heavy galling on roots results in yellowing of leaves which subsequently leads to stunted growth.

Carnation and Gerbera: *M. incognita* is one of the serious limiting factors in commercial cultivation of carnation and gerbera under polyhouse conditions. Most of the highly fetching exotic cultivars of carnation and gerbera from Europe have shown 40 to 60% mortality in polyhouse beds due to root-knot nematode infection in and around Bangalore (Nagesh and Parvatha Reddy, 1996a). The nematode interacts with soil-borne fungi such as *Fusarium* spp. and *Rhizoctonia solani* in causing wilt/root-rot complex in carnation and gerbera.

(iv) Aromatic Crops

Mint: Root-knot disease of Menthol/Japanese mint, Spearmint, Scotch Spearmint, Peppermint and Bergamot mint caused by *M. incognita* and *M. javanica* was observed for the first time in Lucknow and Terai region of Uttar Pradesh which reduce the herbage yield and oil content (Haseeb and Pandey, 1989b). *M. incognita* dominates over *M. javanica* in mixed infection. Leaves turn yellow and thin, scorching easily and eventually turning brown, whereas the veins remain green. The below-ground symptoms are the large numbers of galls on the roots (Haseeb and Pandey, 1989b). Highest numbers of root-knot nematodes were found in the rhizosphere of *Mentha arvensis* and the lowest in *M. citrata*.

Patchouli: In patchouli, the root-knot nematode caused chlorotic leaves, premature drying and shedding of leaves (Prasad and Reddy, 1984).

***Solanum khasianum*:** *M. incognita* infected *Solanum khasianum* plants exhibited severe leaf yellowing and wilting symptoms (Shetty and Reddy, 1984).

Henbane: Henbane (*Hyoscyamus niger* and *H. muticus*) were severely infested with *M. incognita* and *M. javanica*. Up to 60-70% plants were chlorotic and stunted showing patchy appearance with fewer leaves and flowers (Haseeb and Pandey, 1989a).

Basil: Balasubramanian and Rangaswami (1964) reported *Ocimum sanctum* as a host of *M. javanica*. Krishnamurthy and Elias (1967) reported that *O. basilicum* was severely affected by *M. incognita*. Recently, *M. incognita* and *M. javanica* have been found to be the major problems in the cultivation of *O. basilicum, O. canum, O. sanctum, O. gratissimum* and *O. kilmandeschcaricum* (Haseeb *et al.*, 1988; Haseeb and Pandey, 1987).

Davana: Haseeb and Pandey (1990) reported that the root-knot nematode is a major problem in the cultivation of davana (*Artemisia pallens*).

Geranium: *M. hapla* was found to be widely distributed on geranium in Nilgiri Hills of Tamil Nadu (Khan and Parvatha Reddy, 1992).

Jasmine: *M. incognita* was pathogenic to *Jasminum sambac* and *J. flexile* (Rajendran and Rajendran, 1979).

Chrysanthemum: *M. arenaria* and *M. javanica* have been reported on chrysanthemum (Sen and Dasgupta, 1977; Chandwani and Reddy, 1967).

Rose: *M. incognita* was recorded on rose (*Rosa indica*) from Aligarh (Alam *et al.*, 1973b) and *M. javanica* from Delhi (Prasad and Dasgupta, 1964).

(v) Medicinal Crops

Dioscorea: Raveendran and Nadakal (1975) reported severe infestation of *Dioscorea alata* and *D. spinosa* with *M. incognita* in Kerala. The economic threshold and economic injury levels of *M. incognita* were found to be 250 and 1250 nematodes per plant of *D. rotundata* cv. Igave, respectively.

Aswagandha: *M. incognita* (Race 2) infected aswagandha (*Withania somnifera*) plants were chlorotic, stunted, less branched with fewer and smaller leaves. Roots of such plants were severely galled.

Chamomile: *M. incognita* causes considerable reduction in growth, flower buds and essential oil yield of chamomile (*Matricaria recutita*) (Pandey *et al.*, 1999).

(vi) Plantation Crops

Coffee: *M. exigua* occurs in southern India (Anon, 1966). *M. exigua* is a pest of *Coffea arabica* but not of *C. canephora* (robusta coffee) which is very resistant (Schieber and Sosa, 1960). Affected trees look unthrifty and have chlorotic, wilting and dying leaves.

Tea: *M. brevicauda* is the only root-knot nematode that is a serious pest of mature tea. Affected bushes are unthrifty, have severely galled main roots and lack feeder roots. *M. brevicauda* occurs in southern India at altitudes of 3000 to 3500 ft. in the Wynaad area (Venkata Ram, 1963). It can attack both seedlings and mature tea.

(vii) Spice Crops

Cardamom: *M. incognita* and *M. javanica* are widely occurring in cardamom nurseries and plantations in Kerala, Karnataka and Tamil Nadu. Mature plants show stunting, yellowing, reduced tillering, delay in flowering, immature fruit drop and reduction in yield.

Ginger and Turmeric: The root-knot nematodes are important in ginger and turmeric. Affected plants show stunted growth, yellowing, marginal drying of leaves and reduced tillering. Galling and rotting are exhibited on the root system.

2.2.2. The Reniform Nematode, Rotylenchulus reniformis

(i) Fruit Crops: *R. reniformis* is a major problem on grapevine (cv. Anab-e-Shahi), papaya, pineapple and passion fruit. It is responsible for decline in papaya and pineapple.

(ii) Vegetable Crops: The reniform nematode infects tomato, brinjal, okra, cowpea, dolichos, French beans, parwal and other vegetables. Two races of this nematode have been reported from India (Dasgupta and Seshadri, 1971). The emergence of cowpea seedlings was delayed by 7 to 9 days and seedling stand was reduced to the tune of 6 to 11 per cent due to *R. reniformis* at one nematode per one g of soil (Nanjappa *et al.*, 1978).

(iii) Aromatic Crops: *R. reniformis* was observed to cause stunted growth of mint plants, withering of branches with chlorotic leaves at Hessaraghatta near Bangalore (Khan and Parvatha Reddy, 1992).

2.2.3. Burrowing Nematodes, Radopholus similis, R. citrophilus

(i) Fruit Crops

Banana: *R. similis* causes **'rhizome rot'** or **'toppling'** or **'black head'** disease of banana and is becoming a serious problem in Tamil Nadu, Kerala, Karnataka and Andhra Pradesh (Parvatha Reddy and Singh, 1980). It is also spreading slowly to Maharashtra, Gujarat, Orissa (Mohanty *et al.*, 1972), Madhya Pradesh (Tewari and Dave, 1985), Goa, Assam, Tripura and Bihar. It causes retarded growth and extensive root and rhizome necrosis resulting in serious economic consequences viz., fertilizers are not effectively utilized, the period from planting to harvesting is lengthened, maximum bunch weights are not attained, the quality of fruit is impoverished and fields have to be replanted every two to three years because of drastic reduction in plant numbers. Root population of *R. similis* is indirectly correlated with the yield (Charles *et al.*, 1985).

Citrus: Perhaps the most spectacular disease with which the burrowing nematode, *R. citrophilus* has been associated is that of **'spreading decline'** of citrus in Florida, USA. This nematode is of great economic importance to citrus in Florida and causes reduction in top and root growth.

(ii) Medicinal Crops: The burrowing nematode causes considerable damage to betel vine.

(iii) Plantation Crops: The burrowing nematode is serious problem on plantation crops such as coconut, arecanut, oil palm, coffee, tea, cocoa, etc. Among 37 genera of plant parasitic nematodes recorded in India from the root zone of coconut palm, *R. similis* is the most important one and is responsible for considerable amount of root rotting. *R. similis* seems to be a common nematode in coffee lands in southern India (Somasekhar, 1959). It infects both arabica and robusta coffee.

(iv) Spice Crops: The burrowing nematode causes considerable damage to spice crops such as black pepper, turmeric, ginger, etc. *R. similis* causes **'slow wilt'** disease of black pepper resulting in a loss of 10-20% vines every year in Kerala and Karnataka. *R. similis* is responsible for **'yellows'** disease of black pepper in Bangka islands, Indonesia. Out of 22 million pepper vines planted on this island, only 2 million survived. This nematode made a severe dent in the economy of Indonesia, which was based on the export of pepper.

(v) Tuber Crops: The burrowing nematode causes considerable damage to sweet potato.

2.2.4. The Citrus Nematode, Tylenchulus semipenetrans

T. semipenetrans causes **'slow decline'** disease of citrus and is considered to be one of the factors responsible for "dieback" disease. This nematode also infects grapevine.

2.2.5. Lesion Nematodes, Pratylenchus spp.

(i) Fruit Crops: *P. coffeae* is responsible for root-rot or **'slump disease'** of citrus in Uttar Pradesh (Siddiqi, 1964) and root necrosis in banana. An annual loss of Rs.20 million was assessed in coffee due to *P. coffeae* in an area of about 1,000 ha in Karnataka alone. *P. penetrans* was responsible for **'peach replant problem'** in Canada. It also infects temperate fruits. *P. vulnus* infects temperate stone fruits and grapes. *P. brachyurus* infects pineapple.

(ii) Vegetable Crops: *P. delattrei* reduces the vigour of chillies and produce downward cupping and crinkling of leaves. *P. delattrei* was also found to be highly pathogenic to okra (Ahmed, 1970). *P. brachyurus* infects potato.

(iii) Ornamental Crops

Chrysanthemum: Rashid and Khan (1975) reported that *P. coffeae* is responsible for heavy root damage which subsequently leads to poor growth of chrysanthemum. Stunting of plants with premature yellowing and drying of leaves are common above ground symptoms. Flower size is also reduced.

Crossandra: *P. delattrei* was frequently associated with crossandra crop exhibiting stunting, chlorosis of leaves and wilting. The leaves showed mottled appearance which turned to brown and becoming pinkish eventually. Later, the leaf tips dried and entire leaves turned yellow and shed. The plants were devoid of side shoots and remained completely defoliated. The roots exhibited brown to black lesions. Heavily infested plants did not produce tertiary spike and thereby flower yield is reduced (Srinivasan and Muthukrishnan, 1975).

Rose: Sundarababu and Vadivelu (1988a) encountered *P. zeae* associated with poor growth of rose plants. The affected plants showed chlorotic symptoms, poor and stunted growth with necrotic lesions on roots.

(iv) Aromatic Crops

Patchouli: *P. brachyurus* was a major problem on patchouli in Annamalai Hills of Tamil Nadu and Hessaraghatta in Karnataka (Khan and Parvatha Reddy, 1992). The infected plants exhibited wilting symptoms invariably.

Mint: High population of *P. thornei* was found to be associated with different mints and caused wilting and chlorosis of the plants and lesions on the root system (Haseeb, 1992). *Mentha arvensis* favours the high population build up of the nematode followed by *M. cardiaca, M. piperita, M. spicata* and *M. citrata*. *P. minyus* infects peppermint.

Citronella: Very high populations of *Pratylenchus* sp. were encountered from the rhizosphere of apparently diseased plants of citronella which were stunted, chlorotic with less number of tillers (Haseeb, 1992).

(v) Plantation Crops: *P. loosi* infects tea.

2.2.6. Spiral Nematodes, Helicotylenchus spp.

(i) Fruit Crops: *H. multicinctus* is an important pest of banana throughout India and is responsible for 20 per cent loss in fruit yield. It causes extensive root necrosis, dieback and dysfunction leading eventually to debility of the entire plant.

(ii) Vegetable Crops: *H. dihystera* causes perceptible reduction in root growth of chillies. Okra, tomato, brinjal and onion were also found to be good hosts for this nematode (Muthukrishnan *et al.*, 1975).

(iii) Ornamental Crops: *H. dihystera* causes heavy damage to crossandra and patchouli roots and is responsible for decline.

2.2.7. Cyst Nematodes, Heterodera spp., Globodera spp.

(i) Fruit Crops: *H. oryzicola* feeds on fibrous roots of banana and causes necrosis. The infected banana plants are progressively debilitated leading to substantial yield losses in Kerala (Charles and Venkitesan, 1984).

(ii) Vegetable Crops: The potato cyst nematode (*G. rostochiensis* and *G. pallida*) is the most important pest of potatoes. This nematode has been reported from Nilgiris and Kodai Hills of Tamil Nadu and Munar Hills of Kerala (Ramana and Mohandas, 1988). Out of 9,000 ha under potato, 3,000 ha are infected by this nematode in Nilgiris. In Kodai Hills, about 200 ha are infected (Thangaraju, 1983). Tomato and brinjal are also attacked by this nematode. Total failure of the crop has been reported under severe infestation conditions.

The other cyst nematode, *Heterodera cajani* is known to infect cowpea, guar and *Dolichos lablab*.

2.2.8. The Lance Nematode, Hoplolaimus indicus

(i) Fruit Crops: *Hoplolaimus indicus* causes damage to citrus in Punjab. The nematode at 1,000 per 500 g of soil caused some growth reduction of citrus, but marked reduction occurred at or above an initial population level of 2,000 (Gupta and Atwal, 1972).

(ii) Vegetable Crops: *H. indicus* also causes growth reduction in tomato.

2.2.9. The Stunt Nematode, Tylenchorhynchus brassicae

(i) Vegetable Crops: *T. brassicae* is quite common in northern India where it is mainly associated with cruciferous crops such as cabbage, cauliflower, knol-khol and tomato. It has been found to reduce the yields of cabbage and cauliflower in Uttar Pradesh.

2.2.10. The Stubby Root Nematode, Trichodorus allius

(i) Vegetable Crops: *T. allius* infects onion.

2.2.11. Stem and Bulb Nematodes, Ditylenchus destructor, D. dipsaci

(i) Vegetable Crops: The stem and bulb nematode, *D. destructor* causes **'dry rot'** of potato tubers in Shillong. *D. dipsaci* is responsible for **'bloat'** disease in onion. *D. dipsaci* is a serious problem of onion, garlic, celery, etc. in temperate areas. The nematode attacks inflorescence in beans.

(ii) Ornamental Crops: *D. dipsaci* is a serious problem of bulbous ornamental plants in temperate areas. The nematode attacks inflorescence in teasels.

2.2.12. Bud and Leaf Nematodes, Aphelenchoides spp.

(i) Fruit Crops: *A. fragariae* is considered as a problem on strawberry. Nematode feeding on buds results in a 'blind' plant. *A. fragariae* causes **'cauliflower'** or **'spring dwarf'** or **'crimp'** or **'strawberry bunch'** or **'red plant'** disease in strawberry in the presence of bacterium *Clavibacter fasciens*.

(ii) Ornamental Crops: *A. ritzemabosi* causes considerable damage to the foliage of chrysanthemum, zinnia, salvia, aster and dahlia (Gill and Sharma, 1976). The veins hinder nematode movement within leaves and thus the characteristic symptoms take the form of interveinal discolouration. These are pale green at first, gradually becoming yellow and then dark brown or black as the leaf dies.

A. besseyi is responsible for **'summer crimp'** or **'summer dwarf'** in chrysanthemum. *A. besseyi* has also been reported to be pathogenic to tuberose from West Bengal and Orissa (Khan *et al.*, 2004).

2.2.13. The Needle Nematode, *Longidorus africanus*

(i) Ornamental Crops: In crossandra, *L. africanus* causes stunting of plants, branching, swelling and darkening of root tips (Muthukrishnan *et al.*, 1977).

2.2.14. The Dagger Nematode, *Xiphinema diversicaudatum*

(i) Ornamental Crops: *X. diversicaudatum* population had a correlation with stunting of rose plants (Prasad and Dasgupta, 1964).

2.2.15. The Yam Nematode, *Scutellonema bradys*

(i) Tuber Crops: *S. bradys* is economically the most important and causes **'dry rot'** of yam in several African and Latin American countries. It is also prevalent in south India.

2.2.16. Ring Nematodes, *Mesocriconema xenoplax, M. curvata*

(i) Fruit Crops: *M. xenoplax* and *M. curvata* have been implicated as one of the causal agents for peach tree short life (PTSL) disease in USA. *M. xenoplax* predisposes peach trees to bacterial canker caused by *Pseudomonas syringae*.

2.2.17. The Pin Nematode, *Paratylenchus projectus, P. bukoiensis*

(i) Vegetable Crops: *P. projectus* is widely distributed in USA, Canada, Europe and India. It is of economic importance on cabbage. *P. bukoiensis* is a problem in cabbage and members of Umbelliferae in Poland.

2.2.18. The False Root-knot Nematode, *Nacobbus batatiformis*

(i) Vegetable Crops: *N. batatiformis* was found to infect sugar beet, cabbage, turnip, radish, lettuce, pumpkin, cucumber, tomato, carrot and peas.

2.2.19. The Sheath Nematode, *Hemicycliophora arenaria*

(i) Fruit Crops: *H. arenaria* is more important on citrus species and causes terminal galls on roots. It may cause up to 30% retardation in plant growth with comparable reduction in yield.

2.2.20. The Sessile Nematode, *Cacopaurus pestis*

(i) Fruit Crops: It feeds as sedentary ectoparasite on roots of citrus. *C. pestis* has been implicated as causal agent of dieback and ultimate death of walnut in California, USA.

2.2.21. The Red Ring Nematode, Rhadinaphelenchus cocophilus

(i) Plantation Crops: The **'red ring'** of coconut caused by *R. cocophilus* is a very destructive disease on some islands of Caribbean which threatens the existence of this crop. The extensive losses caused by this nematode are a major catastrophe to natives of these areas.

Chapter 3

INTERACTION WITH OTHER MICROORGANISMS

Nematode interactions are important biological phenomena and of great significance in horticulture. It is a fascinating subject which is multidisciplinary by nature and concerns any scientist involved with plant health. Nature does not work with pure cultures and that many plant diseases are influenced by associated organisms. Different parasites on the same plant interact, which result in disease complexes, and these interactions may lead to susceptibility by predisposition or resistance through preinduction of resistance against a particular parasite. Plant parasitic nematodes often play a major role in disease interactions. The fungus-nematode and the bacterium-nematode interactions are numerous, and weakly parasitic fungal and bacterial parasites can cause considerable damage once they gain entry into plant roots in the presence of feeding nematodes.

3.1. Bacteria

Nematodes may predispose plants to bacterial diseases. Nematodes in association with bacteria induce different symptoms in plants from those found when either pathogen is present alone. The association between gall-inducing nematodes and a species of bacteria results in toxin production in the host plant. Bacteria can enter only plants through wounds or through natural openings or by vectors. The role of nematodes in relation to bacterial pathogens has usually been regarded as providing wounds in the roots through which bacteria may enter the plant.

3.1.1. Vegetable Crops

Both capsicum and eggplant are prone to many soil borne diseases among which the bacterial wilt (*Ralstonia solanacearum*) in combination with root-knot nematode (*M. incognita*) takes heavy toll every year all over the world (Naik, 2004).

Potato: Jatala *et al.* (1990) reported the interactions between *R. solanacearum* and nematodes, especially *Meloidogyne*, *Pratylenchus* and *Globodera* spp., in potato in relation to breeding for combined resistance. The potato cyst nematode, *Globodera rostochiensis* in association with the wilt bacterium, *R. solanacearum* causes wilt complex in potatoes.

Tomato: Pani and Das (1972) have reported the association of root-knot nematode with bacterial wilt of tomato.

Brinjal: The root-knot nematode, *M. incognita*, present along with *Ralstonia solanacearum*, greatly increases the incidence of bacterial wilt of brinjal. The root-knot nematode is responsible for breaking bacterial wilt resistance in "Pusa Purple Cluster" cultivar of brinjal (Parvatha Reddy *et al.*, 1979). *M. incognita* interacts with *R. solanacearum* in increasing the speed of development and severity of wilt disease of eggplant. Maximum disease occurred when the bacterium and nematode were inoculated simultaneously. Inoculation with *R. solanacearum* alone resulted in less severe disease. It was concluded that root-knot nematodes modify the plant tissues facilitating bacterial colonization (Nayar *et al.*, 1988).

French bean: In French bean, the root-knot nematode, *M. incognita* reduced the number of *Rhizobium* nodules in the root system (Singh and Parvatha Reddy, 1981a). Nodules containing root-knot nematodes deteriorate more rapidly than those free of nematodes. Thus the concurrent nematode infection reduces the total benefit from the nitrogen-fixing bacteria.

Cowpea: *M. incognita* and *Heterodera cajani* adversely affected the root nodulation and nitrogen content of cowpea (Sharma and Sethi, 1976b). The root-knot nematode also reduced the leghaemoglobin content of the cowpea root nodules (Sharma and Sethi, 1975).

3.1.2. Spice Crops

Bacterial wilt of ginger, caused by *R. solanacearum* is influenced by *M. incognita* (Samuel and Mathew, 1983).

Table 3.1. Nematode – bacterial disease interactions

Crop	Nematode	Bacteria	Reference
Grapevine	*R. reniformis*	*A. tumefaciens*	Lele *et al.*, 1978
Peach	*M. incognita*	*A. radiobacter var. tumefaciens*	Zutra & Orion, 1982
	M. javanica	*A. tumefaciens*	Nigh, 1966
Strawberry (Cauliflower disease)	*A. ritzemabosi* / *A. fragariae*	*Rhodococcus (Corynebacterium) facians*	Pitcher & Crosse, 1958
Tomato	*M. incognita*	*R. solanacearum*	Napiere & Quinio, 1980
	M. hapla	*R. solanacearum*	Libman *et al.*, 1964
	Meloidogyne spp.	*Clavibacter michiganense*	De Moura *et al.*, 1975

	H. nannus	*R. solanacearum*	Libman *et al.*, 1964
Brinjal	*M. incognita*	*R. solanacearum*	Parvatha Reddy *et al.*, 1979; Naik, 2004 Nayar *et al.*, 1988
	M. javanica	*R. solanacearum*	Sitaramaiah & Sinha, 1984
Capsicum	*M. incognita*	*R. solanacearum*	Naik, 2004
Potato	*M. incognita acrita*	*R. solanacearum*	Lucas *et al.*, 1955
	G. rostochiensis	*R. solanacearum*	Pani & Das, 1972
	G. pallida	*R. solanacearum*	Jensen, 1978
Beans	*M. incognita*	*Curtobacter (Corynebacterium) flaccumfaciens*	Schuster, 1959
French bean	*M. incognita*	*Rhizobium sp.*	Singh & Parvatha Reddy, 1981a
Cowpea	*Heterodera cajani*	*Rhizobium sp.*	Sharma & Sethi, 1976b
Garlic	*D. dipsaci*	*P. fluorescens*	Caubel & Samson, 1984
Gladiolus	*M. incognita*	*P. marginata*	El-Goorani *et al.*,1974
Carnation	*Meloidogyne* spp.	*P. caryophylli*	Stewart & Schidler, 1956
	H. nannus	*P. caryophylli*	Stewart & Schindler, 1956
	Ditylenchus spp.	*P. caryophylli*	Schuster, 1959
Rose	*P. vulnus*	*A. rhozogenes*	Munnecke *et al.*, 1963
Rhubarb	*D. destructor*	*Clavibacter michiganense* sub sp. *insidiosum*	Hawn, 1963
Ginger	*M. incognita*	*R. solanacearum*	Samuel & Mathew, 1983

3.2. Fungi

Much experimental evidence indicates a biological interaction between nematodes and certain soil-borne fungi. In most interactions the nematodes are not essential for establishment and development of fungal pathogens. However, the nematodes usually assist and enhance the pathogenicity mechanism of the fungus towards modifications in the host plants.

3.2.1. Fruit Crops

Banana: The incidence of panama wilt in banana caused by *Fusarium oxysporum* f. sp. *cubense* was doubled in the presence of *R. similis*. The onset of wilt was also considerably shortened. *R. similis* caused the breakdown of resistance to panama wilt in "Robusta" bananas.

Citrus: In the presence of *T. semipenetrans,* significantly more citrus plants become infected by *Fusarium oxysporum* and *F. solani.*

3.2.2. Vegetable Crops

M. incognita and *Rhizocotnia bataticola* or *Sclerotium rolfsii* when inoculated simultaneously in soil, reduce the germination of seeds in okra, brinjal and tomato (Chhabra and Sharma, 1981; Shukla and Swarup, 1970).

The incidence and severity of root rot of brinjal, tomato and okra, caused by the soil-borne fungi such as *R. solani, R. bataticola* and *Phomopsis vexans* was increased in the presence of *M. incognita* (Hazarika and Roy, 1974; Chahal and Chhabra, 1984; Chhabra *et al.*, 1977; Sharma *et al.*, 1980).

Potato: The potato cyst nematode, *Globodera rostochiensis* in association with the fungus *Rhizoctonia solani* caused root rot complex in potatoes.

Tomato: Jenkins and Coursen (1957) induced wilting in *Fusarium* wilt – resistant tomato variety 'Chesapeake' only when root-knot nematodes were present along with fungal inoculum. Furthermore, when *M. hapla* was combined with the fungus, only 60% of the plants wilted, whereas *M. incognita acrita* promoted wilt in 100% of the plants.

Bhagawati and Goswami (2000) reported the interaction of *M. incognita* and *F. oxysporum* f. sp. *lycopersici* on tomato (*Lycopersicon esculentum*). It was found that when either or both pathogens were inoculated simultaneously or nematode was inoculated 10 days prior to inoculation of the fungus, the symptoms were visible by 20 days as against no such symptoms in plants where fungus was inoculated 10 days prior to nematodes. The intensity of wilt after 40-60 days of inoculation was significantly higher when simultaneously inoculated or prior inoculation of nematodes compared to prior inoculation of fungus. The number of galls and egg masses and nematode population in soil was significantly reduced when simultaneously inoculated or prior inoculation of nematodes.

M. javanica increased the extent of damage by pre and post-emergence phases of damping off caused initially by *R. solani* and *Pythium debaryanum* in tomato (Nath *et al.*, 1984).

Brinjal: *M. incognita* and *Ozonium texanum* var. *parasiticum* together acted synergistically by reducing the germination of brinjal to 28 per cent (Nath *et al.*, 1976).

French bean: *M. incognita* in association with soil-borne fungi such as *R. solani* and *F. solani* was responsible for root-rot/wilt disease complex in French bean. In the presence of nematode, the root rot due to *R. solani* and wilt due to *F. solani* increased considerably (Parvatha Reddy *et al.*, 1979; Singh *et al.*, 1981).

Cauliflower: Maximum wilt score due to *F. oxysporum* f. sp. *conglutinans* (4.0) and wilt symptom index (80.0) was observed in cauliflower plants which received *M. incognita* 14 days before the inoculation of wilt fungus (Pathak and Keshari, 2004).

Tylenchorhynchus brassicae alone did not affect the percentage emergence of cauliflower seedlings. *R. solani* was highly destructive by itself, but the adverse effect on the emergence of seedlings was greater when nematode infestation occurred along with fungal infection (Khan *et al.*, 1971).

Peas: In peas, early yellowing and root rot disease is dependant upon the presence of both *Hoplolaimus uniformis* and *F.o.*f.sp. *pisi* race 3.

Okra: Okra plants succumbed to wilt at an early stage when the reniform nematode, *R. reniformis* and *R. solani* were present (Kumar and Sivakumar, 1981).

(iii) Ornamental Crops

Crossandra: *M. incognita* interact synergistically with wilt fungi *F. oxysporum* and *F. solani* and results in premature death of crossandra plants (Khan and Parvatha Reddy, 1992). *F. solani* and *F. oxysporum* are invariably associated with the lesions caused by *P. delattrei* in crossandra (Khan and Parvatha Reddy, 1992a; Srinivasan and Muthurksihanan, 1975). Srinivasan (1974) found that wilt symptoms appeared earlier when nematodes were added two weeks prior to *F. solani*.

Chrysanthemum: The presence of *P. coffeae* has been found to aggravate the disease severity of *Pythium aphanidermatum* and *Rhizoctonia solani* in chrysanthemum (Hassan, 1988). Chrysanthemum roots inoculated with *Belonolaimus longicaudatus* and *Pythium aphanidermatum* develop symptoms of root rot earlier and more extensively than those inoculated with the fungus alone.

Carnation: Ferraz and Lear (1976) studied the interaction of four plant parasitic nematodes (*Circonemoides curvatum, Pratylenchus dianthus, Rotylenchus robustus* and *Meliodogyne hapla*) and *F. oxysporum* f. sp. *dianthi* in carnations. They reported that plants wilted earlier and more severe symptoms were present with the nematode-fungus combination than with the fungus alone. All nematodes except *R. robustus* showed synergistic interaction with *Fusarium*.

In an experiment carried out to study the role of root-knot nematode, *M. incognita* in predisposing carnation (*Dianthus caryophyllus*) to *Fusarium* wilt, it was observed that the appearance of the wilt symptoms were accelerated when *M. incognita* was inoculated two weeks prior to *F. oxysporum* f. sp. *dianthi*. The rate of wilting was observed to be 2.1 and 4.6, respectively during 12th and 25th week of observation, while the root galling index (RGI) values were recorded to be 4.15 and 4.75, respectively. It was observed that maximum mortality was recorded when *M. incognita* was inoculated 2 weeks prior to *F. oxysporum* f. sp. *dianthi* which was 33.4% and 79.2%, respectively, at 12th and 25th week of observation. The plant growth was also reduced significantly due to prior inoculation of *M. incognita* (Shylaja, 2004).

Tuberose: In an experiment undertaken to assess the impact of *Fusarium* wilt– root knot disease complex on the flower yield in tuberose (*Polyanthes tuberosa*), it was observed that the disease complex drastically reduced the yield of flowers in *P. tuberosa,* thereby bringing down the production of flowers in this commercial crop. It was found that the damage was maximum in presence of both pathogens *viz; Meloidogyne incognita* and *Fusarium oxysporum* f. sp. *dianthi*, when compared to the presence of either one of these pathogens (Shylaja, 2004).

(iv) Medicinal Crops:

Dioscorea: A positive interaction between *M. incognita* and *F. oxysporum* f. sp. *dioscorea* occurs in *Dioscorea rotundata* cv. Habanero.

Coleus: Simultaneous inoculation of *M. incognita* and *Macrophomina phaseolina* as well as nematode followed by fungus 15 days later, caused 100% root rot disease and significant reduction in plant growth parameters in *Coleus forskohlii* (Senthamarai *et al.*, 2006a).

(v) Aromatic Crops

Mint: Bergeson (1963) reported that when *Pratylenchus penetrans* infected *Mentha piperita* plants were inoculated with *Verticillium albo-utrum,* diagnostic symptoms of *Verticillium* wilt appeared approximately two weeks earlier than the plants inoculated with the fungus in the absence of nematode.

Presence of *Pratylenchus miniyus* increased both incidence and severity of *Verticillium* wilt (*Verticillium dahliae* f. sp. *menthae*) disease, reduced the period for the development of wilt by 2 to 3 weeks, increased the reproduction of nematode and reduced the dry weight of peppermint plants up to 68% when both the pathogens were present (Faulkner and Skotland, 1965).

(vi) Spice Crops

Ginger: *M. incognita* had synergistic effect in rhizome rot/ soft rot (*Pythium aphanidermatum)* of ginger, particularly when nematodes were inoculated to the plants prior to fungal inoculation. The damage to the crop was more and the onset of the disease was early when *M. incognita* was inoculated 40 days prior to fungal inoculation (Ramana *et al.*, 1999).

Cardamom: In cardamom, *M. incognita* is the predisposing factor for *Rhizoctonia solani* infection, damping off and rhizome rot, prevalent in nurseries (Ali and Venugopal, 1992, 1993).

Table 3.2. Nematode – Fusarium wilt disease interactions

Crop	Nematode	*Fusarium* spp.	Reference
Banana	*M. incognita acrita*	*F. o.* f. sp. *cubense*	Loos, 1959
	R. similis	*F. o.* f. sp. *cubense*	Newhall, 1958
Papaya	*Meloidogyne* spp.	*Fusarium* spp.	Saeed *et al.*, 1972
Tomato	*M. incognita acrita* & *M. hapla*	*F. o.* f. sp. *lycopersici*	Jenkins & Coursen, 1957

	M. incognita & *M. javanica*	*F. o.* f. sp. *lycopersici*	Bowman & Bloom, 1966 Harrison & Young, 1941 Chahal & Chhabra 1984 Hasan & Khan, 1985
	M. incognita	*F. o.* f. sp. *lycopersici*	Bhagawati & Goswami, 2000
	P. coffeae & *P. penetrans*	*F. o.* f. sp. *lycopersici*	Hirano & Kawamura, 1972
Eggplant	*M. incognita*	*F. oxysporum*	Smits & Noguera, 1982
Pea	*M. incognita acrita* *M. incognita*	*F. o.* f. sp. *pisi*	Davis & Jenkins, 1963
	P. penetrans	*F. o.* f. sp. *pisi*	Oyekan & Mitchell 1971
Bean	*M. incognita*	*F. o.* f. sp. *phaseoli*	Singh *et al.*, 1981
	M. javanica		Ribeiro & Ferraz 1983
Cowpea	*M. javanica*	*F. oxysporum*	Thomason *et al.*, 1959
Watermelon	*M. arenaria* *M. incognita*	*F. oxysporum*	Sumner & Johnson 1973
Muskmelon	*M. incognita*	*F. oxysporum*	Bergeson, 1975
Cucumber	*M. incognita*	*F. o.* f. sp. *cucumerinum*	Fritzsche *et al.*, 1983
Squash	*M. incognita*	*F. oxysporum*	Caperton *et al.*, 1986
Cabbage	*M. incognita acrita*	*F. oxysporum*	Fassuliotis & Rau, 1969
Cauliflower	*M. incognita*	*F. o.* f. sp. *conglutinans*	Pathak & Keshari, 2004
Crossandra	*M. incognita*	*F. oxysporum*, *F. solani*	Khan & Parvatha Reddy, 1992
	P. delattrei	*F. oxysporum*, *F. solani*	Srinivasan, 1974
Chrysant-hemum	*Meloidogyne* spp.	*F. oxysporum*	Johnson & Littrell, 1969
Tuberose	*M. incognita*	*F. o.* f. sp. *dianthi*	Shylaja, 2004
Carnation	*Meloidogyne* spp.	*F. o.* f. sp. *dianthi*	Schindler *et al.*, 1961 Shylaja, 2004
	Criconemella curvata *M. hapla*, *Paratyl-enchus dianthus*	*F. o.* f. sp. *dianthi*	Ferraz & Lear, 1976
Coffee	*M. incognita*	*F. o.* f. sp. *coffeae*	Negron & Acosta, 1989

Table 3.3. Nematode – Verticillium wilt disease interactions

Crop	Nematode	*Verticillium* spp.	Reference
Strawberry	*P. penetrans*	*V. dahliae*	McKinley & Talboys, 1979
Cherry	*Meloidogyne* spp.	*V. dahliae*	Ndubizu, 1977
Tomato	*P. penetrans*	*V. alboutrum*	Conroy *et al.*, 1972
	T. christiei	*V. alboutrum*	Conroy & Green, 1974
	G. tabacum	*V. alboutrum*	Miller, 1975
	T. capitatus	*V. alboutrum*	Overman & Jones, 1970
	M. incognita	*V. alboutrum*	Overman & Jones, 1970
	M. javanica	*V. dahliae*	Orion & Krikun, 1976
Potato	*P. penetrans*	*V. dahliae*	Martin *et al.*, 1982
	P. thornei	*V. dahliae*	Krikun & Orion, 1977
	P. neglectus / *Meloidogyne* spp.	*V. dahliae*	Scholte & s' Jacob, 1989
	M. hapla	*V. alboutrum*	Jacobson *et al.*, 1979
	G. rostochiensis	*V. dahliae*	Corbett & Hide, 1971
Eggplant	*P. penetrans*	*V. alboutrum*	McKeen & Mountain, 1960
Gerbera	*M. arenaria*	*V. dahliae*	Schlang & Sikora, 1978
Balsam	*P. penetrans*	*Verticillium* sp.	Muller, 1973
Mentha piperata	*P. penetrans*	*V. alboutrum*	Bergeson, 1963
Peppermint	*P. neglectus*	*V. dahliae*	Faulkner & Skotland, 1965

Table 3.4. Nematode – Fungal disease interactions

Crop	Nematode	Fungi	Reference
Banana	*Meloidogyne* spp.	*R. solani*	Sikora & Schlosser, 1973
Tomato	*M. incognita*	*R. solani*	Chahal & Chhabra, 1984
	M. incognita	*Sclerotium rolfsii*	Shukla & Swarup, 1970
Brinjal	*M. incognita*	*R. bataticola*	Chhabra & Sharma, 1981
	M. incognita	*R. solani*	Hazarika & Roy, 1974
Potato	*M. incognita*	*R. solani*	Sharma & Gill, 1979
Chilli	*M. incognita*	*Pythium aphanidermatum* / *R. solani*	Hasan, 1985
Okra	*M. incognita*	*R. solani*	Golden & Van Gundy, 1975 Chhabra *et al.*, 1977
	M. incognita	*R. bataticola*	Chhabra & Sharma, 1981 Sharma *et al.*, 1980
	R. reniformis	*R. solani*	Kumar & Sivakumar, 1981
Cauliflower	*Tylenchorhynchus brassicae*	*R. solani*	Khan *et al.*, 1971

French bean	*M. incognita*	*Macrophomina phaseolina*	Al-Hazmi, 1985
	M. incognita	*Rhizoctonia solani*	Parvatha Reddy *et al.*, 1979
Celery	*M. hapla*	*Pythium polymorphon*	Starr & Mai, 1976
Chrysanthemum	*M. incognita / B. longicaudatus*	*P. aphanidermatum*	Littrel & Johnson, 1969
	P. coffeae	*R. solani, P. aphanidermatum*	Hassan, 1988
Gerbera	*M. arenaria*	*Phytophthora cryptoge*	Schlang & Sikora, 1978
Betel vine	*M. incognita*	*S. rolfsii*	Acharya *et al.*, 1987
	Meloidogyne spp.	*Phytophthora* spp.	Venkata Rao *et al.*, 1973
Solanum khasianum	*M. incognita*	*Corticium rolfsii*	Prasad *et al.*, 1980
Coleus forskohlii	*M. incognita*	*M. phaseolina*	Senthamarai *et al.*, 2006a
Areca nut	*R. similis*	*Cylindrocarpon obtussisporum*	Sundararaju & Koshy, 1982
Cardamom	*M. incognita*	*R. solani*	Ali & Venugopal, 1992
Ginger	*M. incognita*	*Pythium myriotylum*	Lanjewar & Shukla, 1985
	M. incognita	*P. aphanidermaum*	Ramana *et al.*, 1998

3.3. Viruses

Some species of nematodes belonging to the families Longidoridae and Trichodoridae have a unique and important role as pathogens of plants, because they also transmit certain viruses to their host plants. In the Longidoridae, two genera – *Xiphinema* and *Longidorus* – transmit viruses, the nepoviruses, which are polyhedral, isodiametric particles about 28 nm in diameter. Nepoviruses are pathogens of economic importance, as they affect and damage major crops. Grapevine fan leaf and strawberry and raspberry ring spot viruses are transmitted by the Longidoridae (Table 3.5). In the Trichodoridae, 13 species of *Trichodorus* and *Paratrichodorus* are known to transmit rod-shaped, short and long (45-115 nm and 180-210 nm) particles classified as tobraviruses. Tobacco rattle (infects bulbous flower crops and potato) and pea early browning viruses are transmitted by the Trichodoridae (Table 3.5).

Table 3.5. Nepoviruses transmitted by *Xiphinema* and *Longidorus*.

Nematode	Virus	Host range of virus
X. americanum sensu lato	Tomato ring spot	Fruits, vegetables, ornamentals
	Cherry rasp leaf	Fruits
X. americanum sensu stricto	Peach rosette mosaic	Fruits
X. californicum	Tomato ring spot	Fruits, vegetables, ornamentals

	Grapevine yellow vein strain	Grapevine
X. diversicaudatum	Arabis mosaic	Fruits, vegetables, ornamentals
	Strawberry latent ring spot	Fruits, vegetables, ornamentals
X. index	Grapevine fan leaf	Grapevine
X. italiae	Grapevine fan leaf	Grapevine
X. rivesi	Tomato ring spot	Fruits, vegetables, ornamentals
L. apulus	Artichoke Italian latent (Italian strain)	Vegetables
L. attenuatus	Tomato blacking	Fruits, vegetables, ornamentals
L. diadecturus	Peach rosette mosaic	Fruits
L. elongatus	Raspberry ring spot (Scottish strain)	Fruits
	Tomato blacking (Beet ring spot strain)	Sugar-beet
	Peach rosette mosaic	Fruits
L. fasciatus	Artichoke Italian latent (Greek strain)	Vegetables
L. macrosoma	Raspberry ring spot (English strain)	Fruits
L. martini	Mulberry ring spot	Mulberry

Table 3.6. Tobraviruses transmitted by *Trichodorus* and *Paratrichodorus*

Nematode	Virus	Host range of virus
T. cylindricus	Tobacco rattle	Wide
T. hooperi	Tobacco rattle	Wide
T. primitivus	Tobacco rattle	Wide
	Pea early browning	Leguminosae
T. similis	Tobacco rattle	Wide
T. viruliferus	Tobacco rattle	Wide
	Pea early browning	Leguminosae
P. allius	Tobacco rattle	Wide
P. anemones	Tobacco rattle	Wide
	Pea early browning	Leguminosae
P. minor	Tobacco rattle	Wide
	Pepper ring spot	Vegetables

P. nanus	Tobacco rattle	Wide
P. pachydermis	Tobacco rattle	Wide
	Pea early browning	Leguminosae
P. porosus	Tobacco rattle	Wide
P. teres	Tobacco rattle	Wide
P. tunisiensis	Tobacco rattle	Wide

3.3.1. Vegetable Crops

The growth of tomato plants was significantly reduced in simultaneous inoculations with *M. incognita* and tobacco mosaic virus or when nematode preceded the virus inoculation (Goswami and Chenulu, 1974). In case of brinjal and bottle gourd, root-knot development has been enhanced by the presence of mosaic virus (Mahmood *et al.*, 1974).

Combined effect of *M. incognita* and mycoplasma-like organisms (little leaf) on growth of brinjal was more than their individual effects (Dhawan and Sethi, 1977).

3.4. Management of Disease Complexes

3.4.1. Physical Methods

Soil solarization with 0.03 mm thick polyethylene transparent sheets in moist and well-cultivated soil continuously for 30-50 days has been found effective against soil-borne fungi and nematodes. *Ditylenchus dipsaci, Pratylenchus thornei, P. penetrans, Rotylenchus incultus* and *Parartrichodorus lobatus* among nematodes and *Fusarium* spp., *Verticillium dahliae* and *Phytophthora cinnamoni* among fungi were suppressed. In such a soil the growth of tomato and grapevine was significantly enhanced. Significant reduction in root galling on tomato by *Meloidogyne* spp. occurred in soil exposed to summer heat for about 9 weeks.

3.4.2. Cultural Methods

(i) Fruit Crops: In France, to save grapevines from attack by *Xiphinema index* and subsequently infection by grapevine fan leaf virus, a 7-year rotation with cereal and alfalfa crop has been recommended (Lamberti, 1981).

(ii) Vegetable Crops: In fields infested with root-knot nematodes and *V. dahliae,* Scholte (1990) found that a five-year rotation of maize-sugar beet-barley-barley-potato had significantly higher potato yields than continuous potato.

The beneficial effects of neem cake were observed not only against *M. incognita* and the fungus *Macrophomina phaseolina* individually, but also when both the pathogens formed a disease complex on French bean.

(iii) Plantation Crops: Phytosanitation is currently the best method of red ring disease management. This strategy is directed at reducing the vector population (*Rhynchophorus palmarum*) as well as the number of sources of nematode inoculum. As soon as coconut palms with red ring disease or *Bursaphelenchus cocophilus*-induced little leaf symptoms have been detected, they should be destroyed.

3.4.3. Chemical Methods

Dibromochloropropane (DBCP) in addition to nematicidal action possesses fungicidal properties also. Its twin potential in combination with fungicides like sodium azide is synergized.

(i) Fruit Crops: Pentachloronitrobenzene (PCNB) has been found to effectively check the population increase of nematodes like *Longidorus elongatus* and *Xiphinema diversicaudatum*. This efficacy would eventually influence the spread of viruses transmitted by them (raspberry ring spot, strawberry latent ring spot, peach rosette mosaic) (Murant and Taylor, 1965; Taylor and Gordon, 1970).

Grapevine: Dichloropropane-Dichloropropene (DD) at 250 gallons/acre was a suitable recommendation for heavy soils against *Xiphinema index*, a vector of grape fan leaf virus, while only 200 gallons/acre was found to be effective in lighter soils (Rankin *et al.*, 1971).

Strawberry: Soil application of ethylene dibromide (EDB) exhibited dual efficacy against *M. hapla* – *V. dahliae* disease complex in strawberry (Meagher and Jenkins, 1970).

(ii) Vegetable Crops

Potato: Methylisothiocyanate has been successfully used for soil fumigation to reduce the populations of soil-borne fungus, *Verticillium dahliae* and alleviate the effects of disease complex involving wilt-inducing fungus (*V. dahliae*) and the lesion nematode (*Pratylenchus penetrans*) on potato (Rowe *et al.*, 1987).

Soil fumigation with methyl bromide resulted in yield increases of 68% for potato fields infested with *Globodera* spp. and *V. dahliae* (Hide and Corbett, 1973).

In *V. dahliae* – *Globodera rostochiensis* disease complex on potato, a mixture of methyl bromide and chloropicrin was found to have comparatively more fungicidal and nematicidal efficacy (Hide and Corbett, 1974).

Tomato: Application of metham sodium through drip irrigation was found to be effective against root-knot, *Pythium ultimum* and *Fusarium* sp. complex of tomato (Roberts *et al.*, 1988).

Jones and Overman (1976) noticed an improvement in yield of tomato affected by nematodes like *Belonolaimus longicaudatus, Trichodorus christiei, M. acrita* and wilt fungi such as *F. o.* f. sp. *lycopersici* race 2 and *Verticillium albo-atrum*, when carbofuran was applied at a higher soil pH (6.5 to 7.5).

Pentachloronitrobenzene (PCNB) has been found to effectively check the population increase of nematodes like *Longidorus elongatus* and *Xiphinema diversicaudatum*. This efficacy would eventually influence the spread of tomato blacking (Murant and Taylor, 1965; Taylor and Gordon, 1970).

Okra: Significant reduction in root-knot – *Fusarium* wilt disease complex of okra was also obtained by soil fumigation with methyl bromide (Good, 1964).

Carrot: Application of metham sodium through drip irrigation was found to be effective against root-knot, *Pythium ultimum* and *Fusarium* sp. complex of carrot (Roberts *et al.*, 1988).

Cucumber: Benomyl application inhibited feeding of *X. americanum* on cucumber, which was reflected in a marked reduction of its transmission ability of tobacco ring spot virus (McGuire and Good, 1970).

3.4.4. Biological Methods

(i) Vegetable Crops: *Paecilomyces lilacinus* can be effectively used against *Rhizoctonia solani* – *M. incognita* complex of okra (Shahzad and Ghaffar, 1989).

3.4.5. Integrated Methods

(i) Fruit Crops

Banana: Soil application of neem cake + *T. viride* + carbendazim was found effective in reducing the burrowing nematode (*R. similis*) and wilt (*F. o.* f sp. *cubense*) disease complex and in increasing the banana fruit yield (15,147 kg/ha). This treatment also gave minimum lesion index (1.1) and root-knot index (1.0) as compared to control (4.0) with cost: benefit ratio of 1: 2.72 (Ravi *et al.*, 2001).

Grapevine: Treatment of grapevines with *P. fluorescens* at 100 g/vine + FYM at 20 kg/vine gave highest reduction in final soil nematode population of *M. incognita* (56.9%), least root gall index (1.8) and least wilt disease (*Fusarium moniliforme*) incidence (24.8%). The bunch weight of grapevine increased by 155.4% compared to untreated control (Senthil Kumar and Rajendran, 2004).

Papaya: Khan *et al.* (1997) have reported that *P. lilacinus* gave better over all protection of papaya plants against *M. incognita* – *F. solani* disease complex than *T. harzianum*. They reported that application of both the biocontrol agents further limited the damage caused by *M. incognita* or *F. solani* and gave a 35% increase in plant growth compared to individual applications.

(ii) Vegetable Crops: *Pseudomonas aeruginosa* and *Paecilomyces lilacinus* when used together significantly reduced infection of *Meloidogyne javanica* and root infecting fungi (*Macrophomina phaseolina*, *Rhizoctonia solani*, *F. solani* and *F. oxysporum*) on pumpkin (*Cucurbita pepo*), guar (*Cyamopsis tetragonoloba*), chilli (*Capsicum annum*) and watermelon

(*Citrullus lanatus*). *P. aeruginosa* was more effective than *P. lilacinus* in reducing nematode infection. Combined use of *P. lilacinus* and *P. aeruginosa* was more effective in reducing nematode infection in guar; *M. phaseolina* and *F. oxysporum* on pumpkin; and *F. solani* on guar and watermelon, than either used alone (Perveen *et al.*, 1998).

Brinjal: Results related to root-knot and wilt disease complex management in eggplant under field conditions revealed that the combination formulation of *T. harzianum* and *P. chlamydosporia* performed very well. The yield was 2024 g/3m^2 followed by combination of *P. fluorescens* + *P. chlamydosporia* where the yield was 1782.33 g and *T. harzianum* + *P. fluorescens* (1680.33 g). There was a significant difference between these two treatments. Root gall index (RGI) and wilt disease incidence due to *Ralstonia solanacearum* (WDI) was minimum in combination treatment of *T. harzianum* + *P. chlamydosporia* followed by *P. fluorescens* + *P. chlamydosporia* and *T. harzianum* + *P. fluorescens* where the RGI was 1.8, 1.81 and 1.93, respectively and that of WDI was 22.18%, 25.13% and 27.28%, respectively (Naik, 2004).

Capsicum: In capsicum, based on the field performance of bio-agents, Naik (2004) found that combination of *T. harzianum* along with *P. fluorescens* increased the yield (3320g/ 3m^2 plot) followed by combination of *T. harzianum* and *P. chlamydosporia* (2960 g/3m^2). The least gall index was present where the combination of *T. harzianum* and *P. fluorescens* was used (1.5) followed by *T. harzianum* + *P. chlamydosporia* (1.7) and *P. fluorescens* + *P. chlamydosporia* (1.7). But among combination treatment of bio-agents, all the treatments were on par (*P. fluorescens* + *T. harzianum*, *P. fluorescens* + *P. chlamydosporia*, *T. harzianum* + *P. chlamydosporia*). Results related to the percent bacterial wilt disease (*Ralstonia solanacearum*) incidence also showed similar trend in which *T. harzianum* + *P. fluorescens* performed very well and the percent incidence was 16.90 followed by *T. harzianum* + *P. chlamydosporia* (17.20).

Field application of neem based formulation of fungal (*P. chlamydosporia*) and bacterial (*P. fluorescens*) bioagents resulted in significant reduction of the root-knot (*M. incognita*) and bacterial wilt (*Ralstonia solanacearum*) disease complex to the tune of 70% in capsicum and increased the crop yield by 37% (Rao *et al.*, 2002).

Cabbage and Cauliflower: The bioformulation mixture of *P. fluorescens*, *T. viride* and chitin effectively reduced club root and root-knot disease complex in cabbage and cauliflower both under greenhouse and field conditions (Samiyappan, 2003).

(iii) Ornamental Crops

Gladiolus: Gladiolus plants treated with *P. lilacinus* + *T. harzianum* + neem cake and *P. lilacinus* + *T. viride* + neem cake combinations not only controlled *M. incognita* infection but also *Fusarium* wilt (*F. oxysporum* f. sp. *gladioli*) till the harvest of flower spikes. The corms and cormels obtained from plants treated with these combinations were free from *Fusarium* infection. Fungal colonization of galled roots was maximum in *P. lilacinus* + *T. viride* combination (94%) followed by *P. lilacinus* + *T. harzianum* combination (69%). Similarly, the parasitization of eggs was maximum in *P. lilacinus* + *T. viride* combination (44%) followed by *P. lilacinus* + *T. harzianum* combination (38%), while egg mass parasitization was same in both the combinations (58%) (Nagesh *et al.*, 1999).

Carnation: The studies carried out to evaluate combination of bio-agents for the biological control of wilt (*F.o.*f. sp. *dianthi*) and root-knot nematode (*M. incognita*) disease complex in carnation, revealed that a combination of *P. chlamydosporia* + *P. lilacinus* gave the best control of wilt and root knot disease complex (Shylaja, 2004).

Tuberose: In course of the experiments carried out to evaluate combination of bio-agents for the biological control of wilt (*F.o.*f. sp. *dianthi*) and root-knot nematode (*M. incognita*) disease complex in tuberose, the best result was found in plants treated with *P. chlamydosporia* + *T. harzianum* (Shylaja, 2004).

(iv) Spice Crops

Black pepper: Integrated management of foot rot (*Phytophthora capsicii*) and nematodes (*Meloidogyne incognita* and *Radopholus similis*) on black pepper was achieved by (i) mixing AMF and *Trichoderma harzianum* in solarized nursery mixture to raise healthy and robust seedlings, (ii) application of *T. harzianum* and FYM in planting pit, (iii) field application of neem cake at 1 kg/vine mixed with 50 g of *T. harzianum* during August.

Cardamom: *P. lilacinus* in combination with *Trichoderma* spp. suppressed *Meloidogyne* spp. and rhizome rot disease (*Rhizoctonia solani*) complex when incorporated in solarized cardamom nursery beds (Eapen and Venugopal, 1995).

Table 3.7. Integrated management of disease complexes in horticultural crops.

Crop	Disease complex	INM Components	Reference
Banana	*R. similis* + *F.o.* f. sp. *cubense*	*T. viride* + Neem cake + Carbendazim	Ravi *et al.*, 2001
Papaya	*M. incognita* + *F. solani*	*P. lilacinus* + *T. harzianum*	Khan *et al.*, 1997
Grapevine	*M. incognita* + *F. moniliforme*	*P. fluorescens* + FYM	Senthil Kumar & Rajendran, 2004
Tomato	*M. incognita* +*F.o.* f. sp. *lycopersici*	*T. harzianum* + Vydate / Nemacur	Stephen *et al.*, 1996
	M. incognita + *Pythium aphanidermatum*	*T. harzianum* + *T. viride* + Press mud / Carbofuran	Sundaram & Tangaraju, 2001
Brinjal	*M. incognita* + *R. solanacearum*	*T. harzianum* + *P. chlamydosporia*	Naik, 2004

Potato	*Meloidogyne* spp. + *V. dahliae*	5 yr. rotation of maize – sugarbeet – barley – barley – potato	Scholte, 1990
Capsicum	*M. incognita* + *R. solanacearum*	*P. chlamydosporia* + *P. fluorescens*	Rao *et al.*, 2002
		T. harzianum + *P. fluorescens*	Naik, 2004
Cabbage & Cauliflower	*M. incognita* + Club root	*T. viride* + *P. fluorescens* + Chitin	Samiyappan, 2003
Pumpkin, Cluster bean, Chilli, Water melon	*M. javanica* + *M. phaseoli* / *R. solani* / *F. solani* / *F. oxysporum*	*P. aeruginosa* + *P. lilacinus*	Perveen *et al.*, 1998
Tuberose	*M. incognita* +*F.o.* f. sp. *dianthi*	*P. chlamydosporia* + *P. lilacinus*, *P. chlamydosporia* + *T. harzianum*	Shylaja, 2004
Gladiolus	*M. incognita* +*F.o.* f. sp. *gladioli*	*P. lilacinus* + *T. harzianum* / *T. viride* + Neem cake	Nagesh *et al.*, 1999
Carnation	*M. incognita* +*F.o.* f. sp. *dianthi*	*P. chlamydosporia* + *P. lilacinus*, *T. harzianum* + *P. lilacinus*	Shylaja, 2004
Black pepper	*M. incognita* + *R. similis* + *P. capsici*	*T. harzianum* + Endomycorrhizae	Anandraj *et al.*, 2001
Cardamom	*M. incognita* + *R. solani*	*P. lilacinus* + *Trichoderma* spp.	Eapen & Venugopal, 1995

Chapter 4

NEMATODE MANAGEMENT OPTIONS

Realizing that nematodes cannot be eradicated, and that we must live with them, the overall goal is to keep the population density as low as possible. The economic threshold is the density of nematodes where the losses incurred exceed the cost of nematode management.

Horticultural ecosystem provides a congenial ecological niche for one or the other nematodes, the population of which if unchecked, may reach alarming proportions resulting in serious losses. Nematode management is therefore important for high yields and quality that are required by the high cost of modern crop production. The direct and indirect benefits of nematode management include – increased quantity and quality of the produce, improved health of plants thereby reducing their susceptibility to plant pathogens and increasing the ability to withstand adverse growing conditions, and a better utilization of nutrients and moisture.

In the field, several nematode management methods like regulatory, physical, cultural, chemical, biological and host resistance can be effectively employed to keep nematode population to a minimum level. It is not advisable to depend on a single method to control nematodes. Efficient management requires the carefully integrated combinations of several practices.

4.1. Regulatory Methods

If a harmful species of nematode is not present in an area, legal quarantine is the most effective method for prevention. Numerous attempts have been made to prevent the introduction of nematodes into countries or provinces by means of plant quarantine. Quarantines are established by legislative or other government actions and usually give quarantine authorities power to develop and enforce regulations to accomplish the purpose. Such regulations prohibit bringing infected plants into protected areas where similar crops are vulnerable to infection.

4.2. Physical Methods

Heat treatment is one of the oldest methods of nematode control. Plant parasitic nematodes can be killed by heat, desiccation, irradiation, high osmotic pressure, etc., but it is more difficult to employ physical methods in killing nematodes in soil and planting materials. It is virtually impossible to kill nematodes when they are in growing plants. Field scale treatment of soil is also difficult. Nematodes can be killed at 44 to 48°C. The nematode enzymes are inactivated at short time exposure to about 50°C *in vitro*. Bare root dip in hot water has to be specifically determined for different nematode species. Solarization is a modern technique to be developed for each situation.

4.3. Cultural Methods

Basically, cultural management involves depriving the nematode of a suitable host and thereby reducing the nematode population by starvation. There are many cultural practices used in nematode management. Probably the most common of these practices is crop rotation. Other cultural methods used for nematode management include fallowing, flooding, mulching, trap cropping, time of planting, use of organic amendments to the soil, intercropping with antagonistic plants and influence of fertilizers.

Botanicals and their products play an important role in the reduction of nematode population in soil. A large number of reports have been published showing that growing of antagonistic plants (marigold, mustard, sesame, asparagus, etc.) or addition of organic materials (green manures, oil cakes, crop residues, cellulosic soil amendments, etc.) to the nematode infested soils results in definite reduction of several plant parasitic nematodes.

4.4. Chemical Methods

The use of chemicals on a field scale for control of plant parasitic nematodes was not possible until early 1940's when effective and economical soil fumigants like D-D and EDB were discovered which made it possible to provide growers with spectacular differences in growth and yield through the effective control of nematodes and other soil pests. Several effective nematicides belonging to organophosphate and carbamate groups were developed and improvements in methods of application provided more economic control. In many parts of the world today, soil treatment for nematode control is an established farm practice.

The primary advantage of chemical control over other methods is the quick reduction of nematode population within days after the chemical is applied. The grower is able to plant a crop soon after treatment or, in some cases, at the time of treatment. Most crops are especially vulnerable to nematode attack during the seedling stage when the young root system is becoming established. Crops planted in treated soil develop extensive root systems, and the crop, especially annuals, mature before residual population of nematodes can increase to a damaging level.

4. 5. Host Resistance

The use of resistant and tolerant cultivars is economically attractive management tactics because most infested fields have a dominant or *target* nematode. Furthermore, nematodes spread slowly, persist in the soils for long periods of time, and management by physical and chemical means is relatively expensive. Available information indicates that plant resistance to nematodes is genetically controlled by one gene in some cases, and by more than one gene in other plant-nematode combinations. Practical results have been obtained in breeding for resistance to a number of major nematodes.

4.6. Biological Methods

Biological suppression of plant parasitic nematodes with antagonistic fungi, mycorrhizae, bacteria and Actinomycetes is gaining increasing importance as a result of realization that many environmental and health hazards are linked with the use of chemicals. Standardization of methods for effective utilization of biocontrol agents is very important for evolving ecologically sound integrated nematode management strategies.

4.7. Integrated Approach

Integrated nematode management (INM) includes research, development, technology transfer and implementation needed to integrate two or more control measures to manage one or more nematode species. The aim of INM is to reduce and stabilize the nematode pest population below damaging levels, resulting in favourable long term socio-economic and environmental consequences. The components of INM systems include (a) biological monitoring; (b) environmental monitoring; (c) agricultural production systems; (d) nematode crop ecosystem models; (e) system design; and (f) implementation. The proceedings available for INM system design and implementation are based on the principles of exclusion, population reductions and tolerance.

4.8. Biointensive Integrated Nematode Management (BINM)

Biointensive integrated nematode management is a systems approach to pest management based on an understanding of nematode ecology. It begins with steps to accurately diagnose the nature and source of nematode problems, and then relies on a range of preventive tactics and biological controls to keep nematode populations within acceptable limits. Reduced-risk nematicides are used if other tactics have not been adequately effective, as a last resort, and with care to minimize risks.

An important difference between conventional and biointensive INM is that the emphasis of the latter is on proactive measures to redesign the agricultural ecosystem to the disadvantage of a nematode and to the advantage of its parasite and predator complex.

BINM options may be considered as proactive or reactive.

4.8.1. Proactive Options

Proactive options, such as crop rotations and creation of habitat for beneficial organisms, permanently lower the carrying capacity of the farm for the nematode. The carrying capacity is determined by the factors like food, shelter, natural enemies complex and weather, which affect the reproduction and survival of a nematode species. Cultural control practices are generally considered to be proactive strategies. Proactive practices include crop rotation, resistant crop cultivars including transgenic plants, disease-free seed and plants, crop sanitation, spacing of plants, altering planting dates, mulches, etc.

The proactive strategies (cultural controls) include:

- Healthy, biologically active soils (increasing below-ground diversity).
- Habitat for beneficial organisms (increasing above-ground diversity).
- Appropriate plant cultivars.

4.8.2. Reactive Options

The reactive options mean that the grower responds to a situation, such as an economically damaging population of nematodes, with some type of short-term suppressive action. Reactive methods generally include inundative releases of biological control agents, mechanical and physical controls, botanical pesticides and chemical controls.

Chapter 5

REGULATORY METHODS

5.1. Plant Quarantine

A plant quarantine is a restriction, imposed by duly constituted authorities, whereby the production, movement or existence of plants and plant products is brought under regulation, in order that the introduction or spread of a nematode may be prevented or limited, or in order that a nematode already introduced may be controlled or eradicated, thereby reducing or avoiding losses that would otherwise occur through damage done by the nematode or through a continuing cost of control measures. The principle involved in enacting quarantines is exclusion of the nematode from entering into an area which is not infested.

Quarantine is an important eco-friendly strategy for management of nematodes. Quarantines are traditionally used to restrict the movement of infected plant materials and contaminated soil into a state or country. If practiced with due care, it can prevent to a great extent emergence of new nematode problems in clean areas, as many of the nematode problems all over the world are the creation of man himself. Many countries maintain elaborate organizations to intercept plant shipments containing nematodes and other pests. Contaminated materials may be treated to kill nematodes, or entry may be barred. Quarantines are also maintained to prevent the movement of infected plants and soil out of areas known to have nematodes. Although quarantines are useful especially when a nematode is first discovered, they are very expensive and usually cannot prevent the spread of a nematode induced disease.

To be considered for quarantine, a nematode must be a threat to substantial interests. The measures implemented to control, contain or eradicate the nematode must be reasonable and effective. The benefits expected from quarantine programme must justify the expense and inconvenience of the quarantine, i.e., a quarantine must have favourable cost/benefit ratio.

A quarantine should be repealed after either the objectives have been met, or the nematode has been eradicated or it is found not to be a threat in the light of further information.

Quarantine examination of plant materials received for use in crop improvement programmes by National Bureau of Plant Genetic Resources (NBPGR), New Delhi, has resulted in over 500 interceptions of plant parasitic nematodes comprising of about 60 species belonging to 33 genera. Some of the notable interceptions are: *Heterodera humili, H. goettingiana, Ditylenchus dipsaci, D. angustus, D. destructor, Rhadinaphelenchus cocophilus, Radopholus similis, Pratylenchus brachyurus, P. vulnus, P. penetrans, Scutellonema bradys, Aphelenchoides fragariae, A. besseyi, A. subtenius, Meloidogyne incognita, Xiphinema* spp., *Longidorus* spp., *Gracilacus* spp., *Bursaphelenchus* spp. (Sethi *et al.*, 1972; Sanwal and Mathur, 1975).

5.1.1. Nematode Quarantines in USA

- Stem and bulb nematode, *Ditylenchus dipsaci* (enacted in 1926, revoked in 1935)
- Golden nematode, *Globodera rostochiensis* (enacted in 1948)
- Burrowing nematode, *Radopholus similis* (enacted in 1953)
- Soybean cyst nematode, *Heterodera glycines* (enacted in 1957, revoked in 1972)

5.1.2. Nematode Quarantines in India

As per the latest "Destructive Insects and Pests Act, 1914" which was revised w.e.f. 27th October 1989, the plant species imported into India shall be accompanied by an official phytosanitary certificate containing additional declaration that the material is free from nematode pests specified against it (Table 5.1).

Table 5.1. Special conditions for the import of plants, fruits and seeds with respect to nematodes under the revised Regulation Order, 1989.

Sl. No.	Plant species	Nematode species	Special condition for import
1	All species of *Allium* (onion, garlic, leek, shallot, etc.)	Stem and bulb nematode, *Ditylenchus dipsaci*	Nil
2	All species of *Cocos* (coconut, seeds and plants)	Red ring nematode, *Rhadinaphelenchus cocophilus*	Consignment can only be imported by Director, CPCRI, Kasargod, Kerala
3	All species of *Solanum* (potato)	Cyst nematodes, *Globodera rostochiensis, G. pallida*	Consignment can only be imported by Director, CPRI, Shimla (H.P.)

4	All species of *Vigna* (beans)	Cyst nematode, *Heterodera glycines*	Nil
5	Beet root	Cyst nematode, *Heterodera schachtii*	Nil
6	Broad bean (*Vicia faba*)	Cyst nematode, *Heterodera glycines*	Nil
7	Chrysanthemum	Foliar nematodes, *Aphelenchoides fragariae, A. ritzemabosi* Stem and bulb nematode, *Ditylenchus dipsaci*	Nil
8	*Narcissus daffodil*	Stem and bulb nematode, *Ditylenchus dipsaci*	Nil
9	Pea	Cyst nematode, *Heterodera goettingiana*	Nil

In India, there is a Quarantine Act against the cyst nematode of potato (*Globodera rostochiensis*) in Nilgiris. Infected potatoes are not allowed to be transported from Nilgiris to other parts of India for seed and table purposes in order to prevent the spread of potato cyst nematode from Tamil Nadu to other states and Union Territories.

From the imported germplasm material, economically important nematodes like *Heterodera goettingiana, Meloidogyne incognita, Ditylenchus dipsaci, Radopholus similis* (Sethi *et al.*, 1972) and *Globodera rostochiensis* (Renjhen, 1973) have been intercepted.

Table 5.2. Exotic nematode species of quarantine significance to horticultural crops in India.

Nematode	Common name	Major host
Aphelenchoides fragariae	Strawberry leaf nematode	Strawberry
A. ritzemabosi	Foliar nematode	Flower bulbs
Ditylenchus dipsaci	Stem and bulb nematode	Flower bulbs, onion, garlic
D. destructor	Potato rot nematode	Potato, flower bulbs
Heterodera goettingiana	Pea cyst nematode	Pea
H. schachtii	Sugar-beet cyst nematode	Sugar beet
Nacobbus aberrans	Potato false root-knot nematode	Potato
Radopholus citrophilus	Burrowing nematode – citrus race	Citrus
Rhadiaphelenchus cocophilus	Coconut nematode	Coconut and other plants

5.1.3. Prevention of Spread

Nematodes do not move more than a few centimeters a year of their own accord and most spread is passive. They may be spread in soil or in roots of a host plant (*Heterodera, Globodera, Meloidogyne, Pratylenchus, Radopholus, Tylenchulus*) or in aerial parts of plants including the seed (*Aphelenchoides, Ditylenchus dipsaci*).

Spread in Soil

Any means whereby soil is moved will result in soil-inhabiting nematodes being spread and some species of *Heterodera* and *Globodera* will remain viable even in soil adhering to farm implements, labour's feet, seed potatoes and transplants grown in infested soil. Hence farm implements should be thoroughly cleaned before taking them to other fields. Labourers should wash their feet after working in an infested field. Seed potatoes should be freed of adhering soil before transporting them elsewhere. Seedlings should not be raised in infested fields for transplanting elsewhere.

Spread with Seed

Several nematodes are known to be seed – borne *e.g. D. dipsaci* on onion. Many nematodes spread through vegetatively propagated plant material *e.g. T. semipenetrans* on citrus, *Globodera* spp. on potato, *R. similis* on banana, coconut, arecanut, ginger, turmeric, betelvine, black pepper, *etc.* Efficient seed cleaning methods can considerably reduce the amount of nematode infested debris with seed. Ideally there are 3 steps which should be taken to ensure the production of clean seed:

- Only nematode-free seed should be used for growing crop plants.
- The clean seed should be sown or planted in land where the nematode is not known to exist.
- The crop should be inspected periodically for the presence of pathogenic nematodes.

Seed-borne nematodes can be controlled by fumigation with methyl bromide. Hot water treatment of bulbs and other planting material is another method that can be effectively employed to reduce the spread of nematodes.

5.1.4. Prevention of Multiplication

Nematode multiplication can be prevented by prohibition of the cultivation of host plants and by directly controlling the nematode by chemical means. Crop rotation can also prevent the nematode multiplication.

Nematode multiplication can also be prevented by removing the nematode-infested farm from agricultural use. This land can be used for construction of buildings.

5.1.5. Suggestions for the Transfer of Plant Genetic Resources

Tissue culture (wherever possible) is an important safeguard that may be used to lower nematode risk. Firstly, size of the consignment is reduced (an imported variety or clone may be represented by meristem tip or excised buds or embryos instead of several cuttings or rooted plants, scions, tubers, seed, etc.). Secondly, the aseptic plantlet system has built in pest and pathogen detection capabilities. Cultures which remain uncontaminated are free from pests and pathogens.

Scientists who happen to receive the germplasm material directly or brought by them in person from other countries should send them for quarantine checking to Director, NBPGR, New Delhi. The material shall be returned to them after quarantine clearance.

5.2. Seed Certification

The production of vegetatively propagated plant material that is certified to be nematode-free is accomplished by growing clean plants in clean soil. Strict sanitation must be practiced and both the plant material and rooting medium must be periodically checked for the presence of pathogenic nematodes.

Vegetative seed of banana free of the burrowing nematode, potato seed pieces free of the cyst nematode and garlic cloves free of the stem nematode can be produced commercially by seed certification.

Despite failures and short comings, regulatory action against some plant parasitic nematodes has undoubtedly reduced the rate of their spread if not actually confined them.

Chapter 6

PHYSICAL METHODS

Plant parasitic nematodes can easily be killed in the laboratory by the application of heat, irradiation, osmotic pressure, *etc.* However, it is more difficult to employ such methods to large quantities of soil and plant material. Physical treatments may involve hazards for the plants or the operators but, except possibly with irradiation treatments, there are unlikely to be residual effects. Controlling nematodes by killing them with heat is one of the oldest methods and is probably the most successful physical control measure developed so far. It is used for killing the nematodes in the soil and within the plant tissues.

6.1. Heat Treatment of Soil

Sterilization of soil by steam is a long established practice in greenhouse beds, seed beds and other small areas. Insects, weed seeds, nematodes, bacteria and fungi are killed by steam sterilization. The most important nematode pests controlled by greenhouse steaming are cyst nematode of potato attacking tomato and root-knot nematodes on tomato, cucumber and lettuce.

Steam is generally introduced through perforated pipes buried in the soil and heating takes place by condensation of the soil particles. It is essential that the soil surface should be covered during steaming. Usually plastic sheets are employed and the formation of a bubble of steam under them indicates that treatment is complete. Small quantities of soil can be sterilized in an autoclave at 30 lb pressure per square inch for 30 minutes.

Rabbing, i.e., burning of crop residues over the soil surface, has been a traditional practice to reduce weeds, insects and pathogens. Turning of soil prior to rabbing can improve efficiency in killing some nematodes. Krishnamurthy and Elias (1969) found that burning of 20 kg paddy husk per sq. m. is adequate to obtain nematode control. Rabbing was most effective for the management of root-knot nematodes on turmeric which gave maximum plant height and rhizome yield with lowest root-knot index (Patel *et al.*, 2001).

6.2. Hot-Water Treatment of Planting Material

Hot water treatment is widely used for the killing of nematodes within plant tissues before planting and has proved useful with nematode-infested bulbs, tubers, seeds and roots of plants such as banana, citrus, strawberry, chrysanthemum, *etc.* (Table 6.1). First, the time – temperature – mortality curves of the nematodes have to be determined. This entails immersing the nematodes into thermostatically controlled water baths at different temperatures for different lengths of time. Information must also be obtained on the plant's reaction to temperature because if the treatment has to be successful the plants must be able to tolerate temperatures lethal to nematodes. Since most enzymes are inactivated after short exposure to temperatures near 50°C, death can probably be explained on this basis.

Table 6.1. Time and temperature recommendations for hot-water treatment for denamatizing planting stock.

Nematode	Planting stock	Temp. °C	Time min.
Tylenchulus semipenetrans	Citrus rooted cuttings	46.7	10
		45.0	25
Radopholus similis	Banana suckers	55.0	25
Meloidogyne incognita		43.0	30
Meloidogyne spp.	Grape rooted cuttings	52.7	5
		50.0	10
		47.8	30
	Papaya seedling roots	50	10
	Peach & Cherry rootstocks	50-51	5-10
	Potato tubers	46.0-47.5	120
	Chinese Potato tubers	53	10
	Sweet potatoes	46.7	65
	Yam tubers	51	30
	Ginger rhizomes	50	10
	Rose roots	45.5	60
	Begonia tubers	48	30
		45	60
	Tuberose bulbs	49	60
	Caladium tubers	50	30
	Gladiolus corms	57.8	30
	Mentha spicata	48	30
Aphelenchoides fragariae	Strawberry runners	46	10
	Easter lily bulbs	44	60
	Lily bulbs	41	60
		36	360
A. ritzemabosi	*Heliconia* spp.	50	30

		48	60
	Chrysanthemum stools	47.8	15
		43.0	30
Ditylenchus dipsaci	Tulip bulbs	43.3	180
	Narcissus bulbs	43	240
		44.4	180
	Onion	43.5	120
D. destructor	Irish bulbs	43.3	180
Pratylenchus penetrans	Dormant strawberry plants	49.4	17.5
		51.0	7.5
		52.8	3
		54.4	1
	Peach seedlings	50.0-51.1	10
		52	15
	Cherry	50.0-51.1	10
	Lily	46	20
P. vonvallariae	Lily of the valley *Convallaris majalis*	43.3	60
P. zeae	Rose seedlings	50	5
		48	10
P. vulnus	Apple rootstocks	46.1-46.7	30
		50	15
	Rose plants	48.3	35
		2.2	3 weeks (cold hardiness)
		37.8	24 hours (heat hardiness)
Rotylenchulus reniformis	Sweet potato	50	3-5

Incubation of potato tubers at 45°C for 48 hours has been shown to kill about 98.9 per cent *M. incognita* without affecting tuber viability (Nirula and Bassi, 1965).

6.3. Soil Solarization

Another method of killing nematodes with heat is by soil solarization. This technique consists of covering moist soil with a plastic film during periods of intense sunshine and heat, thereby capturing radiant heat energy from the sun, causing physical, chemical and biological changes in the soil. The soil temperature is increased (by 5-15°C) to levels lethal to many soil-borne fungi, bacteria, nematodes and weeds. The short wave incident solar radiation penetrates the polyethylene sheet but the long wave radiation is prevented from soil, thus trapping the heat resulting in thermal inactivation and production of heat shock proteins and irreversible heat injury.

Solarization also improves plant nutrition by increasing the availability of N and other essential elements. This practice can be profitably used on a small scale and is especially useful for nursery beds.

Tarping with Linear Low Density Poly Ethylene (LLDPE) 100 gauge film during summer for 15 days (in between April 15th to June 15th) in tomato nursery at Anand gave 66 and 92% reduction of root-knot nematodes and weeds, respectively. This practice produced 615% more seedlings over control giving net profit of Rs. 7,661 from 1000 m^2 area with cost: benefit ratio of 1:5.65.

Soil solarization for 40-45 days using 400 gauge transparent LDPE was proved ideal pre-sowing treatment for the management of nematodes and other soil-borne diseases in cardamom nurseries. Mean soil temperature during the solarization period increased by 8.7°C in solarized beds and the germination was enhanced by 25.5%, while weed growth was suppressed by 82%. *Pythium vexans* was totally eliminated, while populations of *Rhizoctonia solani, Phyllosticta elettaria* and nematodes were suppressed to varying levels. Solarization was also found to stimulate the growth of cardamom seedlings (Eapen, 1995).

White polyethylene mulching for 15 days on beds prepared for planting betelvine revealed high reduction in nematode populations due to increased soil temperature to 44.1°C (Sivakumar and Marimuthu, 1987).

Soil solarization using thin transparent polyethylene mulches for six weeks caused as much as 72.6 and 88.5% decrease in the population densities of *Tylenchorhynchus vulgaris* and *Hoplolaimus indicus*, respectively in nursery beds. Application of mahua cake before solarization usually caused slightly greater reduction in nematode population (Kamra and Gaur, 1998).

6.4. Barriers

Barriers on a smaller scale are being used to confine *Radopholus citrophilus* within limited areas of infestation (Poucher *et al.*, 1967). These man made barriers, surrounding an infested area, consists of a strip of land 3 to 8 m wide, free of citrus trees and treated with a nematicide at 6 month intervals to keep the soil in the barrier zone free of roots as well as burrowing nematodes. The surface of the soil is also treated with herbicides to keep the barrier free of all plant growth.

Small-scale production of nematode-free seedlings of perennial plants can be accomplished by planting the seedlings in plastic bags of about 100 cm^3 capacity filled with uninfested or sterilized soil. The bags should not be placed on the soil surface, but on boards supported about 20 cm above the soil over a polyethylene sheet. In this way, reinfestation can be minimized.

6.5. Irradiation

There is a good evidence that nematodes are killed or seriously affected by irradiation. Fassuliotis and Sparrow (1955) demonstrated that irradiation of potato tubers with x-rays inhibited sprouting and also development of *Globodera rostochiensis*. Cysts of this nematode exposed to 20,000 r contained only brown and dead eggs, while at 40,000 r the eggs lost their contents completely. Wood and Goodey (1957) found that r-rays in between 48,000 and 96,000 r inactivated *Ditylenchus myceliophagus* in mushroom compost. Ultraviolet light rays will also kill nematodes.

6.6. Washing Process

Plant parasitic nematodes are often spread through soil adhering to potato tubers, bulbs and other planting material and such spread is especially important with cyst forming species. Careful washing can considerably reduce the risk. Commercial vegetable, sugarbeet and potato tuber washing apparatus have been developed.

Chapter 7

CULTURAL METHODS

Cultural methods are attempts to adapt crop husbandry practices so as to minimize the effects of nematodes especially on low value crops. Usually the operation will be a normal one carried out at little or no extra expense.

7.1. Crop Rotation

Crop rotation is an effective and most widely used land management practice for reducing nematode population in the soil. This practice was being followed by early civilization long before its significance as a means of nematode control was recognized. Crop rotation attempts to keep nematode populations to a level at which crop damage is reduced to a minimum. By growing non-host crops, the numbers of plant nematodes are reduced, the number of years for this to occur depending on the initial population and the rate of population decrease. Rotations must also provide economically useful crops. In choosing rotations one must be careful not to promote a new set of pathogens in place of the ones to be controlled.

Some plants appear to be antagonistic to nematodes for they reduce populations to a greater extent than non-hosts. Ochse and Brewton (1954) found that *Crotalaria spectabilis* and *C. striata* root systems have a toxic effect on root-knot nematodes and that, planted in advance of a crop, they may reduce subsequent damage. Meijneke and Oostenbrink (1958) showed that, *Tagetes* spp. reduce populations of *Pratylenchus, Tylenchorhynchus, Paratylenchus* and *Rotylenchus robustus*, giving better growth of crops. Hence, these antagonistic plants can also be used in crop rotations to control certain plant nematodes.

Crops which have been shown to reduce root-knot populations in the soil include varieties of cereals (sorghum, bajra, maize, wheat, rice, sugarcane), cruciferous crops (cabbage, cauliflower, kohl rabi, mustard), onion, garlic, groundnut, cotton, sesame, *Hibiscus sabdariffa*, chrysanthemum, cosmos, sunnhemp, dhaincha, dahlia, marigold, balsam, lentil, radish, spinach, fenugreek, zinnia and pigeon pea.

A number of crops and other plants are reported resistant to the reniform nematode, *Rotylenchulus reniformis* which include sugarcane, *Coffea robusta*, jack bean (*Canavalia ensiformis*), velvet bean (*Stizolobium deeringianum*), chillies, maize, sorghum, *Leucaena glauca*, finger millet (*Eleusine coracana*), *Crotalaria, Capsicum fruitescens*, carrot, *Anethum graveolus*, coriander, *Foenicutum vulgare, Spinacea oleracea, Beta vulgaris,* onion, garlic, mustard, cauliflower, pigeon pea, chrysanthemum, coriander, cosmos, sunnhemp, dhaincha, cluster bean, dahlia, marigold, horse gram, balsam, linseed, radish, sesame, zinnia and *Raphanus sativus* (Khan and Khan, 1973).

Two successive crops of maize or sorghum preceding susceptible crops are effective in controlling *R. reniformis*. Rotations with sugarcane and pangola grass are recommended for control of the reniform nematode.

7.1.1. Fruit Crops

Banana: Planting of sugarcane or pangola grass (*Digitaria decumbens*) following the destruction of banana infected with *R. similis* eliminates the nematode after 10 weeks. Rice or green gram grown after banana suppressed the population of *R. similis, Pratylenchus coffeae* and *Helicotylenchus multicinctus* (Rajendran *et al.*, 1979).

Pineapple: Planting of pineapple after pangola grass reduced the population of *M. incognita, Criconemoides* spp., *Helicotylenchus* spp. and *R. reniformis* and increased the fruit yield in the planted crop (23.5 MT/acre as against 15.1 MT/acre in control) and first ratoon crop (18.4 MT/acre as against 9.3 MT/acre in control) (Ayala *et al.*, 1967).

7.1.2. Vegetable Crops

Out of 15 different cropping sequences studied, tomato-onion-resistant tomato cv. Hisar Lalit-okra was found to be the best in terms of root-knot nematode reduction and economic returns (Kanwar, 1990). Crop rotations having okra-garlic-cluster bean and okra-coriander-resistant tomato cv. Hisar Lalit effectively managed the root-knot nematode population (Anon, 1993b). Likewise, decrease in root-knot nematode population on tomato, brinjal, okra, chillies and sponge gourd occurs following marigold, spinach and bottle gourd (Khan *et al.*, 1975).

Potato: Continuous cropping with potatoes increases the cyst population of *Globodera rostochiensis*. More than three year rotations with wheat, strawberry, cabbage, cauliflower, peas, maize and beans reduces the nematode population to a safe level.

Tomato: Tomato rotated with cotton-wheat-cotton resulted in low gall index due to *M. javanica* (root-knot index 1-2) and gave maximum yield (38% increase over continuous tomato) (Hashmi and Hashmi, 1990).

Brinjal: The population levels of *M. incognita* on highly infested land dropped to zero after 18 months of continuous cultivation of *Panicum maximum*. Brinjal growing after *Panicum maximum* was completely free of root-knot nematodes (Netscher, 1983).

Sheela *et al.* (2002) conducted crop rotation trials in farmer's fields in brinjal for the management of root-knot nematodes (*M. incognita*). When brinjal was raised in heavily root-knot nematode infested field after sweet potato (cv. Sree Bhadra) crop, the population of nematodes in soil was reduced by 47% and the fruit yield was increased by 22%.

Considerable reduction in population of *Hoplolaimus indicus* was noted in brinjal (Alam *et al.*, 1980).

Chilli: Considerable reduction in population of *Hoplolaimus indicus* was noted in chilli (Alam *et al.*, 1980).

Cabbage and Cauliflower: There was sudden decline in population of *Tylenchorhynchus brassicae* when cabbage and cauliflower were rotated with wheat (Siddiqi *et al.*, 1973). The mustard (rabi), radish (summer), sesame (kharif) sequence considerably reduced the population of *Tylenchorhynchus* spp. (Haque and Gaur, 1985). Considerable reduction in population of *T. brassicae* was noted in potato and tomato, of *Hoplolaimus indicus* in cabbage (Alam *et al.*, 1980).

Okra: Sheela *et al.* (2002) conducted crop rotation trials in farmer's fields in okra for the management of root-knot nematodes (*M. incognita*). When okra was raised in heavily root-knot nematode infested field after sweet potato (cv. Sree Bhadra) crop, the population of nematodes in soil was reduced by 21% and the fruit yield was increased by 21%.

Sugar beet: Sugar beet rotated with cotton-barley-cotton resulted in low gall index (root-knot index 1-2) and gave maximum yield (56% increase over continuous sugar beet) (Hashmi and Hashmi, 1990).

Water Spinach: Pre-planting *Cosmos bipinnatus, Gaillardia pulchella* (blanket flower), *Tagetes erecta, Tithonia diversifolia* or *Zinnia elegans* reduced root galling and J2 of *M. incognita* in and around water spinach (*Ipomoea reptans*). Amendment of soils with roots, stems or leaves of *G. pulchella* was effective in controlling *M. incognita* on *I. reptans*. Root exudates of *G. pulchella* were lethal to J2 of *M. incognita* and were inhibitory to the hatch of eggs at the concentration of 250 ppm or higher. *G. pulchella* could be used to manage *M. incognita* as a rotation crop, a co-planted crop or as a soil amendment for control of root-knot nematodes (Tsay *et al.*, 2004).

7.2. Trap Cropping

Cowpea causes the root-knot nematode eggs to hatch; the larvae enter the roots and develop to immobile stage. Then the crop is destroyed before the nematodes mature. *Crotalaria* is highly susceptible to invasion by the root-knot nematode but is resistant to the development of larvae into adults. Specific resistance to nematodes which operates on a gene for gene basis, may involve a hypersensitive reaction to prevent invasion as occurs in sunnhemp (*Crotalaria spectabilis*), which survives invasion by *Meloidogyne* juveniles by failing to create the giant cells vital for the nematode survival (Cook and Baker, 1983).

Sree Bhadra, a high yielding released variety of yam by Central Tuber Crops Research Institute, Trivandrum, is identified as a resistant trap crop for the root-knot nematode. The nematodes were able to penetrate the root but giant cell formation was not induced and as a result the nematodes die inside the roots. Planting this variety in root-knot nematode infested field helped in clearing the field free of nematodes and giving good tuber yield. Subsequently growing of susceptible crops like okra, coleus and African yam escaped nematode damage (Mohandas, 2001).

7.3. Intercropping

7.3.1. Fruit Crops

Banana: The nematode population both in soil and roots were considerably reduced when banana was intercropped with nematode antagonistic plants. *Crotalaria juncea* was most effective in reduction of soil nematodes and *Coriandrum sativum* was more effective in reduction of root nematodes. *C. juncea* gave maximum bunch weight (16.7 kg) recording 51% more than control (11.03 kg) followed by *Tagetes erecta* (16.2 kg), *Sesamum indica* (15.87 kg), *Acorus calamus* (15.67 kg) and carbofuran (15.23 kg) (Charles and Venkitesan, 1993).

Citrus: Interculture of acid lime with marigold and mustard markedly reduced the rate of multiplication of *Tylenchulus semipenetrans* to 10.03 and 8.59, respectively (Mani, 1988a).

7.3.2. Vegetable Crops

Interculture of tomato, brinjal, cabbage and cauliflower with neem seedlings reduced the rate of multiplication of *Tylenchorhynchus brassicae* by 59, 48, 50 and 68%, respectively compared to control (Siddiqi and Saxena, 1987b).

Interculture of tomato and brinjal with neem and Persian lilac seedlings reduced the rate of multiplication of *M. incognita* and *R. reniformis* and improved plant growth significantly (Siddiqi and Saxena, 1987a).

Potato: Intercropping of onion and maize reduce galling by *M. incognita* on potato roots.

Tomato: Marigold intercropped with tomato at 1:4 and 1:6 ratios and mustard at 1:2 ratios was found to be effective in reducing root galls, egg masses and nematode population at harvest and gave cost:benefit ratio of 1:8.36, 1:7.88 and 1:3.31, respectively (Rangaswamy *et al.*, 1999). Intercropping of tomato with marigold, margosa and Persian lilac have been recorded to reduce nematode population and root galling.

Interculture of onion with tomato reduced gall index due to *M. javanica* significantly and improved tomato fruit yield. Intercropping with 2 rows of onion with one row of tomato was the most effective treatment in reducing the nematode infection and population and enhancing tomato yield (Ram and Gupta, 2001).

In pots having *Zinnia elegans* with tomato, the population of *M. incognita* and *R. reniformis* declined. The root-knot index on tomato was reduced to 1.08 compared to 3.75 in control. The nematode reproduction factor of *M. incognita* and *R. reniformis* was reduced to 0.88 and 0.80, respectively compared to 3.86 and 2.43 in control (Tiyagi *et al.*, 1986).

Cultivation of tomato with castor results in considerable reduction in number of root-knot galls on tomato (Hackney and Dickerson, 1975).

Brinjal: The number of root-knot galls on the susceptible brinjal is reduced when it is planted in association with marigold (Choudhury, 1981). Intercropping of brinjal with marigold, margosa and Persian lilac have been recorded to reduce nematode population and root galling.

Intercropping of sweet potato with brinjal was found to reduce the final nematode population of *M. incognita* by 14.41%. Even number of galls, egg masses and root-knot index were significantly lower. Further, the yield of brinjal was also enhanced when intercropped with sweet potato (Shivaprasad *et al.*, 2001).

Chilli: Intercropping of chillies with marigold, margosa and Persian lilac have been recorded to reduce nematode population and root galling.

Okra: Intercropping of okra with sesame decreased root-knot nematode penetration in okra and also delayed the life cycle of *M. incognita* (Tanda and Atwal, 1988). Similarly, intercropping of chillies, brinjal and tomato with marigold, margosa and Persian lilac have been recorded to reduce nematode population and root galling.

Lettuce: Planting 4 and 5 seedlings of *Tagetes erecta* in a pot around a lettuce plant reduced root galling by 56 and 72% and nematode population by 73 and 94%, respectively (Pervez *et al.*, 1988).

(iii) Medicinal Crops: In pots having 5 seedlings of *Tagetes erecta* around a betel vine plant, the population of *M. incognita* was reduced by 41.4% and root galling by 54% (Medhana *et al.*, 1985).

(iv) Ornamental Crops: Intercropping of crossandra with enemy plant like marigold significantly reduced the root-knot and lesion nematode population both in soil and roots (Khan and Parvatha Reddy, 1994).

(v) Plantation Crops: Growing of intercrop like cocoa that enriches the soil with sizeable quantities of shed foliage which help in the build up of beneficial organisms may inhibit *R. similis* multiplication on coconut.

7.4. Green Manure Cropping

Green manuring is a conventional practice in which green plants are ploughed into the soil to rot and provide nutrients for the succeeding crop. Green manuring is a low cost but effective technology for the management of plant nematodes. Leguminous green manure crops can be included in rotations and they can be ploughed in during flowering. Growing of leguminous green manure crops such as sunnhemp (*Crotalaria juncea*), dhaincha (*Sesbania aculeata*), green gram (*Phaseolus* sp.), cowpea (*Vigna unguiculata*), khesari (*Lathyrus sativus*), berseem (*Trifolium alexandrium*) and *in situ* mulching adds 8 to 28 tonnes/ha of organic manure.

Glyricidia rooted cuttings may be planted at every 2 m interval on the existing bunds. About 120 to 130 plants will produce sufficient biomass after 2 years. About 15 kg of biomass is produced from each plant. This production will be 3 times in a year.

7.4.1. On Farm Generation of Organic Matter

***In situ* Generation of Green Manure Crops:** The limitations on transport of bulky organic materials produced elsewhere to the field where they have to be utilized, there is an urgent need to identify strategies for the generation of organic matter on the farm itself in a non-competitive manner without interfering with the crop production in order to make sustainable horticulture – a practical reality.

Raising a fast growing, high biomass producing, drought tolerant, herbaceous legume like horse gram, can be incorporated in the soil at flowering. After about 75 days of growth, horse gram can be ploughed in. By ploughing in, about 1.5 tonnes of dry biomass could be added to the soil contributing about 35 kg N/ha.

Bund Farming with Trees and Grasses: A strategy to generate organic matter on farm is to plant fast growing N-fixing trees or bushes on single or both sides of bunds which are used for conserving soil and water. *Glyricidia maculata* cuttings can be planted on both sides of bunds at 1 m interval across the slope. *Stylosanthus hamata*, a pasture legume can be seeded on the flat bund surface. *Cenchrus* slips can be planted in between *Glyricidia* cuttings on the upstream side of the bunds. This strategy shall assure regular supply of organic matter from loppings of *Glyricidia*. Every running meter of live bunds can yield 2.88 kg fresh biomass of *Cenchrus* + *Stylosanthus* and 2.96 kg of *Glyricidia* leaves.

Some of the common leguminous green manure plant species are as follows:

Legumes: Greeen manure – Dhaincha, Sunnhemp, Kolinji, Sesbania, Cowpea, Wild Indigo, Pillipesara, Cluster bean, Green gram, Black gram, Horse gram, Field bean, Berseem, Madras Indigo.

Green leaf manure – Glyricidia, Cassia, Pongamia.

Non legumes: Greeen manure – Sunflower, Buck wheat.

Green leaf manure – Calotropis, Adathoda, Thespesia.

The crops such as black gram, horse gram, green gram, field bean, and cowpea can also be grown as green manures. They should be ploughed in 8 to 9 weeks after sowing, when they flower. They rot in about a month's time and become manure. After monsoon crops are harvested, sowing of green manure crops can be taken up.

The organic matter applied through ploughing in of green manure crops at flowering, brings about important physical, chemical and biological changes during the decomposition in the micro sites where micro flora and micro fauna are active. Patel and Desai (1964) also found that application of chopped leaves of *Pongamia glabra* and *Crotalaria juncea* to root-knot infested soil reduced the disease severity. The decomposition of green leaves of neem, *Melia azadirachta* and *Sesbania aculeata* in the infested soil was also very effective (Singh and Sitaramaiah, 1967, 1970).

7.5. Mulching

Watson (1937) demonstrated that heavy organic mulches reduced nematode damage under Florida conditions. Linford and Yap (1940) experimented with five nemic trapping fungi in Hawaiian soils high in organic matter. Root-knot injury was reduced when *Dactylella ellipsospora* was introduced. Loos (1953) applied 20 to 30 tonnes of Guatemala grass, *Tripsicum laxum*, around tea bush severely infested with *Pratylenchus coffeae*, and the results were most satisfactory.

7.6. Organic Amendments

The use of natural plant products to manage nematodes has a number of advantages over synthetic chemicals which cause environmental pollution, phytotoxicity, ground water contamination and adversely affect the soil and its biotic environment. The plant products are cheaper, easy to apply, biodegradable, ecofriendly, more readily available in bulk, effective in management of nematodes and improve soil health structurally and nutritionally. Application of organic amendments such as farm yard manure, compost, oil cakes, oils, mature crop residues, cellulosic soil amendments to the soil helps in reducing the nematode population. Most of the chemicals produced during decomposition of organic amendments are toxic to nematodes. The population of predators and parasitic fungi is stimulated which leads to reduction in numbers of phytonematodes.

7.6.1. Organic manures

(i) Fruit Crops: Reduction in citrus nematode populations by amendments with sheep and chicken manure has also been reported.

(ii) Vegetable Crops

Potato: Addition of FYM and compost to nematode infested soil reduced the infestation level of *Globodera rostochiensis* in potato roots. These manures hampered the development of nematodes in plant roots, where they developed more slowly than in plants grown in unmanured plots. Obviously there was some type of biochemical resistance imparted to the plants by nutrients from these manures.

Tomato: Deep ploughing during hot summer months followed by application of FYM at planting spots increased the yield of tomato in root-knot infested soil.

(iii) Ornamental Crops: Organic N in the form of sewage sludge was more effective than ammonium nitrate for reducing populations of *Belonolaimus longicaudatus* and *Hypsoperine graminis* in turf grass.

7.6.2. Green Manures

Green manuring is a conventional practice in which green plants are ploughed into the soil to rot and provide nutrients for the succeeding crop. The organic matter thus applied during the decomposition brings about important physical, chemical and biological changes in the micro sites where micro flora and micro fauna are active.

Linford and his associates as early as in 1938 had demonstrated that the incidence of root-knot could be significantly reduced by incorporating into soil chopped pineapple leaves. Patel and Desai (1964) reported that application of chopped leaves of *Pongamia glabra* and *Crotalaria juncea* when applied to root-knot infested soil reduced the disease severity. The decomposition of green leaves of neem, Persian lilac and sesbania in infested soil was also very effective (Singh and Sitaramaiah, 1970).

7.6.3. Cellulosic Soil Amendments

Excellent success in the control of plant nematodes has been obtained with the use of cellulosic soil amendments.

(i) Fruit Crops: Reduction in citrus nematode, *Tylenchulus semipenetrans* in cellulose amended soil was attributed to increase in the number of microbivorous and fungivorous nematodes.

(ii) Vegetable Crops: Singh and Sitaramaiah (1973) reported field reduction of root-knot nematodes of okra and tomato by amending the nematode infested soil with saw dust supplemented with N fertilizers. Supplemental N in saw dust amended soil caused several fold increase in yield.

7.6.4. Mature Crop Residues

Application of mature crop residues such as dried straw is not a conventional manuring practice. However, a substantial quantity of the dry crop debris is ploughed into the soil at the end of every cropping season. Every type of dried crop residue introduced into the nematode infested soil reduced root-knot severity and the greatest control was obtained when the crop residues were allowed to decompose for at least 30 days (Johnson, 1963). The dried crop residues of alfalfa, oats, flax, rye straw, timothy hay, buck wheat hull, cocoa bean hull, lucerne hay, peanut hull were all effective in reducing several kinds of plant nematodes.

7.6.5. Oil Cakes

fertilization of soil with oil cakes is a common and routine practice in many parts of the world. Lear (1959) reported that a significant reduction in population of plant nematodes when large quantities of castor bean pomace was mixed with soil. Populations of lesion nematodes are reduced by amendments with soybean and corn meal. The microbial decomposition of N substances especially the process helps in decreasing the nematode population. Oil cakes of neem, groundnut, castor, linseed, mustard, mahuva are effective on a field scale in controlling root-knot nematodes in vegetables. Oil cakes of neem and karanj which are not used as animal feeds have potential and proved to be economical.

Various parts of neem and its seed cake have suppressed the multiplication of *M. incognita, R. reniformis* and *T. brassicae.* The commonly occurring chemicals in neem include azadirachtin, kaemfor, nimbin, nimbicidin, nimbicidic acid, querctin and thionemone are believed to be the toxic factors.

Neem and its derivatives (especially neem cake) are eco-friendly and quite effective in the management of nematode pests infecting horticultural crop plants. The residual effect of neem derivatives persist in the subsequent crop. Besides nematode management, the neem derivatives supplied soil nutrients for the crop plants. The use of neem for the management of nematodes through seed dressing, nursery bed treatment or seedling bare root-dip treatment is highly economical. Incidentally, neem products check leaching of nematicides and nitrogen from the fertilizers, which prevents ground water contamination and nitrogen loss, respectively.

Neem and Chinaberry products are the most favoured materials used for the management of nematodes as seed treatment, nursery bed treatment, seedling bare root-dip treatment, soil incorporation and as a component in integrated management.

7.6.6. Oils and Sugars

Use of oils and sugars has been shown to be effective against several plant nematodes. Several vegetable oils such as mustard, corn, cotton seed, groundnut and soybean oils when applied at 1% (w/w) reduced the population of lesion nematodes, corn oil being most effective. Use of sugar containing substances such as sugarcane bagasse and molasses with 0.2% sucrose concentration was effective giving 50% larval mortality of *M. javanica* in about 9 days.

(i) Fruit Crops: Spraying of 1.5% water soluble hydrogenated fish oil on surface of soil reduced the population of *Radopholus citrophilus* in citrus roots.

(ii) Vegetable Crops: Ellenby (1945) reported adverse effect of black mustard oil (containing allyl isothiocyanate) on potato cyst nematode. In field experiments, application of 0.25 cwt oil/ha increased potato yields by 100% and decreased the emergence of larvae from cysts.

7.6.7. Methods of Application

(i) Seed treatment: Neem derivatives such as azadirachtin, nimbin, neem bitter, Repelin, Wellgro, seed kernel extract, cake extract, oil, leaf extract and Chinaberry derivatives have been effectively employed for seed treatment of tomato, brinjal, okra, cowpea, cabbage and cauliflower against *Meloidogyne incognita, Rotylenchulus reniformis* and *Tylenchorhynchus* *brassicae* (Table 7.1). The effect of various neem products and formulations might be due to the direct nematode toxicity of the seed coatings and making the environment unfavourable for nematode activity. Plants grown from coated seeds possibly acquire resistance or tolerance to the nematodes (Siddiqi and Alam, 1989b). These might have influenced the metabolism of the germinating seeds rendering the seedlings unfavourable for nematode multiplication and stimulating plant growth (Siddiqi and Alam, 1988b).

Table 7.1. Effect of neem/Chinaberry derivatives as seed dressers for the management of plant parasitic nematodes in horticultural crops

Nematode spp.	Crop	Neem/Chinaberry derivatives	Reference
Meloidogyne incognita	Tomato, brinjal, okra	Neem/Chinaberry leaf, flower, fruit, bark, root, gum extract	Siddiqi & Alam, 1987 Singh *et al.*, 1980
		Azadirachtin	Siddiqi & Alam, 1988b
		Nimbin	Siddiqi & Alam, 1990
	Tomato	Neem bitter, Repelin, Wellgro, Seed kernel extract	Das & Padhi, 1990
	Tomato, chilli	Neem oil	Singh *et al.*, 1980 Vijayalakshmi & Goswami, 1986 Akhtar & Mahmood, 1993, 1996
	Tomato, okra	Achook, Suneem, Neem oil	Akhtar & Mahmood, 1995a,b; 1997a
	Cowpea	Neem leaf extract	Khan & Hussain, 1988
Rotylenchulus reniformis	Tomato, brinjal, okra	Neem/Chinaberry leaf, flower, fruit, bark, root, gum extract	Siddiqi & Alam, 1987 Singh *et al.*, 1980
		Azadirachtin	Siddiqi & Alam, 1989b
		Nimbin	Siddiqi & Alam, 1990
	Cowpea	Neem leaf extract	Khan & Hussain, 1988
Tylenchorhynchus brassicae	Cabbage	Azadirachtin	Siddiqi & Alam, 1989a
	Cauliflower	Nimbin	Siddiqi & Alam, 1990

Vegetable Crops

Tomato: Neem bitter, Repelin, Wellgro, neem oil and seed kernel extracts gave satisfactory control of *M. incognita* infecting tomato. Neem cake extract efficiently controlled mixed population of nematodes on tomato.

Cowpea: Seed treatment of cowpea with neem leaf extract not only controlled *M. incognita* and *R. reniformis* but also the disease complexes involving these nematodes and the root rot fungus, *Rhizoctonia solani* (Khan and Hussain, 1988).

(ii) Nursery bed treatment: Application of neem cake to the nursery beds effectively controlled nematodes.

Fruit Crops: Nursery bed treatment proved very effective against fruits like mango, apricot, lemon, pear, peach and walnut (Alam, 1990).

Vegetable Crops: Nursery bed treatment was found effective against vegetables like tomato, brinjal, chilli, cabbage, cauliflower, knol khol and onion (Khan *et al.*, 1976).

Incorporation of neem leaves at 500 g/m^2 or 250 ml of neem leaf extract (5%) or neem cake extract (5%) per m^2 in nursery beds of tomato and brinjal gave effective control of *M. incognita* and *M. javanica* (Jain *et al.*, 1988; Rao and Parvatha Reddy, 1992a, 1992b).

Ornamental Crops: Nursery bed treatment proved very effective against ornamentals like calendula, rose, corn flower and bougainvillea (Khan *et al.*, 1976).

(iii) Seedling Bare Root-dip Treatment: Various derivatives of neem and Chinaberry have been used in small amounts as bare root-dip treatment for the management of *M. incognita, M. javanica, R. reniformis* and *T. brassicae* on tomato, brinjal, chilli, cabbage and cauliflower (Table 7.2). In many instances systemic activity has been noticed.

Table 7.2. Effect of neem/Chinaberry derivatives as bare root dips for the management of plant parasitic nematodes in horticultural crops

Nematode spp.	Crop	Neem/Chinaberry derivatives	Reference
Meloidogyne incognita	Tomato, brinjal	Azadirachtin, nimbin, nimbidic acid	Alam, 1993
		Neem leaf extract	Siddiqi & Alam, 1988a Akhtar & Mahmood, 1994 a,c; 1996a; Aziz *et al.*, 1995; Vats & Nandal, 1995 Mukhtar *et al.*, 1995
		Chinaberry leaf extract	Siddiqi & Alam, 1988a
	Tomato	Neem cake extract	Vijayalakshmi & Goswami, 1987
		Neem oil	Pradhan *et al.*, 1989
	Chilli, tomato	Chinaberry leaf extract	Akhtar *et al.*, 1992
		Neem oil	Akhtar & Alam, 1993

			Akhtar & Mahmood, 1993 a, 1996 b
		Nimin, Neem oil	Akhtar & Mahmood, 1993 a, 1996 b,c
M. javanica	Tomato, brinjal	Neem leaf, cake extract	Abid & Maqbool, 1991
Rotylenchulus reniformis	Tomato, brinjal	Azadirachtin, nimbin, nimbidic acid, Neem/ Chinaberry leaf extract	Alam, 1993 Siddiqi & Alam, 1989b Siddiqi & Alam, 1988a
Tylenchorhynchus brassicae	Cabbage, cauliflower	Neem leaf extract	Siddiqi & Alam, 1989b

Significant control of *M. incognita* on tomato and brinjal has been reported with neem cake extract and neem oil. Neem leaf extract has also been used for the management of *M. incognita, M. javanica* and *R. reniformis* on tomato and brinjal, *T. brassicae* on crucifers.

The mode of action of leaf extracts of neem and Chinaberry used as a root-dip treatment may be due to poor root-knot development resulting from poor penetration and retardation in feeding and/or reproduction of the juveniles of *M. incognita* (Siddiqi and Alam, 1988a). Likewise, the plants did not support the multiplication of *R. reniformis* after root-dip treatment. Some chemicals present in the extracts were either absorbed by the roots or some factor (acting as elicitor/activator) present in the extracts triggered the chain reaction (Cascade mechanism).

(iv) Soil Incorporation: Khan *et al.* (1966) and Singh and Sitaramaiah (1966) were the first to report effectiveness of neem cake against *M. incognita* on okra and *M. javanica* on tomato and okra, respectively. Thereafter, neem derivatives (cake, leaves, seed kernel powder, cake extract, nemin, neem oil coated urea) and Chinaberry derivatives have been used as soil amendments for the management of plant parasitic nematodes (Table 7.3).

Table 7.3. Effect of neem/Chinaberry derivatives as soil treatment for the management of plant parasitic nematodes in horticultural crops

Nematode spp.	Crop	Neem/Chinaberry derivatives	Reference
Meloidogyne incognita	Papaya	Neem cake	Routaray & Das, 1988
	Tomato, brinjal	Neem cake	Alam, 1987; Gowda, 1972
	Tomato, chilli, potato, okra, radish, betel vine	Neem cake	Singh & Sitaramaiah, 1966, 1971a; Akhtar & Mahmood, 1996 a; Alam, *et al.* 1980; Goswami & Swarup, 1971; Jagdale *et al.*,

			1985; Acharya & Padhi, 1988; Bhattacharya & Goswami, 1988; Bhatnagar & Goswami, 1987; Akhtar & Mahmood,1996 a; Prasad *et al.*, 1994; Jacob & Illaque, 1998
	Chilli, okra	Neem cake	Parvatha Reddy & Khan, 1991
	French bean	Neem cake	Srivastava & Singh, 1990
	Davana	Neem cake	Haseeb & Butool, 1991
	Basil	Neem cake	Haseeb *et al.*, 1988
	Okra	Neem cake extract	Khan *et al.*, 1991
	Tomato, chilli	Neem leaves	Akhtar & Alam, 1990
	Tomato, brinjal, okra, potato	Neem leaves	Singh, 1965; Akhtar, 1993; D'Errico & Maio, 1980; Akhtar & Alam, 1991; Akhtar & Mahmood, 1993b, 1996a
		Nimin, neem oil coated urea	Alam, 1993
	Brinjal	Neem leaves	Verma, 1986
	Tomato	Chinaberry leaf, cake, seed power	Verma, 1986
		Neem oil	Pradhan *et al.*, 1989
	Brinjal	Chinaberry leaf, cake, seed power	Ahuja & Mukhopadyaya, 1987
	Chilli	Chinaberry leaf, cake, seed power	Akhtar & Alam, 1990
	Cowpea	Hind-O-Meal	Goswami *et al.*, 1988
	Tomato, chilli	Nimin	Akhtar & Alam, 1993; Akhtar & Mahmood, 1994d, 1997b
	Tomato, betel vine	Neem saw dust	Singh *et al.*, 1967; Singh & Sitaramaiah, 1967, 1971a,b; Akhtar, 1997; Jagdale *et al.*, 1985; Acharya & Padhi, 1988; Kushwaha *et al.*, 1983; Siddiqi & Alam, 1990
M. javanica	Tomato, brinjal	Neem cake	Srivastava *et al.*, 1971

	Okra	Neem cake	Singh & Sitaramaiah, 1971a
	Acid lime	Neem cake	Mani & Murti, 1986
	Tomato, okra	Neem/Chinaberry leaves	Singh & Sitaramaiah, 1967
M. arenaria	Tomato	Neem leaf, seed kernel powder	Rossner & Zebitz, 1987
Rotylenchulus reniformis	Okra	Neem cake	Rao *et al.*, 1987
	French bean	Neem cake	Padhi & Mishra, 1987
	Garden bean	Neem cake	Sundaram & Velayutham, 1988
	Brinjal	Neem leaves	Lal *et al.*, 1977
	Tomato, chilli	Nimin (urea coated)	Alam, 1993
Tylenchorhynchus brassicae	Cabbage, cauliflower	Neem cake	Alam, 1989
	Tomato, chilli	Nimin (urea coated)	Alam, 1993
Radopholus similis	Banana	Neem cake Neem coated urea	Ravichandra & Krishnappa, 1985; Vadivelu *et al.*, 1987
Tylenchulus semipenetrans	Acid lime	Neem cake	Parvatha Reddy *et al.*, 1993
Helicotylenchus dihystera	Garden bean	Neem cake	Sundaram & Velayutham, 1988
H. indicus	Tomato, chilli	Nimin (urea coated)	Alam, 1993
H. multicinctus	Banana	Neem cake	
Hopololaimus indicus	Tomato, chilli	Nimin (urea coated)	Alam, 1993
Pratylenchus penetrans	Tomato	Neem leaf, seed kernel powder	Alam, 1989
P. zeae	Chilli	Neem leaves	Khan, 1990
P. coffeae	Banana	Neem cake	Vadivelu *et al.*, 1987
Aphelenchoides composticola	Mushroom	Neem leaves	Grewal, 1989
Tylenchus filiformis	Tomato, chilli	Nimin (urea coated)	Alam, 1993

Neem cake gave effective control of *M. incognita* on tomato, brinjal, chilli, okra, French bean, papaya, davana and basil; *M. javanica* on tomato, brinjal and okra; *R. reniformis* on okra, French bean and garden bean; *T. brassicae* on crucifers; *R. similis* on banana; *Tylenchulus semipenetrans* on acid lime and *H. dihystera* on garden bean. Similar studies were also made with neem cake extract as soil amendment for the management of *M. incognita* on okra. Seed kernel powder of neem was used for controlling *M. arenaria* and *Pratylenchus penetrans* on tomato.

Proper application of neem cake and adequate irrigations are necessary for optimum nematode management. Thorough mixing of cake with soil proved better than its application on the soil surface. Similarly, temperature also influences the efficacy of neem cake as the population of *Tylenchorhynchus elegans* was reduced at and above 20°C. Penetrability of *M. incognita* J2 was retarded by neem cake and seed extracts, thus fewer galls were produced on tomato roots (Mishra and Prasad, 1977). Duration of decomposition of neem cake also influence the nematoxicity of extracts. Mortality *of M. incognita, R. reniformis* and *T. brassicae* progressively increased with increase in the decomposition period up to 15 days (Alam *et al.*, 1982). In another study, toxicity of extracts from soil amended with neem cake to *M. incognita* increased up to 3 weeks of decomposition (Goswami and Vijayalakshmi, 1987). The increase in nematoxicity of different extracts with decomposition occurred either due to the release of more toxic substances (allelochemicals) during degradation of complex substances or due to accumulation of toxic metabolites of microorganisms which multiply during decomposition (Kirmani *et al.*, 1978). In view of this, neem cake is more efficacious against *M. incognita* in unsterilized soil than in sterilized soil (Bhattacharya and Goswami, 1987).

Singh *et al.* (1979) reported that neem cake reduced the movement of dazomet, oxamyl, dichlofenthion in soil through surface absorption and thereby prevented their leaching into deeper soil and thus prevented ground water contamination with the nematicides.

The beneficial effects of neem cake were observed not only against *M. incognita* and the fungus *Macrophomina phaseolina* individually, but also when both the pathogens formed a disease complex on French bean (Srivastava and Singh, 1990).

7.6.8. Mechanism of Control

Explanations for the possible mechanisms involved in the control of nematodes by the use of organic amendments are as follows:

- The volatile fatty acids (formic, acetic, propionic and butyric acids), ammonia and hydrogen sulphide that are released during the microbial decomposition of organic amendments are directly toxic to the nematodes.
- Microbivorous and fungivorous nematodes rapidly reproduce on the abundant bacteria and decomposition products, which in turn kill plant nematodes.
- Changes in the physical and chemical properties of the soil may alter the host-nematode relationship.
- Parasitism and predation of plant nematodes by nematophagous fungi, bacteria and soil micro fauna may be enhanced.
- Organic amendments improve soil conditions for more rapid root growth, which enhances the utilization of soil nutrients and masks the effects of nematode damage.

- Amino acids released during the decomposition of organic amendments may be responsible for reducing the nematode population in the soil (Parvatha Reddy *et al.*, 1975b). Yamashita and Akiya (1968) have shown that the amino acids are released from the organic amendments applied to the soil. Application of amino acids has been shown to control root-knot nematodes in tomato (Parvatha Reddy *et al.*, 1975a).

It appears from the above information that nematode suppression by organic amendments of soil may not be due to one mechanism but a combination of several mechanisms including that of host nutrition and vigour.

7.7. Antagonistic plants

Many plants and plant products are known to cause reduction in the nematode population below the damaging threshold. In few cases, plants have been found to be antagonistic towards nematodes and some have been shown to produce toxic materials inhibitory to nematodes. Antagonistic plants also serve as useful tool for nematode biocontrol. Many of these plants have good nematode control potential. There is an age old practice to interplant mustard, marigold, neem etc. with crop plants.

7.7.1. Mustard

Morgan (1925) accidentally discovered that potatoes grown with white mustard in a pot of infested soil were less heavily attacked by nematodes (*Globodera rostochiensis)* than potatoes growing alone. Triffit (1930) showed that potato root diffusate was ineffective in the presence of leachings from the roots of mustard seedlings. Ellenby (1951) showed that, in field trials, mustard oils increased the yield of potatoes by reducing the severity of nematode attack. Mustard has also inhibiting effects on *Tylenchorhynchus brassicae, Hoplolaimus indicus, M. incognita, R. reniformis, Helicotylenchus indicus, Tylenchus filiformis* and *T. mirzii* (Alam *et al.*, 1977). The active principle involved in mustard is allyl isothiocyanate which is toxic to the nematodes.

7.7.2. Marigold

The root-knot development on tomato and okra was low when interplanted with *Tagetes erecta*. The population of *Tylenchorhynchus, Helicotylenchus, Hoplolaimus, Rotylenchulus* and *Pratylenchus* was also markedly reduced (Khan *et al.*, 1971a; Alam *et al.*, 1977). Marigold has effectively reduced *Pratylenchus penetrans* on narcissus and reduced root rot disease. Different varieties of *T. patula* and *T. erecta* reduced population of lesion nematode up to 90% besides *Tylenchorhynchus* spp. and *Pratylenchus crenatus*.

Alpha-terthienyl is the active principle in *Tagetes* spp. which is toxic to these nematodes. Non-specific resistance usually involves several genes, for e.g. those that stimulate the secretion of nematicidal toxins in marigold,

Tagetes patula to repel juveniles of *Meloidogyne* spp. and other soil-inhabiting nematodes (Kyo *et al.*, 1990).

7.7.3. Asparagus

Asparagus officinalis would not support populations of *Trichodorus chiristiei* for more than 40 to 50 days. Tomato, normally a good host of this nematode, supported only a low population when asparagus was growing in the same pots. Asparagus contain nematicidal glycosides with systemic nematicidal action to reduce activity of *Paratrichodorus minor, T. chiristiei*, and completely inhibited emergence of *G. rostochiensis* juveniles from cyst and produced toxic effect on *M. hapla, P. penetrans* and *P. curvitatus.* A glycoside (asparagusic acid) is the active principle involved which is toxic to *T. christiei* and several other nematode species.

7.7.4. Sesame

Atwal and Mangar (1969) showed that root exudates from sesame (*Sesamum orientale*) have nematicidal properties against *M. incognita*. When okra was grown in *M. incognita* infested soil it was only slightly attacked and there were fewer nematodes compared with when okra was grown in the absence of sesame.

Non-specific resistance usually involves several genes, for e.g. those that stimulate the secretion of nematicidal toxins in sesame to repel juveniles of *Meloidogyne* spp. and other soil-inhabiting nematodes (Atwal and Mangar, 1969).

7.7.5. Sunnhemp

Crotalaria spectabilis and *C. juncea* are known to suppress *M. incognita, H. indicus, T. mirzii, R. reniformis, T. brassicae* and *T. filiformis* (McBeth and Taylor, 1944; Alam *et al.*, 1979). Monerotatum, a pyrrolizidine ester is believed to be the chief toxic factor (Fassuliotis and Skucus, 1969).

7.7.6. Pangola Grass

Another reported naturally occurring nematicide is pyrocatechol. This compound accumulates in high concentration in roots of pangola grass (*Eragrostis curvula*) and is a potent nematicide against *M. javanica, M. hapla, M. arenaria thamsei* and *M. incognita acrita.*

7.7.7. Bitter Cucumber

Cucurbitacins are triterpenoids that accumulate as aglycones or glycosides in cucumbers carrying a dominant gene for bitterness. Extracts from bitter cucumber attracted fewer *M. incognita* larvae than did a comparable extract from non bitter cucumbers which do not accumulate cucurbitacin. The cucurbitacins are repellant to *M. incognita.* Once larvae penetrate the roots, it is possible that cucurbitacin (an antibiotic) acts as a toxin that either kills the invading larvae or repels them from the root.

7.8. Deep Summer Ploughing

Two or three deep ploughings with a soil turning device, in hot summer months expose the nematodes in soil and infected tissue to solar heat and dehydration. The population densities of most nematodes are apparently brought down. This practice also suppresses weeds, soil fungi and bacteria. The efficacy can be increased by polyethylene mulching to tarp and retain more heat than the exposed soil. Exposure of moderately moist soil to solar heat is more efficient since the active nematodes are more vulnerable to moist heat.

7.9. Fallowing

Fallowing is the common practice of keeping the land free of all vegetation for a certain period by occasional ploughing of soil during hot and dry weather. This method ensures that the plant parasitic nematodes will have no host plants to feed and there will be no reproduction. In the absence of host plants, the nematodes will die after reserved food in the nematode body is depleted.

Crops that have been damaged by root-infecting nematodes of any kind should not be allowed to stand in the field for a long time after harvest but should be destroyed immediately, preferably by ploughing to expose the roots to the drying action of wind and sun. Weeds which are susceptible to nematodes should not be allowed to grow during fallow.

Fallowing for a period of three months after banana effectively suppressed the burrowing nematode population (Rajendran *et al.*, 1979). Decrease in *M. incognita* population on tomato, brinjal, okra, chillies and sponge gourd occurs when the field was left fallow (Khan *et al.*, 1975).

7.10. Flooding

Keeping the land submerged in water decreases oxygen content of the soil and kills the nematodes by asphyxiation and also toxicity of toxins produced as a result of anaerobic bacteria. Flooding can be employed to reduce the incidence of phytoparasitic nematodes if sufficient amount of water is available. Nematodes are killed due to asphyxiation and production of chemicals lethal to nematodes such as butyric acid, propionic acid and production of H_2S and NH_3 gases because of low pH as a result of decomposition of organic matter.

Flooding for three months after banana effectively suppressed the burrowing nematode (Rajendran *et al.*, 1979). This method of control is also effective against root-knot and stunt nematodes on vegetables.

7.11. Use of Nematode-free Planting Stock

Use of nematode-free planting stock is sure and effective means of nematode control. Cost of these practices is relatively low. With plants that are propagated vegetatively, it is often possible to eliminate nematodes by selecting uninfected plants or plant parts for propagation. Thus the citrus nematode (*Tylenchulus semipenetrans*), the burrowing nematode of banana (*Radopholus similis*), the potato cyst nematode (*Globodera rostochiensis*) and the coffee lesion nematode (*Pratylenchus coffeae*) can be eliminated by selecting nematode-free planting material. Similarly, onion and garlic seeds infested with stem nematode; strawberry and chrysanthemum plants infested with bud and leaf nematodes can also be controlled by using nematode-free planting stock.

7.12. Influence of Manuring

Application of $CaCN_2$, NaCN and urea cynamide is effective in controlling root-knot nematodes. This may be due to the release of HCN or NH_3 which are toxic to the nematodes.

(i) Fruit Crops: Kirkpatrick *et al.* (1959a) showed that population of *Xiphinema* and *Pratylenchus* were significantly lower under cherry trees receiving high rates of potash (2 and 6 kg of K_2 SO_4 per tree). They also showed that the numbers of these nematodes were correlated negatively with leaf potash (Kirkpatrick *et al.*, 1959b).

(ii) Vegetable Crops

Tomato: Increased levels of potash have significantly reduced the number of galls by *M. javanica* in tomato (Gupta and Mukhopadhyaya, 1971).

The growth of tomato increased (35%) and root galling due to *M. incognita* (root- knot index 1 as against 4 in control), nematode population in soil (80% reduction) and egg mass production (egg mass index 1 as against 4 in control) reduced in soil amended with 20% fly ash (Haq *et al.*, 1985). The increase in plant growth may be ascribed to increased nutrient availability (N, K, Ca, Mg, Na, B, SO_4) and the reduction in nematode population might be due to toxic compounds (polycyclic aromatic hydrocarbons, dibenzofuran and dibenzo-p-dioxin mixtures) present in fly ash.

Okra: Application of potash in combination with phosphorus or nitrogen or potash alone checks the reniform nematode multiplication on okra to a great extent (Sivakumar and Meerazainuddin, 1974).

7.13. Time of Planting

Crops may be planted in winter when soil temperatures are still below for activation of nematodes. Both early potatoes and sugar beets grow in soils too cold for cyst nematode activity. By the time the soil warms sufficiently for the nematodes to become active, the plants are advanced to grow and eventually produce a crop.

(i) Vegetable Crops: Tomato, brinjal and other susceptible vegetables are planted in cooler winter (rabi) months to avoid root-knot nematode damage (Sethi and Guar, 1986).

Planting of potato during third or fourth week of March in Shimla Hills would reduce the damage due to *M. incognita* (Prasad *et al.*, 1983). The yields were maximum which was concomitant with the lowest tuber infestation and lowest larval population in the soil at harvest.

7.14. Sanitation/Removal or Destruction of Infected Plants

Soon after the harvest of vegetable crops, the stubbles along with infected roots are removed and exposed to reduce the root-knot nematode population in soil.

Chapter 8

CHEMICAL METHODS

The primary advantage of chemical control over other methods is that the nematode population is reduced to a very low density within a matter of days after the chemical is applied, enabling the grower to plant a crop soon after treatment or in some cases at the time of treatment. Most crops are especially vulnerable to nematode attack during the seedling stage. Crops planted in treated soil develop good root systems, so that usually, in the case of annuals, the crop is mature before residual population of nematodes has increased to a level which would cause damage. Another important benefit resulting from chemical treatment with soil fumigants (methyl bromide) is that in addition to nematodes, fungi, bacteria and weeds are controlled. Weed control alone results in tremendous savings in labour costs in the production of vegetable transplants and ornamentals.

Nematicides are one of the most important and reliable means of controlling a wide variety of nematodes. The greatest significance of these nematicides lies in the fact that these not only offered the possibilities of nematode control but also served as useful tools for demonstration of crop losses inflicted by nematodes.

Nematicides can directly or indirectly reduce nematode populations. They are used as fumigants, contacts, drenches, root-dips, foliar applicants or seed treatments. Many fumigants are broad spectrum pesticides used as nematicides, fungicides, insecticides or herbicides. This may be detrimental to non-target or potential biocontrol organisms. DBCP, EDB and DD have been detected in ground water and are banned for use.

Non-fumigant nematicides (organophosphates and carbamates) are relatively non-phytotoxic and they do not usually bioaccumulate in the environment, but they have high mammalian toxicities. Some non-fumigant nematicides are translocated in plant tissue as systemics. There are several alleged cases of nematode resistance and a number of significant environmental concerns. All current non-fumigant nematicides have been detected in ground water.

In 1977, dependence on nematicides received a serious set back. Many of the employees at a chemical plant at Lathrop, California, USA, who handled DBCP, a soil fumigant, were found to be either azoospermic or oligozoospermic. Low sperm counts were found among formulators, applicators, farmers and salesmen. The US EPA banned production and sale of DBCP. Several regulatory measures followed as most of the nematicides were found to be the cause of environmental concerns and human health risks.

Nematicides cause soil pollution and adversely affect useful microflora and fauna which interact with pathogens at the soil-root interface. They are harmful to endomycorrhizae, rhizobia, actinorhyza and other micro-organisms normally contributing to the nutritional economy of the plants and possibly bring about disturbance in the microbial equilibrium. Nematicides pollute water and deteriorate water quality. Most of them have been detected in ground water. The key message is the need-based use of nematicides as a last resort.

8.1. Groups of Nematicides

8.1.1. Halogenated Hydrocarbons

Nematodes are sensitive to halogens and most of our successful nematicides are halogenated hydrocarbons. Of these that have been tried, the most effective compounds are: Chloropicrin (Trichloronitromethane) (tear gas), Methyl bromide (Bromomethane), DD (1,2–Dichloropropane + 1,3–Dichloropropene), EDB (Dibromoethane) and DBCP (1,2–Dibromo–3–chloropropane).

The halogenated hydrocarbons can not be used in the standing crops (except DBCP) since they are phytotoxic. They may injure the roots of living plants and stunt their growth. To avoid injury to crops an adequate interval must elapse between application and planting, usually supplemented by tillage.

8.1.2. Organophosphates

The need for a non-phytotoxic nematicide led to the investigation of this group of compounds. The most important compounds are: Parathion [0,0-Diethyl 0-(P-nitrophenyl)-phosphorothioate], Dichlofenthion [0-(2,4-Dichlorophenyl) 0,0-diethyl phosphorothioate], Thionazin (0,0-Diethyl-0,2-pyrazinyl phosphorothioate), Fensulfothion [0,0-Diethyl-0-P (methyl sulfinyl) phenyl phosphorothioate], Fenamiphos [Ethyl-4-(methylthio)-m-tolyl isopropyl phosphoramidate], Ethoprop (0-Ethyl S,S- dipropyl phosphorodithioate) and Phorate [0,0-Diethyl S-(ethylthiomethyl) phosphorodithioate].

8.1.3. Dithiocarbamates

Certain chemicals in this group that are effective in killing nematodes include Metham Sodium (Sodium N-methyldithiocarbamate), Aldicarb [2-

Methyl-2-(methylthio) propionalde-hyde-0-(methylcarbamoyl) oxime], Carbofuran (2,3-Dihydro-2,2-dimethyl 7-benzofuranyl methyl carbamate), Methomyl [S-Methyl N-{(methylcarbamoyl) oxy} thioacetamidate] and Oxamyl [Methyl N, N'-dimethyl - N-{(methylcarbamoyl) oxy}-1-thio oxamimidate].

8.2. Types of Treatment

8.2.1. Nursery Bed Treatment

(i) Fruit Crops: Nursery bed treatment with aldicarb or carbofuran both at 4 kg a.i. per ha was effective for the control of *T. semipenetrans* on citrus.

(ii) Vegetable Crops: Nursery bed treatment with aldicarb and carbofuran both at 2 g a.i. per m^2 were effective in increasing seedling growth and reducing root-knot nematode population on tomato, brinjal and chillies (Jain and Bhatti, 1983; Ramakrishnan and Balasubramanian, 1981). In the main field, the above treatments were also effective and increased the fruit yields.

DBCP at 50 l per ha and metham sodium at 250 l per ha were effective in reducing the reniform nematode population in tomato nursery and in increasing the growth of seedlings (Sivakumar *et al.*, 1977).

(iii) Ornamental Crops: Nursery bed treatment with phorate, aldicarb, fensulfothion and carbofuran each at 5 g per m^2 gave good and healthy crossandra seedlings free from *Pratylenchus delattrei* (Vadivelu and Muthukrishnan, 1979).

8.2.2. Seed Treatment

(i) Vegetable Crops: Fenamiphos, aldicarb and carbofuran all at 1 per cent concentration were effective in controlling the root-knot nematode (*M. incognita*) infecting cowpea, French bean and peas and in increasing their pod yields (Parvatha Reddy, 1984). Seed treatment with aldicarb and carbofuran both at 6 and 12 per cent was effective against the root-knot and reniform nematodes infecting okra (Sivakumar *et al.*, 1976). Aldicarb and carbofuran at 3 per cent were also effective against root-knot nematodes infecting chilli and bottle gourd (Kandasamy and Sivakumar, 1981). Jain and Bhatti (1981) reported effective control of *M. javanica* on okra by seed treatment with fenamiphos at 2, 4 and 6 per cent.

8.2.3. Seedling Bare-Root Dip Treatment

(i) Fruit Crops: Dipping of banana suckers in 0.1 per cent aldicarb solution increased plant height, bunch weight, fingers per bunch, induced early flowering and effectively reduced the burrowing nematode population (Venkitesan and Charles, 1983).

(ii) Vegetable Crops: Carbofuran and oxamyl at 1,000 ppm for 15 to 30 min. (Parvatha Reddy and Singh, 1979); fenamiphos at 250 to 750 ppm for 30 min. (Thakar and Patel, 1985); dimethoate at 500 ppm for 6 hr, phosphamidon and dichlofenthion both at 1000 ppm for 8 hr (Jain and Bhatti, 1978) and thionazin at 500 ppm for 15 min (Reddy and Seshadri, 1975) gave effective control of root-knot nematodes on tomato when used as bare-root dips. Alam *et al.* (1973a) reported that oxamyl was effective for the control of root-knot nematodes infecting brinjal and okra. VC-13 was found effective against *M. incognita* in chillies, brinjal and tomato (Saxena *et al.*, 1974; Alam *et al.*, 1973).

Bare-root dip treatment of brinjal seedlings with aldicarb, carbofuran and turbofos at 500 to 1000 ppm was effective in reducing the reniform nematode population (Prasad and Krishnappa, 1981).

(iii) Ornamental Crops: Dipping of bare-roots of rose in 0.1% solution of fenamiphos for 30 min. gave considerable reduction in root-knot infection. Dipping of gladiolus corms in thionazin or fensulfothion solution (0.5 g a.i. per litre) gave reduced root-knot nematode infestation. Root dip treatment for 8 hr in 0.05 to 0.1% solution of aldicarb, carbofuran, fensulfothion, fenamiphos or phorate protected crossandra seedlings from *Pratylenchus delattrei* for 30 days after transplanting (Vadivelu and Muthukrishnan, 1979).

(iv) Medicinal Crops: Bare root-dip treatment of *Solanum khasianum* seedlings with fenamiphos and aldicarb sulfone both at 1000 ppm for 2 hr was effective against *M. incognita* infection (Shetty and Reddy, 1985a).

8.2.4. Soil Treatment in the Main Field

(i) Fruit Crops: Significant reduction in the citrus nematode, *Tylenchulus semipenetrans* population (76 to 84 per cent) in sweet orange and root-knot nematode, *M. incognita* population (81 per cent) in grapevine was observed with soil application of aldicarb and carbofuran both at 6 kg a.i. per ha. The significant increase in yield was also observed in sweet orange (161 to 224 per cent) and grapevine (37 to 212 per cent) from the above treatments (Bhagel and Bhatti, 1983 a,b). Rajendran and Naganathan (1981) have reported that carbofuran at 2 kg a.i. per ha gave significant reduction in soil population of the reniform nematode, *Rotylenchulus reniformis* and increased the papaya fruit yield by 38.4 per cent over control. Pre-plant soil application of metham sodium at 40 l per ha was effective in controlling *R. reniformis* infecting grapevine and increased the bunch weight by 63 per cent over control. The burrowing nematode on banana, *Radopholus similis* can be controlled by soil application of fenamiphos at 2 kg a.i. per ha (Parvatha Reddy *et al.*, 1991b).

(ii) Vegetable Crops: Aldicarb, carbofuran, ethoprophos and fenamiphos each at 1 to 2 kg a.i. per ha were found effective in reducing the root-knot nematode population and in increasing fruit yields of tomato, cauliflower (Ahuja, 1983b), brinjal (119 to 145 per cent); okra (56 to 87 per cent), French bean (100 to 112 per cent); cowpea (40 to 50 per cent), peas (58 to 85 per cent) (Singh *et al.*, 1978; Rao and Singh, 1978; Singh and Parvatha Reddy, 1981b, 1982b; Parvatha Reddy, 1985a) and chillies (Handa and Mathur, 1981). Aldicarb at 0.5 kg a.i. per ha and carbofuran at 1.5 kg a.i. per ha were effective in reducing *Tylenchorhynchus brassicae* and *T. dubius* population and in increasing yields of cabbage (Varma *et al.*, 1978). Reddy and Seshadri (1972) reported that thionazin at 4 kg a.i. per ha proved effective against *R. reniformis* infecting tomato.

(iii) Ornamental Crops

Crossandra: Carbofuran, phorate, fenamiphos and disulfoton at 1g per plant effectively controlled *P. delattrei* and increased flower yield of crossandra (Vadivelu and Muthukrishnan, 1979). Carbofuran at 2.5 kg a.i. per ha was found to be effective against root-knot and lesion nematodes on crossandra (Srinivasan and Muthukrishnan, 1976).

Chrysanthemum: Aldicarb and phorate each at 1.5 kg a.i. per ha were effective in controlling *Aphelenchoides ritzemabosi* on chrysanthemum (Gill, 1981).

Tuberose: For the control of *M. incognita* on tuberose, aldicarb, fensulfothion and phorate were effective and increased flower yield.

(iv) Aromatic Crops

Patchouli: Aldicarb, carbofuran and phorate each at 2 to 3 kg a.i./ha significantly reduced *Helicotylenchus dihystera* and increased leaf yield of patchouli (Kumar and Nanjan, 1983). Gunasekaran (1979) reported the effectiveness of aldicarb at 1.5 kg a.i. per ha in controlling phytonematodes infecting patchouli.

Geranium: Aldicarb at 1 g per plant gave best control of *M. incognita* and increased shoot height, number of branches and leaf yield in geranium (Doraiswamy *et al.*, 1979).

Mint: Application of carbofuran increased plant growth and oil yield of *Mentha arvensis* and reduced the root-knot nematode population (Haseeb, 1992).

8.2.5. Foliar Application

Foliar application of methyl parathion, chloropyriphos and quinalphos each at 0.5% were effective in the control of *Aphelenchoides ritzemabosi* infecting zinnia and chrysanthemum (Gill, 1981; Gill and Walia, 1980).

8.3. Problems Associated with the Use of Chemical Nematicides

Use of nematicides did help in minimizing the avoidable losses due to nematodes in agriculture, but this benefit is bereft with many problems such as:

- The development of nematode resistance to nematicides.
- Phytotoxic effects of D-D became apparent, particularly in colder regions of Europe, Canada and USA, when applied during spring months in cool (5 to 10°C), wet soils. D-D persisted for weeks and was toxic to seeded crops.

- D-D and Vidden D could affect a number of soil organisms including suppression of nitrifying bacteria with consequent accumulation of phytotoxic levels of ammonia (Mc Kenry, 1980).
- Environmental pollution.
- Disruption in biodiversity.
- DBCP was found to adversely affect the fertility of workers (depressed sperm counts) in a factory manufacturing and packaging DBCP and the compound was a carcinogen (Thomason, 1982).
- Ground water in a number of states in USA has been shown to be contaminated with many organic chemicals, including the namaticides such as Aldicarb, carbofuran, DD, DBCP and EDB.
- Avian toxicity with carbofuran.

Excessive chemicalization of horticultural eco-system with various agro chemicals has affected the soil bio-diversity adversely tilting the balance in soil flora and fauna. This deleterious alteration has increased soil sickness in the absence of natural suppression by antagonistic and useful organisms. This scenario resulted in population increase of nematodes in the soil.

There is an urgent need to shift from the 'Chemical intensive' to 'Bio-intensive' cropping system with a view to restore the viability, sustainability and health of our agro-ecosystems. There is a need for development and adoption of INM in agriculture because of the economic and environmental benefits it brings to us. Major emphasis must be given to non-chemical methods of control, such as regulatory, physical, cultural and biological methods. Nematode resistant varieties of important crops must be evolved as these would be most useful tool of INM for our resource poor farmers.

8.4. Economizing on the Nematicide Usage

Most of the nematicides are imported and hence they are very costly. Besides, there is a need to reduce the quantity of nematicide used. The following practices are recommended to cut down the cost and quantity of nematicide usage:

- The use of nematicides must be limited to fields where they are economically justified.
- The reduction in nematode populations of 50 to 75% gives satisfactory yields with certain host-pathogen combinations. The lower levels of control can often be achieved with 1/3 to ½ the chemicals required for 95% kill. Using tickle irrigation, low dosage rates applied repeatedly through the first half of the growing season can give excellent crop response even when total amount applied is only ½ to 2/3 the recommended rate. These lower concentrations of nematicides minimize the chances of chemicals being leached through a soil profile to ground water.
- Nematologists look for a single chemical (preferably soil fumigant) that could be applied once (normally pre-plant) to manage nematodes

throughout a growing season on annual plants or for 2 to 4 years on perennial plants. A treatment of this type has appeal for both management and economic reasons. Furthermore, experience showed that post-emergence treatments (usually with OP and carbamates) of annuals gave unsatisfactory control because of high water solubility, low sorptive, potential pH stability and resistance to physical and biological degradation.

- To prevent leaching of aldicarb to ground water, use lower rates, applied at potato emergence (greater root mass is available to absorb aldicarb) when soils were warmer and biologically more active, with reduced amounts of irrigation.
- Seed or planting stock treatment (dipping or coating with stickers) with nematicides provide protection to the seedlings from nematode infection for several weeks to several months. Slow release formulations extend the period of activity and reduce the phytotoxicity to seeds and emerging seedlings.
- Nursery bed treatment with carbofuran at 2 kg a.i./ha for transplantable vegetable crops like tomato, brinjal, chilli, capsicum, cruciferous vegetables, etc. to produce nematode-free seedlings.
- Seedling bare-root dip treatment before transplanting in the main field with systemic nematicides like aldicarb, carbofuran, oxamyl and thionazin at 500 ppm for 15 min to protect the seedlings for 6 to 8 weeks from nematode infection after transplanting.
- Row and spot application of nematicides in the main field for widely spaced crops.
- Indigenous manufacturing of nematicides.
- Reduction of customs duty on nematicides which are imported.
- Giving subsidy to the farmers for the purchase of nematicides.

Chapter 9

HOST RESISTANCE

Plant resistance is regarded as an extremely feasible method for controlling nematodes. It is an effective, economical and environmentally safe means of reducing losses from nematodes. Use of resistant plants enables the grower to control nematodes without increasing production costs associated with the purchase of expensive chemicals, applicators and in numerous mechanical operations that go into the production of a crop. Resistance is especially valuable in controlling nematodes in low value crops and minor crops which can increase crop yields equal to that obtained by chemical methods. The use of varieties resistant to nematodes may be the only method of control available, which is economical on a field scale, for certain nematodes. At present, largely because of limited research in this area, a few nematode-resistant varieties of plants are available to the commercial grower.

9.1. Screening of Germplasm

Sources of resistance have been identified in certain horticultural crops and ornamentals (Table 9.1). Nematode-resistant varieties of a very few fruit and vegetable crops have been developed that too against one or two nematodes out of many that exists.

Table 9.1. Nematode-resistant cultivars of horticultural crops.

Crop	Nematode	Resistant cultivars	Reference(s)
Banana	*Radopholus similis*	Kadali, Pedalimoongil, Peykunnan, Ayiramkapoovan, Kunnan, Pisang Seribu, Tongat, Vennettu Kunnan, Anaikomban, Gold Finger	Charles *et al.*, 1983 Rajendran *et al.*, 1979
Citrus	*Tylenchulus semipenetrans*	Trifoliate orange, Swingle Citrumello	Parvatha Reddy, 1983
	Pratylenchus coffeae	*Poncirus trifoliata* *Microcitrus* selections	O'Bannon & Esser, 1975

Grapevine	*Meloidogyne incognita*	Black Champa, Dogridge, 1613, Salt Creek, Cardinal, Banquabad, Zinfundel, Kandharia, Kishmish Chorni, Redvine, Convert Large Black, Luglienga	Parvatha Reddy and Singh, 1984 Darekar & Patil, 1982
	Rotylenchulus reniformis	Dakshi, Joazbeli, Mukchilani	
Papaya	*R. reniformis*	Solo, Washington	Patel *et al.*, 1989
Passion Fruit	*M. incognita*	Yellow, Kaveri	Parvatha Reddy, 1981
Tomato	*Meloidogyne* spp.	Nematox, SL-120, NTR-1, SL-12, Patriot, VFN-8, VFN Bush, Piersol, Radiant, Nemared, Ronita, Anahu, Bresch, Helani, Campbell-25, Punuui, Arka Vardan, Pelican, Hawaii-7746, Hawaii-7747, Hisar Lalit, Punjab NR-7	Rao & Tikoo, 1973 Patel *et al.*, 1979 Jain *et al.*, 1983 Kanwar & Bhatti, 1990
	R. reniformis	Healani, VFN Bush, Radiant, SL-120, Patriot, Kalyanpur selection I, Kalyanpur selection III, Petropride, Peto 95, Peto 108, Nema 1400, ACE 55	Swarup *et al.*, 1989
Brinjal	*M. incognita*	Giant of Banaras, Black Beauty, Gola, SM – 67, Gachha Baigun, Syamala (MR)	Alam *et al.*, 1974 Sheela *et al.*, 2002 Harinath Naidu *et al.*, 2006
Chilli	*M. javanica*	579, CAP-63, Pusa Jwala Wonder Hot, Teja, LCA 206	Jain *et al.*, 1983 Harinath Naidu *et al.*, 2006
Capsicum	*R. reniformis*	Sweet Bell, California Wonder	Sirohi & Mohan, 2002 Swarup *et al.*, 1989
Potato	*M. incognita*	Kufri Dewa	Raj & Gill, 1983
	Globodera rostochiensis	Kufri Swarna	
	Pratylenchus penetrans	Katahdin, Kenebec, Superior	Bernard & Laughlin, 1976
	R. reniformis	La Rouge, Red Aboda	Swarup *et al.*, 1989
Onion	*R. reniformis*	Evergreen	Swarup *et al.*, 1989
Sweet potato	*M. incognita*	Sree Bhadra, Sree Vardhini, Sree Nandini, Kanjangad	Mohandas & Ramakrishnan, 2002
	R. reniformis	Gold Rush	Swarup *et al.*, 1989
Cowpea	*M. incognita*	Barsati Mutant, Iron, Clay, Chinese Red, Groit, New Era, Missisipi Purple, Missisipi Silver,	Sharma & Sethi, 1976a Darekar & Patil, 1981; Singh & Parvatha

		New Selection, Viktor K 798, Broweye 7, Brabham, Moneetta, Catjang, Columbia, Gujarat Cowpea 1 (GAU-1), C-152, 92-1-B, IC 9642-B, TVU 2430-P	Reddy, 1982a Dasgupta & Ganguly, 1986; Singh *et al.*, 1974 Hartman, 1976
	Heterodera cajani	CO 6	Kalaiarasan *et al.*, 2006
French bean	*M. incognita*	Banat, Blue Lake Stringless, Bountiful Flat, Brown Beauty, Cambridge Countess, Gallaroy, Kenya-3, Pinto W5-114, Seafarer, Suttan's Masterpiece	Singh *et al.*, 1981
Okra	*M. incognita*	Parbhani Kranthi (MR)	Harinath Naidu *et al.*, 2006
Muskmelon	*M. incognita*	Scarsol	Khan *et al.*, 1971b
	M. javanica	S-445	Jain *et al.*, 1983
Water-melon	*M. incognita*	Shehjanpuri	Khan *et al.*, 1971b
Ridgegourd	*M. incognita*	Panipati, Meerut Special	Khan *et al.*, 1971b
Ash gourd	*M. incognita*	Jaipuri, Agra	Khan *et al.*, 1971b
Pumpkin	*M. incognita*	Jaipuri, Dasna	Khan *et al.*, 1971b
China aster	*M. incognita*	Shashank-Resistant Poornima-Mod. Resistant	Nagesh *et al.*, 1995
Tuberose	*M. incognita*	Shringar-Resistant Suvasini-Tolerant	Nagesh *et al.*, 1995
Mentha	*M. incognita*	Kukrail, Arka, Neera, SS-1-4, SS-2-7, SS-18	Pandey & Patra, 2001
Black pepper	*M. incognita*	Pournami (tolerant)	George *et al.*, 2001
Yam	*M. incognita*	Sree Latha, Sree Keerthi	Mohandas *et al.*, 1998
Cassava	*M. incognita*	Sree Sahya, H-97, Narayaniyakappa	Sreeja *et al.*, 1998
Coffee	*R. reniformis*	*Coffeae robusta, C. canefora*	Swarup *et al.*, 1989

Among four species of *Solanum* tested for their reaction to *M. incognita, S. torvum* and *S. seaforthianum* gave resistant reaction which was reflected in the reduction in number of galls, egg masses and fecundity of females (Shetty and Reddy, 1985b).

Screening of the four commercially available varieties of carnation (*Dianthus caryophyllus*) for their resistance to root-knot nematode *M. incognita,* revealed that *cv.* Ivonne was the most susceptible while, Randez was moderately tolerant. The other two varieties, Dark Randez and Sunshine were observed to be moderately susceptible (Shylaja, 2004).

Among the six varieties of tuberose (*Polyanthes tuberose*) screened for their resistance to root-knot nematode *M. incognita, cv.* Mexican Single was tolerant under screen house experiments in sterilized soil, while under field conditions it was found to be moderately tolerant. Both, under screen house and field conditions, *cv.* Vybhav and Suvasini were found to be moderately susceptible, while *cv.* Shringar and Pearl Double were found to be moderately tolerant under screen house conditions, but moderately susceptible under field conditions. In case of *cv.* Prajwal, it was observed to be moderately susceptible under screen house conditions, but susceptible under field conditions (Shylaja, 2004).

9.2. Nature of Resistance

Nature of resistance to *M. incognita* in cowpea selections IC 9642-B and TVU 2430-P was investigated by Singh and Parvatha Reddy (1985). The resistance was associated with reduced larval invasion, root galling, egg mass production and fecundity, delayed development of larvae to adult females coupled with high concentration of magnesium.

Histological studies revealed that there was a direct correlation between the number of cork layers and resistance to *M. incognita* in cowpea. Cortical sclereids were noticed in resistant selection (IC 9642-B) while they were absent in the susceptible cultivar (S-488). There were more number of starch grains in the cortex of susceptible cultivar of cowpea. Death of cells (hypersensitive reaction) around infecting *M. incognita* larvae occurred in the roots of resistant selection (Singh *et al.*, 1984).

9.3. Isolation of Nematicidal Plant Products

Methanol extracts of *Catharanthus roseus,* onion, *Gloriosa superba* and scented geranium exhibited nematicidal activity against root-knot nematodes. The active nematicidal compounds identified were citronellol, geraniol and linalool from scented geranium essential oil; colchicine from *G. superba* seeds; serpentine from *C. roseus* and amino acids (Methionine) from onion seeds. The feasibility of utilization of serpentine at 5000 ppm for the management of *M. incognita* in tomato nursery was ascertained (Leela *et al.*, 1992; Nidiry *et al.*, 1992).

9.4. Biochemical Mechanism of Nematode Resistance

The formation of phytoalexin 'Kievitone' was observed in root-knot resistant cowpea cv. C-152 in response to *M. incognita* infection, while it was absent in susceptible cv. S-488. Higher concentrations of phenolic and ortho-dihydroxy phenolic compounds and more active IAA-oxidase and peroxidase systems have been observed in resistant tomato varieties (Bajaj and Mahajan, 1977; Masood and Husain, 1976; Ganguly and Dasgupta, 1981).

9.5. Biotechnological Approaches in Nematode Management

Biotechnology has a role to play in incorporation of resistance against nematodes and biological control of plant nematodes. Embryo rescue technique and protoplast fusion or somatic hybridization have been used in nematology to overcome interspecific plant breeding problems with a view to incorporate nematode resistance.

9.5.1. Embryo Rescue Technique

Embryo rescue technique involves hybridization between economic plant species and wild relatives followed by excision and culturing of hybrid zygotic embryo on nutrient media. One of the classical examples of successful application of embryo rescue technique in plant nematology is the synthesis of tomato cultivars resistant to root-knot nematodes. Cultivated tomato species, *Lycopersicon esculentum,* is highly susceptible to attack by *Meloidogyne* spp. but the related wild species *L. peruvianum* has a high degree of resistance against the nematodes. A cross between these two species yielded a hybrid in which case the endosperm does not develop and the embryo aborts. This problem was solved by carefully isolating and culturing the hybrid embryo on tissue culture media (Smith, 1944). This technique thus provides a mechanism for transfer of nematode resistance gene from *L. peruvianum* into cultivated species of tomato, *L. esculentum*.

9.5.2. Protoplast Fusion / Somatic Hybridization

In somatic hybridization, plant tissues from two parental sources are digested with a mixture of pectinase and cellulase in an osmotically stable solution to produce protoplasts. Protoplasts are naked cells surrounded by a plasma membrane and have the capacity to fuse spontaneously or are induced to fuse by way of chemicals (polyethylene glycol, dextrin and polyvinyl alcohol) or electro fusion. These fused protoplasts are capable of cell wall regeneration. After cell division, a single nucleus may be formed containing the chromosomes of one of the parent cell or more frequently, selective elimination of chromosomes of one of the parental species during cell division and tissue formation.

This technique of protoplast fusion has been used as a means of transferring root-knot nematode resistance factors from *Solanum sisymbrifolium* into eggplant, *S. melongena.* S1 and S2 progenies of *S. sisymbrifolium* plant regenerated from callus not only showed resistance to *M. javanica* but also retained resistance to *M. incognita* (Fassuliotis and Bhatt, 1982). The nematode did not develop beyond second-stage larvae. Transfer of nematode resistance by protoplast fusion between sexually incompatible plant species holds promise in future.

9.5.3. Recombinant DNA Technology

Application of genetic engineering made possible the synthesis of transgenic plants. Transgenic plant is one in which foreign DNA gene (a transgenic) is stably incorporated in the genome early in embryonic development and so is present in both somatic and germ cells, is expressed in one or more tissues, and is inherited to offspring in a Mendalian fashion. The aim of gene transferring technology is to improve a top variety in one additional character without disturbing the rest of the genome. For gene transfer, three methods are available; vector-mediated gene transfer (*Agrobacterium tumefaciens*), microinjection by capillaries, laser micro beams or micro projectiles and direct transfer.

This technique of recombinant DNA technology has been used in tomato to transfer *Mi* gene which shows resistance to *M. arenaria, M. incognita* and *M. javanica*; but not to *M. hapla* (Gilbert and Mc Guire, 1956).

A lot of information is being generated with respect to identification of resistance gene loci against various nematode pests in different crops of importance. Particular interest has been shown in the chromosomal location of resistance to *Globodera rostochiensis* in potato (Pireda *et al.*, 1993), *Heterodera schachtii* in beet (Jung *et al.*, 1992), and root-knot nematode, *Meloidogyne* spp. (Klein-Lankhorst *et al.*, 1991; Messeguer *et al.*, 1991). The identification and cloning of such genes would facilitate their transfer and expression in transgenic crops. In case of potato, a major dominant locus *Gro* 1 conferring resistance to some pathotypes of *G. rostochiensis* was mapped by restriction fragment length polymorphism (RFLP) to potato linkage group IX; the nearest neighbouring RFLP markers are CP 51 and TG 20(a) (Barone *et al.*, 1990). Other groups are developing high resolution RFLP maps around the *Mi* genes in tomato. The *Mi* genes originated from wild tomato species *L. peruvianum* confers resistance to all major species of *Meloidogyne.* The *Mi* gene is located at position 44 on chromosome 6 and is closely linked to the leaf calyx markers of *yv* (yellow virescent) and acid phosphatase-1 (Aps-1). The work of Messeguer *et al.*(1991) led to the production of a more detailed consensus linkage map of chromosome 6. Four co-dominant RFLP markers (TG 297, CD 14, Tom 25 and GP 79) can be used as markers for selection of the *Mi* gene. The extent of *L. peruvianum* DNA segment in different *Mi* gene containing cultivar was also determined. Three groups of cultivars (ABC) with different amounts of *L. peruvianum* DNA were found. Only in the cultivars of group A, there is linkage between *Mi* and Aps-1. For cultivars of group B, the marker Gp 79 can be used and new markers are needed for nematode resistance in group C cultivars.

Many patents have already been made in the area of transgenic plants resistant to nematodes. A transgenic potato with a novel nucleic acid molecule capable of increasing the resistance of a plant to nematode infection comprising a nucleotide sequence encoding a polypeptide with a collagenase (EC-3.4.24.3) activity or its precursor has been patented by Stratford and Sheilds (1995). A lectin from garlic has been patented to confer resistance against wide array of nematode pests (Geoghegan *et al.*, 1995). Similarly, protease from *Paecilomyces lilacinus* (patented information) can be used to construct nematode resistance transgenic plants (Den-Belder *et al.*, 1994). A transgenic plant with nematode resistance may be prepared by introduction of an RNAase gene under control of a promoter which induces RNA production in response to nematode attack of root tissue, resulting in prevention of formation of nematode feeding sites. This strategy has been patented by Monsanto, USA.

Chapter 10

BIOLOGICAL METHODS

Biological control aims at increasing the parasites, predators and pathogens of nematodes in the soil, in order to increase the mortality of plant parasitic nematodes. This can be done by introducing the biocontrol agents directly which would enhance the biotic component of environmental resistance. In practice, this has not been possible on a large scale.

The role of potential biocontrol agents such as bacteria (including Actinomycetes) and fungi (including VAM fungi) in reducing population of parasitic nematodes in horticultural crops along with integration of these agents with eco-friendly components into overall nematode management strategies have been discussed.

10.1. Bacteria

Bacteria destroy nematodes in virtually all soils because of their constant association in the rhizosphere. Bacteria like *Pasteuria penetrans* destroy nematodes by their parasitic behaviour while the non parasitic rhizobacteria reduce nematode populations by colonizing the rhizosphere of the host plant. A large number of rhizobacteria are known to reduce nematode populations and important genera include *Agrobacterium*, *Alcaligenes*, *Bacillus*, *Clostridium*, *Desulfovibrio*, *Pseudomonas*, *Serratia* and *Streptomyces*. Application of some of these bacteria has given very promising results. Practical control systems, formulation, mechanisms of nematode suppression and possible uses of bacteria in nematode bio-control are suggested (Siddiqui and Mahmood 1999).

Depending upon their mode of action, bacteria may be categorized mainly into two groups. The first group includes bacteria which release metabolites that have a killing or inhibitory effect on plant nematodes (*Bacillus, Azatobacter*). The second group includes bacteria which parasitize directly on nematodes, thus affecting the penetration and reproduction of nematodes (*Pasteuria*).

10.1.1. *Bacillus spp:*

(i) *Bacillus licheniformis*: Two bacteria *Bacillus licheniformis* and *Pseudomonas mindocina* and 2 fungi *Acrophialophora fusispora* and *Aspergillus flavus* were evaluated for their efficacy as bio-control agents individually and in various combinations against *Meloidogyne incognita* infecting tomato. Individually, *B. licheniformis* gave best control although when all 4 agents were used, nematode multiplication was reduced by 82.06% (Siddiqui and Husain, 1991).

(ii) *Bacillus thuringiensis*: Various formulations of *B. thuringiensis* are toxic to eggs and larvae of *Meloidogyne incognita* under *in vitro* conditions (Chahal and Chahal, 1991). The exotoxin produced by *B. thuringiensis* was found toxic to the populations of *Meloidogyne*. It also deterred *M. incognita* juveniles forming galls on tomato roots. Dipel and SAN 415 strains of *B. thuringiensis* sub sp. *kurstaki* have been found to suppress populations of both *M. javanica* and *Tylenchulus semipenetrans*. SAN 415 was found as effective as the standard nematicide, fenamiphos. CR 371, another isolate of *B. thuringiensis* was effective in reducing galling by root-knot nematode on tomato and also resulted in reduced population of *Rotylenchulus reniformis*.

B. thuringiensis, besides increasing the dry weight of shoot and root, reduced *M. incognita* infection in pea both in naturally infested soil and sterilized soil by 69.4 and 74.7%, respectively. *B. thuringiensis* was also effective in reducing the number of galls/plant and total nematode population in tomato and brinjal. Due to reduced nematode infection, dry weight of shoot, chlorophyll contents and nitrate reductase activity of brinjal leaves increased (Chahal and Chahal, 2003).

(iii) *Bacillus subtilis*: Multiplication rate of *R. reniformis* was reduced significantly (44.5%) when soil drenching with *B. subtilis* (isolate Bs_t) was done a week before nematode inoculation. But, it was not so in the case of treatments that received simultaneous inoculations of the bacterium and the nematode. These results indicated that *B. subtilis* could induce systemic resistance in tomato against *R. reniformis* (Niknam and Dhawan, 2001a).

10.1.2. *Azotobacter chroococcum*

The heterotrophic bacterium, *A. chroococcum* antagonistic to *M. incognita*, inhibited the hatching of egg masses and prevented the larvae to penetrate into the roots of brinjal to form galls (Chahal and Chahal, 1988).

Chahal and Chahal (2003) reported significant reduction in the gall formation, soil and root population of *M. incognita* both in tomato and brinjal by *A. chroococcum*.

10.1.3. *Pasteuria penetrans*

P. penetrans is known to parasitize 205 nematode species belonging to 96 genera from 10 orders of the phylum Nematoda. It occurs widely (recorded from 51 countries in 5 continents) and is a potential biocontrol agent. It is a mycelial endospore forming bacterium which is an obligate parasite of large number of nematodes with a life cycle remarkably synchronizing with the development of its host. Infection occurs through direct penetration of the cuticle by germinating spores that encumber to the surface of the nematodes. The life cycle begins with the vegetative mycelial colonies which eventually fill the nematode body with a large mass of bacterial endospores. Sporangia are liberated in the soil from the body of the nematode upon decomposition.

A number of strains are reported to exist in *P. penetrans*. An isolate of the parasite infecting *M. incognita* did not encumber *Pratylenchus brachyurus* and two isolates that adhered to *P. brachyurus* could not adhere to *M. incognita*. In India, two different strains of *P. penetrans* have been reported i.e. one infecting *M. incognita* but not *Heterodera cajani* and the other infecting *H. cajani* but not *M. incognita* (Singh and Dhawan, 1998).

P. penetrans reduces root penetration by the nematode juveniles, inhibits egg formation, reduces nematode population in soil and resulted in increased yields. The bacterium being resistant to heat, desiccation, its favourable compatibility with nematicides and its ability to survive for more than 2 years in soil, qualifies it as the most potential bacterial biocontrol candidate which could be used on a large scale under field conditions. However, the obligate nature of the organism, its lack of mobility and dependence on dispersal agents and host specificity are its major drawbacks.

(i) Fruit Crops: *P. penetrans* offers potential for the biocontrol of *Tylenchulus semipenetrans* infecting citrus (Mani, 1988a) and tomato (Rao *et al.*, 1997).

(ii) Vegetable Crops: *P. penetrans* has been demonstrated to exert effective biocontrol of root-knot nematode on tomato (Mankau, 1975). There has been a significant reduction in the root gall index and the number of egg masses of *M. incognita* by about 52 and 60 per cent, respectively due to *P. penetrans* antagonism in brinjal (Nagesh and Parvatha Reddy, 1998).

Significant reduction in *M. incognita* population and increased plant growth of brinjal was found when *P. penetrans* was inoculated in sandy loam soil. The bacterial spores/female were found higher in coarse textured soil compared to clay loam soils (Singh *et al.*, 1998).

(iii) Spice Crops: Black pepper plants inoculated with *P. penetrans* showed reduction in *M. incognita* population and root gall indices and improved plant growth parameters (Sosamma and Koshy, 1995).

10.1.4. *Pseudomonas* spp:

(i) *Pseudomonas fluorescens* belongs to Plant Growth Promoting Rhizobacteria (PGPR), the important group of bacteria, which play a major role in the plant growth promotion, induced systemic resistance, biological control of pathogens etc. PGPR are known to enhance plant growth promotion and reduce severity of many nematode diseases.

Most research is dedicated to improve crop yields and plant growth with microbial inoculants has focused on symbiotic bacteria. However, free-living soil microorganisms, such as *Azospirillum* and various *Pseudomonads,* also might be used to promote plant growth. PGPR are rhizobacteria that have the ability to promote the growth of plants following inoculation onto seeds or subterranean plant parts. PGPR mediate improved plant growth either directly by stimulation of the plant or indirectly through control of pathogens or triggering and inducting of host defense mechanism.

P. fluorescens produce the antibiotics 2,4-diacetylphloroglucinol (DAPG), phenazine-1-carboxylic acid (PCA), phenazine (Phe), antimicrobial metabolites such as pyoluteorin (Plt), hydrogen cyanide (HCN), siderophores pyoverdine (Pvd), salicylic acid (Sal) and pyochelin (Pch) which inhibit a broad spectrum of plant pathogenic fungi, bacteria, nematodes (Raaijmakers *et al.*, 1997; Schmidli *et al.*, 1997; Boruah and Kumar, 2002) and control a variety of root and seedling diseases. There are many evidences substantiating the importance of DAPG in biological control. The production of DAPG is governed by *Phl* gene. The population size of DAPG producers in the rhizosphere correlated with the disease suppressiveness of the soil and *in situ* antibiotic production. The diverse DAPG producing *Pseudomonas* spp. have been isolated from the rhizosphere of various crop plants and their role in promoting plant growth and inhibiting root diseases are the subjects of ongoing investigations worldwide.

(i) Fruit Crops

Banana: Three *P. fluorescens* strains and the type strain *P. putida* inhibited invasion of *Radopholus similis* in banana roots. *P. fluorescens* strain kept *R. similis* numbers significantly lower in banana roots after the initial invasion stage. All strains also showed an *in vitro* repellent effect towards the *R. similis* (Aalten *et al.*, 1998). *P. fluorescens* significantly reduced *R. similis* invasion by 50%.

Grapevine: Application of *P. fluorescens* at 4 g/plant significantly reduced root galling (40%), egg mass production (60%) and increased root colonization by *P. fluorescens* and fruit yield in grapevine (166% over control) (Shanthi *et al.*, 1998).

(ii) Vegetable Crops

Potato: DAPG producing *P. fluorescens* F113 is proposed as a potential bio-control inoculant for the protection of potato crop against the potato cyst nematode, *Globodera rostochiensis* (Cronin *et al.*, 1997; Raaijmakers and Weller, 1998).

Tomato: Shanthi and Sivakumar (1995) reported the increased plant growth of tomato with *P. fluorescens* strain Pf-1. Bacterial treatment also reduced the level of infestation by the nematode, which was concentration dependent. The suppressing ability of the bacterial strains appeared to be related to their root colonizing ability.

Three *P. fluorescens* strains and the type strain *P. putida* inhibited invasion of *Meloidogyne* spp. in tomato roots. All strains also showed an *in vitro* repellent effect towards *Meloidogyne* spp. (Aalten *et al.*, 1998).

Sugar Beet: *P. fluorescens* had induced systemic resistance and inhibited early root penetration of *Heterodera schachtii,* the cyst nematode in sugar beet (Oostendorp and Sikora, 1989).

(iii) Spice Crops: Turmeric rhizome treatment with *P. fluorescens* at 10 g/kg increased plant growth characters (58 and 31% increase in pseudo stem height and number of tillers over control), and decreased nematode population (40%) and gall indices (25%) (Seenivasan *et al.*, 2001).

(ii) *Pseudomonas chitinolytica*: *P. chitinolytica,* a bacterium isolated from crustacean shell amended soil was evaluated for the control of *Meloidogyne javanica* on tomatoes. Greenhouse, screen house and micro plot tests all indicated the improved growth and yield and reduced nematode numbers in treated plants (Spiegel *et al.*, 1991).

Van Gundy and associates at the University of California, Riverside, have found it possible to stimulate growth of rhizosphere bacteria by adding nutrients (e.g. peptone) through a tickle irrigation system or to distribute the rhizosphere bacteria through such a system. There is a need to search for bacteria that would be active in the rhizosphere and produce compounds lethal to nematodes, inhibitory to nematode movement or development and repellent to nematodes on roots.

Table 10.1. Biological control of plant parasitic nematodes in horticultural crops using bacteria.

Bacteria	Crop	Nematode	Reference
Pseudomonas fluorescens	Citrus	*T. semipenetrans*	Shanthi *et al.*, 1998
			Parvatha Reddy & Nagesh, 2000
	Banana	*R. similis* & *M. incognita*	Aalten *et al.*, 1998
	Grapevine	*M. incognita*	Shanthi *et al.*, 1998
	Tomato	*M. incognita*	Verma *et al.*, 1999
			Shanthi & Sivakumar, 1995
		R. reniformis	Nicknam & Dhawan, 2001
	Potato	*M. incognita*	Mani *et al.*, 1998
	Okra	*M. incognita*	Sobita Devi & Dutta, 2002
	Sugar beet	*H. schachtii*	Oostendorp & Sikora, 1989
P. stutzeri	Black pepper	*M. incognita*	Eapen *et al.*, 1997
	Turmeric	*M. incognita*	Seenivasan *et al.*, 2001

	Tomato	*M. incognita*	Khan & Tarannum, 1999
P. chitinolytica	Tomato	*M. javanica*	Spiegel *et al.*, 1991
	Black pepper	*M. incognita*	Sosamma & Koshy, 1995
Pasteuria penetrans	Banana	*R. similis*	Vidya & Reddy, 1998
	Citrus	*T. semipenetrans*	Shanthi *et al.*, 1998
			Parvatha Reddy *et al.*, 2000
			Mani, 1988
	Grapevine	*M. incognita*	Mani *et al.*, 1999
			Stirling & White, 1982
	Tomato	*M. incognita*	Rameshchand & Gill, 2002
			Somasekar & Gill, 1991
			Walia & Dalal, 1994
			Parvatha Reddy *et al.*, 1997
		M. javanica	Giannakiou *et al.*, 1999
			Maheswari & Mani, 1988
			Rao *et al.*, 1997
			Mankau, 1975
	Brinjal	*M. incognita*	Nagesh & Parvatha Reddy, 1998
			Karuna *et al.*, 2001
			Singh *et al.*, 1998
	Black pepper	*M. incognita*	Sosamma & Koshy, 1995
Agrobacterium radiobacter / Bacillus sphaericus	Potato	*Globodera pallida*	Racke & Sikora, 1992
Azatobacter chroococcum	Tomato	*M. incognita*	Chahal & Chahal, 2003
	Brinjal	*M. incognita*	Chahal & Chahal, 1988
		M. javanica	Bansal & Verma, 2002
Bacillus subtilis	Grapevine	*M. incognita*	Santhi & Sivakumar, 1998
	Tomato	*R. reniformis*	Nicknam & Dhawan, 2001
B. thuringiensis	Tomato	*M. incognita*	Khan & Tarannum, 1999
	Pea	*M. incognita*	Chahal & Chahal, 2003
B. licheniformis	Tomato	*M. incognita*	Siddiqui & Hussain, 1991
Streptomyces avermitilis	Tomato, Brinjal, Chilli	*M. incognita*	Parvatha Reddy & Nagesh, 2002

10.2. Actinomycetes

The avermectins are a new class of macro-cyclic lactones isolated from the soil Actinomycete, *Streptomyces avermitilis*. These compounds were reported to be possessing insecticidal, acaricidal and nematicidal properties (Putter *et al.*, 1981). They are commonly distributed in most of the cultivated soils.

While avermectins B1 and B2a have shown high toxicity against *M. incognita* in greenhouse conditions when incorporated into soil, avermectin B2a was found to be more potent than B1 (Putter *et al.*, 1981).

In a micro plot study, avermectin B1 at 0.17 kg a.i. per ha was equal to ethoprop, aldicarb, fenamiphos, oxamyl and carbofuran at 6.7 kg a.i. per ha (Nordmeyer and Dickson, 1981).

The available literature on avermectins indicate that *M. incognita*, *M. javanica*, *M. arenaria* and *T. semipenetrans* were the commonly used test organisms to assess the nematicidal activity of these metabolites on tomato and citrus (Garabedian and Van Gundy, 1983).

The avermectins have a direct action on neurotransmission. They are antagonists of gamma-amino butyric acid.

(i) Fruit Crops: In citrus, a monthly rate of 1.1 kg a.i. per ha of avermectins for 7 months gave maximum increase in yield and reduction of *T. semipenetrans* population.

(ii) Vegetable Crops: The aqueous solution of avermectins (250 ml of 0.001%/m^2) significantly reduced root galls of tomato seedlings raised in root-knot infested nursery beds which yielded robust and healthy seedlings with no root infection (Parvatha Reddy and Nagesh, 2002).

In a root-knot infested micro-plot study, avermectins at 0.015 kg/ha effectively controlled *M. incognita* on tomato and increased yield by 11 and 8% compared to the untreated and carbofuran treated plots, respectively. Similarly, avermectins effectively reduced root-knot nematode infection in chilli and brinjal under field conditions (Parvatha Reddy and Nagesh, 2002).

10.3. Fungi

Interactions between fungal antagonists and nematodes have been known to occur in agricultural soils for many years (Mankau, 1980). Nematode destroying fungi play an important role in regulating nematode population dynamics.

Recent studies have indicated that the existence of opportunistic soil fungi which are capable of colonizing females and eggs of nematodes and deleteriously affecting them. Nematodes belonging to such genera as *Meloidogyne*, *Tylenchulus* and *Rotylenchulus*, at sedentary stages of their life cycle, are vulnerable to attack by fungi either within host plant roots or when females and egg masses are exposed on the root surface or within soil. Fungal growth is known to be enhanced in the rhizosphere, increasing the likelihood of fungal invasion of nematode structures present in the ecological niche.

Once in contact with eggs, larvae or females, the fungi grow rapidly and eventually parasitizes all the eggs that are in the early embryonic developmental stages.

Several fungi such as *Paecilomyces lilacinus*, *Verticillium lecanii*, *Pochonia chlamydosporia*, *Trichoderma harzianum*, *T. viride* and *T. virens* have been reported to parasitize eggs of plant parasitic nematodes.

Nematode antagonistic fungi are capable of solubilizing the insoluble phosphate. *P. lilacinus* gave significantly higher (30%) P solubilization as compared to *Aspergillus awamori,* the fungus commonly used as P solubilizing agent (Somasekhar *et al.*, 1999).

Alginate is a water-soluble polysaccharide gum that has excellent gel-forming properties. Living organisms that are capable of biologically controlling a variety of nematodes can easily be entrapped in a calcium alginate matrix in aqueous form at room temperature. The gel beads, granules or seed coatings that are produced by the process are biodegradable. Antagonistic fungi (*Trichoderma*, *Gliocladium*, *Talaromyces* and *Penicillium* spp.) and bacteria (*Pseudomonas* and *Bacillus* spp.) that can control soil-borne plant pathogens can be formulated in calcium alginate (Connick, 1988).

10.3.1. Paecilomyces lilacinus

P. lilacinus is by far the most promising and practicable biocontrol agent for the management of root-knot and cyst nematodes. The fungus grows abundantly in sterilized paddy seeds, oil cakes, mashed potato, chopped water lily plants, ipil ipil (*Leucaena glauca*) leaves and corn grits.

(i) Vegetable Crops: Eggs of *M. incognita* on potato roots were found to be infected by *P. lilacinus*. This fungus was also capable of invading mature females and cysts of *Globodera pallida* (Jatala *et al.*, 1980).

P. lilacinus increased growth of brinjal plants and reduced root galling, egg masses and number of eggs per egg mass of *M. incognita* (Sharma and Trivedi, 1989). Morgan-Jones *et al.* (1984) showed that hyphae of *P. lilacinus* readily penetrate egg shells of *M. arenaria* through small pores dissolved in the vitelline layer. The effect seems to involve disruption of embryonic development and death of larvae, resulting in reduction of nematode populations.

P. lilacinus was also effective for the management of reniform nematode infecting tomato and brinjal (Parvatha Reddy and Khan, 1988, 1989).

P. lilacinus reduced root-knot nematode egg masses by 91% after 60 days and by 96% after 90 days, while aldicarb reduced egg masses by 90 and 91% after 60 and 90 days after inoculation, respectively in okra. In general, *P. lilacinus* and aldicarb treatments greatly suppressed the numbers of nematode galls and egg masses on okra plants. However, it was evident that *P. lilacinus* was more effective in reducing nematode egg masses than aldicarb (Ibrahim and Rezk, 1988).

The soil incorporation of *P. lilacinus* (in rice hulls and bran soil mix) was effective in reducing the gall index (2.7 as against 5.0 in control), nematode population in soil (77.3% over control) and in increasing the number of okra fruits (121.7 as against 86.0 in control) and fruit weight (3.26 kg as against 2.11 kg in control) (Davide and Zorilla, 1986).

(ii) Medicinal Crops: *P. lilacinus* at 4 and 8g per kg soil significantly reduced *M. incognita* population in soil and root-knot index and increased number of leaves per plant, weight of 100 leaves, shoot and root weight of betel vine (Nakat *et al.*, 1995).

(iii) Spice Crops: *P. lilacinus* was most effective in the management of root-knot nematodes in black pepper vines (Sosamma and Koshy, 1995).

P. lilacinus suppressed *Meloidogyne* spp. population by 58.3-86.9 per cent in cardamom nursery. The number of quality seedlings for transplanting was greater in beds treated with this biocontrol agent (Eapen and Venugopal, 1995).

P. lilacinus was effective in increasing the plant growth parameters and in reducing root galling and egg mass production due to *M. incognita* in coriander (Sobita Devi and Pandey, 2001).

10.3.2. Pochonia chlamydosporia

The fungus *Pochonia chlamydosporia* is found in soil and can colonize roots of certain plants. Isolates of the fungus parasitize the eggs of root-knot nematodes, where they form robust chlamydospores. This fungus is a potential bio-control agent that destroys nematode life cycle. An increase in the number of fungal propagule and chlamydospores has been recorded on the roots of tomato plants that have become suppressive to the root-knot nematode infestation (Rao *et al.*, 2000). Different isolates of *P. chlamydosporia* vary in their bio-control efficacy and both the plant and nematode hosts were implicated in this interaction (Bourne *et al.*, 1994; de Leij *et al.*, 1993). The potential of *P. chlamydosporia* as a nematode bio-control agent will be increasingly important. Hence exploitation of this bio-control agent for the management of nematode diseases is gaining much importance in the recent years.

P. chlamydosporia is a very slow grower and it is very difficult to isolate from other group of fast growing fungi and saprophytes. Semi selective medium (SSM) has been developed (Kerry *et al.*, 1993) for the isolation of *P. chlamydosporia* from soil and roots infected by root-knot nematodes. This SSM contains benzimidazole fungicides- carbendazim and thiabendazole- to which *P. chlamydosporia* has relatively high levels of resistance compared with many other soil fungi.

Morgan-Jones *et al.* (1981) encountered *P. chlamydosporia* parasitizing *M. arenaria* females and the same species has been found in cysts of *Heterodera glycines* (Gintis *et al.*, 1983). *P. chlamydosporia* has been shown to be capable of preventing egg hatching of *M. arenaria* and to colonize eggs by hyphal penetration (Morgan Jones *et al.*, 1983).

(i) Vegetable Crops: Application of *P. chlamydosporia* at 50 g/m^2 in the nursery beds significantly increased the plant growth parameters (42 and 50

% increase in height and weight of seedlings, respectively over control) and root colonization by the bioagent on transplants. When the above seedlings were transplanted in field, there was reduction in root galling index (5.2 compared to 7.8 in control), number of egg masses in 5 g root (20 compared to 49 in control), number of juveniles in 5 g root (62 compared to 162 in control) and number of juveniles in 100 cc of soil (67 compared to 152 in control) and increased plant growth and yield in eggplant at harvest (Naik, 2004).

Bio-agent *P. chlamydosporia* and *Pseudomonas fluorescens* were evaluated under field conditions for their efficacy against root-knot nematode *M. incognita* infecting brinjal and capsicum. Treatment of the nursery bed with the formulation of *P. chlamydosporia* at 50g/m^2 was significantly effective in reducing the galling index, number of nematodes in roots and soil, increasing the percent parasitization of eggs by bioagent and also yields. Seed treatment with *P. fluorescens* alone and the nursery bed treatment with *P. chlamydosporia* alone were effective. However, their individual effect was maximized when both these bio-agents were integrated in the nursery bed stage. This could be because of combined effect of both bio-agents in the management of root-knot nematode (*M. incognita*) on this crop, under field conditions (Naik, 2004). The main purpose of these studies was to produce eggplant seedlings that are colonized by *P. chlamydosporia* before transplanting so that they could carry the bio-agent to the field. During this process the field soil would be enriched with the propagules of eco-friendly component of management (here it is a bio-agent) in two or three seasons.

10.3.3. *Trichoderma spp*

Trichoderma spp. have been described as antagonistic agents against various phytopathogenic fungi and nematodes. This ability makes *Trichoderma* a healthier, safer and environmentally friendlier biological choice, compared with polluting chemical pesticides, usually employed in agriculture. Mycoparasitism is one of the mechanisms involved in antagonistic processes, based on activity of extra cellular lytic enzymes produced by *Trichoderma* and has been proven as a good biological agent.

The main types of destructive activity of opportunistic fungi are thought to be enzymatic disruption of nematode egg shells and larval cuticles and physiological disturbances brought about by biosynthesis of diffusible toxic metabolites (Morgan Jones and Rodriguez Kabana, 1985).

The feasibility of using selected opportunistic soil fungi such as *P. lilacinus*, *P. chlamydosporia*, *V. lecanii*, *Trichoderma harzianum*, *T. viride* and *T. virens* have shown some effect in controlling *M. incognita*, *T. semipenetrans* and *R. reniformis*. Species of *Verticillium*, *Trichoderma* and *P. lilacinus* were recovered from females obtained in galls of plants.

Trichoderma harzianum: *T. harzianum* was found to be a potential bio-agent of root-knot nematodes (Rao *et al.,* 1997b, 1998e). *T. harzianum* suppressed *Meloidogyne* spp. by 58.3-68.9 per cent in cardamom nursery and increased the number of quality seedlings for transplanting (Eapen and Venugopal, 1995).

Table 10.2. Biological control of plant parasitic nematodes using fungi.

Fungi	Crop	Nematode	Reference
Paecilomyces lilacinus	Banana	*M. incognita*	Devarajan & Rajendran, 2002
		R. similis	Devarajan & Rajendran, 2001
	Citrus	*T. semipenetrans*	Manzoor *et al.*, 2002
	Tomato	*M. incognita*	Ekanayake & Jayasundara, 1994
			Cannayane & Sivakumar, 2001
			Sharma *et al.*, 1988
			Gautam *et al.*, 1995
		M. javanica	Zaki, 1998
			Khan & Esfahmi, 1990
		R. reniformis	Alan & Walters, 1994
			Parvatha Reddy & Khan, 1988
	Brinjal	*M. incognita*	Pandey & Dwivedi, 2001
			Saikia & Das, 2001
			Trivedi, 1992
			Rao *et al.*, 1998d
		M. javanica	Zaki & Maqbool, 1991
		R. reniformis	Parvatha Reddy & Khan, 1989
	Potato	*M. incognita* & *G. pallida*	Jatala *et al.*, 1980
		G. rostochiensis & *Pratylenchus* sp.	Davide & Zorilla, 1995
	Chilli	*M. incognita*	Pandey & Trivedi, 1992
	Okra	*M. javanica*	Walia *et al.*, 1991
		M. incognita	Ibrahim & Rezk, 1988
			Davide & Zorilla, 1986
	Carrot	*M. hapla*	Sivakumar, 1998
	Tuberose	*M. incognita*	Shylaja, 2004
	Crossandra	*M. incognita*	Nagesh & Parvatha Reddy, 1995
	Carnation	*M. incognita*	Shylaja, 2004
	Betel vine	*M. incognita*	Jonathan *et al.*, 1995
			Nakat *et al.*, 1995
		M. incognita &	Hazarika *et al.*, 1998

		R. reniformis	Jonathan *et al.*, 1995
		R. similis	Sosamma *et al.*, 1994
	Black pepper	*R. similis*	Sudha *et al.*, 2000 Sosamma & Koshy, 1995
	Cardamom	*Meloidogyne* spp.	Eapen & Venugopal, 1995
	Coriander	*M. incognita*	Midha *et al.*, 2001 Sobita Devi & Pandey, 2001
P. margaundi, Trichoderma harzianum	Crossandra, Tomato, Potato	*M. incognita* *M. incognita* *G. rostochiensis*	Khan & Parvatha Reddy, 1994 Nagesh *et al.*, 2001 Saifullah, 1996
	Okra	*M. incognita*	Lal *et al.*, 2002
	Tuberose, Carnation	*M. incognita*	Shylaja, 2004
T. viride	Banana	*R. similis*	Sosamma *et al.*, 1994
	Tomato	*M. incognita*	Rekha Arya & Saxena, 1998
		M. incognita	Xiujuan *et al.*, 2000 Nagesh *et al.*, 2001
	Okra	*M. incognita*	Lal *et al.*, 2002
T. virens	Tomato, Capsicum	*M. incognita*	Susan *et al.*, 2002a,b Eapen & Venugopal, 1995
Trichoderma spp.	Cardamom	*Meloidogyne* spp.	Gopinath *et al.*, 2002
Pochonia chlamydosporia	Tomato, Tuberose, Carnation	*M. incognita* *M. incognita*	Rao *et al.*, 1998b Shylaja, 2004
Verticillium lecanii	Crossandra	*M. incognita*	Nagesh & Parvatha Reddy, 1995
Cylindrocarpon destructus	Potato	*G. rostochiensis* & *G. pallida*	Morgon Jones & Rodriguez Kabana, 1985
Fusarium solani	Potato	*G. rostochiensis* & *G. pallida*	Morgon Jones & Rodriguez Kabana, 1985
F. oxysporum	Potato	*G. rostochiensis* & *G. pallida*	Morgon Jones & Rodriguez Kabana, 1985

Enrichment of FYM with *Trichoderma harzianum*

Commercial formulation of *T. harzianum* was mixed with moist FYM (1 kg/ton). Mixture was spread 15 cm high heap under shade, covered with polyethylene sheet and incubated for 3 weeks at 25 to 30°C and sprayed with water regularly to maintain moisture.

Colonized FYM can be used at 6 g/kg soil in pot culture experiments (12 g/2 kg soil in 15 cm diameter pots).

T. harzianum colonized compost significantly enhanced germination and growth of both tomato and okra as compared to non-colonized FYM. Population of *T. harzianum* in colonized compost was several folds higher (48.6×10^{12}) than non- colonized FYM (3×10^4).

Colonization of FYM by *T. harzianum* helps to improve the quality of compost. It significantly enhanced total and water soluble humic matter. A number of macro and micro-nutrients like P, K, S, Zn, Cu and Fe (both total and water soluble content) were significantly higher in *T. harzianum* colonized compost. Six times increase in water soluble humic content due to colonization of *T. harzianum* seems quite important as it is reported to enhance the plant growth.

Compatibility of Fungicides with Fungal Biological Control Agents

Copper oxychloride, Metalaxyl, Mancozeb 72 WP and Potassium Phosphonate were found compatible with native isolates of *Trichoderma* spp. and *Paecilomyces lilacinus.*

Carbendazim 50% WP and Captan 50% WP were found compatible with *T. harzianum* and *T. viride* (Gowder and Kulkarni, 1999).

Vydate and Nemacur were found compatible with *T. harzianum* (Stephen *et al.*, 1996).

Benomyl was not compatible with *Trichoderma* spp. (Saifullah, 1996).

T. harzianum was not sensitive to Mancozeb and highly sensitive to Carbendazim.

Carbendazim inhibited *Aspergillus* spp. and *T. harzianum,* while Mancozeb completely inhibited *Aspergillus* spp. but did not affect *T. harzianum*. This gives an indication to use Mancozeb in either liquid fermentation or solid fermentation at 1g/l, which facilitates to get the contamination-free biomass (Naik, 2004).

P. chlamydosporia has relatively high level of resistance to carbendazim and thiabendazole (Kerry *et al.*, 1993).

10.3.4. Multiple Fungi

(i) Ornamental Crops: An experiment carried out for the biological control of root-knot nematode *M. incognita* in carnation (*Dianthus caryophyllus*) showed that, though the growth parameters were found to be at par in plants treated with *P. lilacinus* and *P. chlamydosporia*, significantly higher reduction of root galling was achieved with *P. chlamydosporia.* Thus, it was concluded that under polyhouse conditions, *P. lilacinus* and *P. chlamydosporia* were better bio-agents for the control of root-knot nematode compared to *T. harzianum* (Shylaja, 2004).

The experiment carried out for the biological control of root-knot nematode (*M. incognita*) in tuberose (*Polyanthes tuberosa)* showed that, *P. chlamydosporia* and *T. harzianum* treatments gave significantly higher control of nematodes than *P. lilacinus* (Shylaja, 2004).

In course of the *in vitro* experiments conducted to determine the compatibility of the bio-agents *viz; T. harzianum, P. chlamydosporia* and *P. lilacinus* with the inorganic fertilizers, it was observed that urea encouraged the growth of *T. harzianum, P. chlamydosporia* and *P. lilacinus*. The increase in growth was observed to be directly proportional to the concentration of urea. Single Super Phosphate (SSP) at lower concentrations (below 500 ppm) encouraged the growth of *P. chlamydosporia*, while at concentrations above 500 ppm SSP inhibited the colony growth of *T. harzianum* and *P. lilacinus*. *T. harzianum* and *P. lilacinus* were compatible with Muriate of Potash (MoP), while *P. chlamydosporia* was incompatible with MoP. The growth of the bio-agents *T. harzianum* and *P. chlamydosporia* were highly inhibited with Calcium Ammonium Nitrate (CAN), while the growth of *P. lilacinus* was encouraged at low concentrations (100 ppm) (Shylaja, 2004).

10.4. Arbuscular Mycorrhizal Fungi (AMF)

The word mycorrhiza is derived from two Greek words (Mycos – fungus, rhiza – root). So it is called 'fungus root' which denotes existence of mutualistic symbiotic association between fungal hyphae and the roots of the plants. The fungal hyphae inside the roots of the plant develop some special organs called vesicles and arbuscules in the cortical cells. Vesicles are borne at the tip of hyphae growing intercellularly in the cortical region which serve as food storage organ as well as reproductive structures of the fungus. Arbuscles are dichotomously branched hyphae much like haustoria which grow intracellularly in the cortical cells which are major sites for bidirectional transfer of nutrients between the two symbionts. Since mycorrhiza possess vesicles and arbuscles, they are called Arbuscular Mycorrhizal Fungi (AMF). According to the present classification, AMF belongs to two Orders – Endogonales and Glomales (Morton and Benney, 1990).

Table 10.3. Taxonomy of arbuscular mycorrhizal fungi

Order	Sub-order	Family	Genera
Endogonales	---	Endogonaceae	*Endogone*
			Sclerogone
Glomales	Glomineae	Glomaceae	*Glomus*
			Sclerocystis
		Acaulosporaceae	*Acaulospora*
			Entrophospora
	Gigasporinae	Gigasporaceae	*Gigaspora*
			Scutellospora

Attributes of AMF

- Mobilize increased uptake of P in P-deficient soil saving P fertilizer to the tune of 25-30%.
- Helps in absorption of N, K, Ca, Mg, S, Zn and Cu.
- Extensive network of external hyphae increase absorptive surface area of the root system for nutrients especially P and water transport.
- Helps the plant to withstand draught conditions (high temperature).
- Decreases transplant injury of seedlings.
- Promotes establishment of plants in waste lands.
- Better equipped to withstand stress like draught, heavy metal pollution in industrial sectors and toxicity of Al, Mn or salts in soil.
- Promotes mine waste reclamation.
- Helps in improvement of soil structure and stabilization of organic status of the soil.
- Promotes growth of N fixing bacteria keeping in view the high P requirement by legumes and the improvement in root nodulation and N fixation in legumes in the presence of AMF, twin symbiotic associations within a legume plant from *Rhizobium* as well as AMF hold promise in integrated nutrient management systems.
- Promotes production of plant hormones.
- Plays major role in checking the severity of disease complex.
- Increases host tolerance/resistance by suppressing establishment and development of soil-borne plant pathogens.
- Stimulates the uptake of Zn, Cu, S and K by the plants, enhances nodulation in legumes, decreases rots caused by fungi and root penetration and development of nematode larvae.
- Plays a key role to have healthy plants on earth and sustainable system of agricultural production.

Arbuscular mycorrhizal fungi have shown antagonistic influence on the population of plant parasitic nematodes which may be due to increased amino acid contents (especially arginine) in the mycorrhizal roots (Nemec and Meridith, 1981) or the presence of additional amount of phosphorus in roots. AMF can markedly alter the plant response to nematodes.

(i) Fruit Crops

Banana: *G. mosseae* was found effective in reducing the *M. incognita* and *R. similis* population infesting banana besides enhancing plant height (18%), shoot weight (30.1%), number of leaves (2.5%), leaf area (30.7%), bunch weight (35.7%), number of hands (30.0%) and number of fingers (39.6%) (Sosamma *et al.*, 1998). Umesh *et al.* (1988) recorded low number of *R. similis* in mycorrhizal treated banana plants.

Citrus: When Rough lemon seedlings infected with *Tylenchulus semipenetrans* were transplanted into soil infested with *Glomus mosseae*, the mycorrhizal fungus infection increased seedling growth compared to non-mycorrhizal seedlings.

(ii) Vegetable Crops: The reduction in larval penetration of *R. reniformis* on mycorrhizal (*G. mosseae*) bush bean, cucumber and muskmelon was recorded by Sitaramaiah and Sikora (1982).

Potato: *G. fasciculatum* was found to reduce *Globodera rostochiensis* levels on potato.

Tomato: Tomato roots colonized by *G. fasciculatum* and infected by *M. incognita* or *M. hapla* exhibited fewer and smaller galls than did nematode infected non-mycorrhizal plants (Bagyaraj *et al.*, 1979). Small giant cells with few nuclei and retarded nematode development are observed in mycorrhizal tomato roots infected with *M. incognita* (Sikora, 1978). In addition to adverse effect on nematode penetration and development, there are indications that nematode reproduction is also suppressed on mycorrhizal plants.

Suresh and Bagyaraj (1984) reported that mycorrhizal inoculation caused reduced *M. incognita* infestation in tomato plants which had increased quantities of P, K, Ca, total and reducing sugars and amino acids (Phenyl alanine and Serine) which alone or collectively played a role in suppressing the development of *M. incognita* in mycorrhizal tomato roots.

Sitaramaiah and Sikora (1982) indicated that the penetration of *R. reniformis* decreased on mycorrhizal tomato plants. They also observed delayed development of gelatinous matrix with few eggs per egg sac.

G. fasciculatum was found to increase the resistance of tomato plants to *Rotylenchulus reniformis*.

Okra: Establishment of *G. fasciculatum* two weeks prior to *M. incognita* resulted in increased fresh and dry shoot weight of okra and reduced larval penetration, development, number of galls and egg masses per plant (Sharma and Trivedi, 1994).

Cowpea: The establishment of *G. fasciculatum* and *G. epigaeus* in the root system of cowpea plants caused an increase in plant resistance to *Heterodera cajani* infection (Jain and Sethi, 1988). They suggested that the reduction in penetration may be due to the fact that mycorrhizal roots were undesirable sites for larval penetration. Jain and Sethi (1988) reported reduction in number of galls of *M. incognita* and number of cysts of *H. cajani* on cowpea roots with the prior colonization of *G. fasciculatum*.

(iii) Ornamental Crops: *G. mosseae* was the most effective AMF fungi in reducing *M. incognita* population in roots and promoting growth of crossandra (Nagesh *et al.*, 1996b). The AMF colonization of roots mechanically may prevent nematode penetration and establishment, increased nutritional status, increased sugar content of roots (Smith, 1987), increased levels of P in roots or increased host tolerance.

(iv) Plantation Crops: A mixture of AMF consisting of multiple endophytes *viz. Acaulospora bireticulata, G. fasciculatum, G. macrocarpum,*

G. mosseae, G. versiforme, Sclerocystis rubiformis and *Scutellospora nigra* was found effective in improving plant growth and in reducing the *R. similis* population infesting coconut seedlings (Koshy *et al.*, 1998).

(v) Spice Crops

Black Pepper: The plant height, dry weight of shoot and root of black pepper plants were increased due to *G. etunicatum* inoculation (82, 60 and 104%, respectively over *M. incognita* infested plants). *G. mosseae* was more effective in suppressing nematode population in root (60%) and soil (33%) (Sivaprasad and Sheela, 1998).

Ginger and Turmeric: Root-knot nematode infestation and multiplication was generally low in arbuscular mycorrhizal fungus inoculated ginger and turmeric plants. *G. fasciculatum* was most effective followed by *G. mosseae* (Sivaprasad, 1995).

Although most of the research conducted under greenhouse conditions indicated that mycorrhizal fungi have the potential to nullify the damage due to phytonematodes, but intensive investigations need to be carried out under field conditions to assess their potential utilization as biocontrol agents. Since AMF fungi are obligate parasites, they can not be cultured on large scale which hinders their effective utilization in integrated nematode management programmes under field conditions.

Table 10.4. Biological control of plant parasitic nematodes using arbuscular mycorrhizal fungi

Arbuscular Mycorrhizal Fungi	Crop	Nematode	Reference
Glomus fasciculatum	Banana	*M. incognita*	Prakash Babu, 2000
			Suma *et al.*, 2002
			Vidya & Reddy, 1998
			Umesh *et al.*, 1988
		R. similis	Shanthi, 2003
			Prakash Babu *et al.*, 2001, 2002
			Prakash Babu, 2000
			Channabasappa *et al.*, 1995
	Grapevine	*M. arenaria*	Atilano *et al.*, 1982
	Potato	*G. rostochiensis*	Sikora, 1981
	Tomato	*M. incognita*	Mishra & Shukla, 1997
			Hemavathi, 2002
			Kantharaju *et al.*, 2002
			Bagyaraj *et al.*, 1979
			Suresh & Bagyaraj, 1984
			Nagesh *et al.*, 1999

		M. javanica	Sundarababu *et al.*, 1993
		R. reniformis	Sitaramaiah & Sikora, 1982
	Brinjal	*M. incognita*	Sharma & Trivedi, 1989
	Okra	*M. incognita*	Sharma & Trivedi, 1994
	Onion	*M. hapla*	Mac Guidwin *et al.*, 1985
	Cowpea	*H. cajani*	Jain & Sethi, 1988
		M. incognita	Jain & Sethi, 1988
			Pramila & Goswami, 1992
		R. reniformis	Lingaraju & Goswami, 1993
	Henbane	*M. incognita*	Pandey *et al.*, 1999
	Coconut	*R. similis*	Koshy *et al.*, 1998
	Black pepper	*M. incognita*	Anandraj *et al.*, 1991
G. mosseae	Banana	*M. incognita* &	Vidya & Reddy, 1998
		R. similis	Parvatha Reddy *et al.*, 1997
	Citrus (Rough lemon)	*T. semipenetrans*	O'Bannon *et al.*, 1979
	Peach	*Pratylenchus vulnus*	Pinochet *et al.*, 1995
	Tomato	*M. incognita*	Sikora, 1978
		M. incognita, M. hapla	Sikora & Schonbeck, 1975
	Brinjal	*M. incognita*	Borah & Phukan, 2000
			Rao & Parvatha Reddy, 2001
	Cucumber	*R. reniformis*	Sitaramaiah & Sikora, 1981
	Musk melon	*R. reniformis*	Sitaramaiah & Sikora, 1981
	Crossandra	*M. incognita*	Nagesh & Parvatha Reddy, 1996b
	Henbane	*M. incognita*	Pandey *et al.*, 1999
	Coconut	*R. similis*	Koshy *et al.*, 1998
	Black pepper	*M. incognita*	Sosamma & Koshy, 1995
			Shivaprasad & Sheela, 1998
			Anandraj *et al.*, 1991
G. macrocorpus	Coconut	*R. similis*	Koshy *et al.*, 1998
G. versiformae	Coconut	*R. similis*	Koshy *et al.*, 1998

G. aggregatum	Henbane	*M. incognita*	Pandey *et al.*, 1999
G. etunicatum	Citrus (Rough lemon)	*R. similis*	O'Bannon & Nemec, 1979
	Tomato	*M. incognita*	Bhagwati *et al.*, 2000
	Bean	*M. javanica*	Olivera & Zambolim, 1987
G. margarita	Peach	*M. incognita*	Strobel *et al.*, 1982
Acaulospora bireticulata	Coconut	*R. similis*	Koshy *et al.*, 1998
A. laevis	Black pepper	*M. incognita*	Anandraj *et al.*, 1991
Sclerocystis rubiformis	Coconut	*R. similis*	Koshy *et al.*, 1998
Scutellospora nigra	Coconut	*R. similis*	Koshy *et al.*, 1998
Gigaspora margarita	Black pepper	*M. incognita*	Anandraj *et al.*, 1991

Table 10.5. Commercially available AMF cultures.

Country	Commercial product
Canada	Mycorimix
	Nutria-link
Philippines	Mykovam 1
	Mykovam 2
Japan	Dr. Kinkon

Chapter 11

INTEGRATED NEMATODE MANAGEMENT

The aim of sustainable agriculture is to produce sufficient affordable food and fiber, economically and in an environmentally and socially sensitive manner, maintaining farmer's natural resource base for future generations. The demands of a growing world population for food and fiber require world agriculture to produce higher yields from cultivated land. Feeding future populations with today's yields is not viable; it would require a drastic expansion of planted acreage. In many parts of the world, additional land is unavailable. In others, an expansion of cropped area would be environmentally and socially unacceptable. To increase yields from existing land requires good crop protection against losses before and after harvesting. The challenge will be to do this without harming the environment and the resource base for future generations of farmers and consumers.

Integrated nematode management (INM) is an economically viable and socially acceptable approach to crop protection. INM is a system that, in the context of the associated environment and the population dynamics of the nematode species, utilizes all suitable techniques and methods in as compatible manner as possible and maintains the nematode populations below those causing economically unacceptable damage or loss. The criteria for judging the effectiveness of INM are productivity, stability, sustainability and equity. Sound knowledge of nematodes, cropping systems and the environment are important for the implementation and success of INM.

INM is an important principle on which sustainable crop protection can be based. The goal of INM is to achieve long-term sustainable systems of crop protection and production. The ideal objective of INM should be to maximize profit, while minimizing human health and environmental risks.

INM is the best combination of cultural, biological, chemical and regulatory measures that provides the most effective, environmentally sound and socially acceptable method of managing nematodes under the circumstances in which they work. Conserving the natural enemies of nematodes is recognized as an important part of INM programmes.

There is an urgent need to shift from the 'chemical intensive' to 'bio-intensive' cropping system with a view to restore the viability, sustainability and health of our agro-ecosystems. There is a need for development and adoption of INM in agriculture because of the economic and environmental benefits it brings to us. Major emphasis should be given to non-chemical methods of control, such as the biological, cultural, physical and regulatory. Nematode-resistant varieties of important crops must be evolved as these would be most useful tools of INM for resource poor farmers.

To alleviate nematode problems, INM procedures are implemented based on the principles of prevention, population reduction and tolerance. INM seeks to stabilize populations of target nematodes at acceptable levels resulting in favourable long term socio-economic and environmental consequences. In practice, integration of different methods of nematode control is the rule than the exception. In most cases, practical control of a nematode disease involves integration of several diverse control measures. In this context, biological control agents form ideal components in the INM along with botanicals and to limited extent chemicals.

11.1. Components / Tools of INM

INM strategies are based on 3 main components/tools:

- Prevention
- Monitoring/Observation/Regulation
- Intervention

11.1.1. Prevention

Prevention of nematodes through indirect measures is an important part of INM. The aim of prevention is to reduce the initial severity of nematode infestation. Most of the INM activities emphasize heavily on preventive measures especially before the crop is sown/planted *i.e.* at the field preparation or seed treatment stages since the initial nematode inoculum is largely responsible for the subsequent high nematode build up. Agronomic practices such as deep summer ploughing, soil solarization, cleansing of crop refuse, elimination of weed hosts, crop rotation, water management, intercropping/mixed cropping, trap cropping, nutrition management and genetic sources of resistance along with habitat management are the major tools of prevention.

Site Selection: Avoid growing a crop in nematode-infested area/sick plot.

Crop Sanitation: Reduces the nematode load.

Removal of Alternate Hosts: Breaks the nematode life cycle.

Land Preparation: Deep summer ploughing or soil solarization kills vulnerable stages of nematodes due to high temperatures.

Crop Rotation: Growing of non-host crops reduces the pest build up.

Use of Organic Amendments: Use decomposed FYM/vermicompost which enriches natural population of bioagents.

Timely Sowing: Avoidance of peak activity of the nematode e.g. planting during winter helps to avoid nematode attack.

Need Based Irrigation: Avoid excessive irrigation as it favours nematodes.

Biodiversity or Habitat Management: Crop mixes and variety mixes should be adopted in the cropping system. Follow intercropping/mixed cropping/trap cropping.

Host Resistance: Use nematode-resistant cultivars for planting.

11.1.2. Monitoring/Observation/Regulation

The aim of monitoring/observation/regulation is to determine when and what action to take. Regular monitoring is very crucial to nematode management as it is one of the most important decision making tools. An efficient monitoring programme can pay big dividends in lowering nematode management costs. An Agro-Eco-System Analysis (AESA) based on crop health at different stages of growth, population dynamics of nematodes and natural enemies, soil condition and farmer's past experiences are considered for decision making. Soil sample analysis for nematodes is usually employed as monitoring tool. Economic threshold levels (ETL) are now available to assist in proper timing of INM interventions.

- Quarantine of the pest against introduction of exotic pests.
- Domestic quarantine against further distribution e.g. potato cyst nematode in Nilgiri Hills of Tamil Nadu.
- Regulation on supply of quality bioagents.

11.1.3. Intervention

Once it is determined that intervention must be made, decisions must be made on the type of intervention which will reduce the effects of economically damaging nematode populations to acceptable levels. Mechanical, biological, cultural and chemical control measures are applied individually or in combination, taking into consideration costs, benefits, timing, available labour force, as well as ecological and environmental effects. Some of the principal INM interventions include cultural and physical measures, monitoring of nematodes, biocontrol agents (predators, parasitoids and pathogens), host plant resistance, botanicals and target specific, less hazardous chemical pesticides.

Mechanical and Physical Methods: a number of cultural and physical methods such as collection and destruction of diseased plants continue to be used. Recommendations should be practicable. The possibility of integrating appropriate cultural methods with judicious use of crop protection products should be explored (instead of broadcast application of nematicide, use bare-root dip of seedlings before transplantation, spot/row application).

Biological Methods: Use of bioagents for seed treatment or soil application with *Trichoderma viride, T. harzianum, Paecilomyces lilacinus, Pochonia chlamydosporia, Pseudomonas fluorescens.*

Monitoring and Surveillance: Nematode surveillance and decision making ought to be done either visually or through soil sampling. Similar monitoring of natural enemies is also desirable.

Botanicals: Use of neem/karanj cake for nematode management.

Chemical Methods: Nematode monitoring should be done for scheduling the chemical nematicide application. Nematicides should be used in right dosages, time and through appropriate equipment.

11.2. Integrated Nematode Management Methods

11.2.1. Bio-agents and Botanicals

(i) Fruit Crops

Banana: The root-knot and the burrowing nematode on banana were effectively managed by integration of neem cake at 500 g and FYM enriched with *T. harzianum* at 2 kg/plant. The above treatment increased the fruit yield to 45 kg/plant and bunches came to harvest 65 to 75 days earlier.

Integration of neem cake at 400 g per plant with *Pasteuria penetrans* at 100 g soil (300 spores/g soil) was found effective in reducing *R. similis* population both in soil and roots of banana by more than 50 per cent and increased plant growth parameters (Channabasappa *et al.*, 1995).

Significant reduction in population of *Helicotylenchus multicinctus* was observed in banana plants treated with *Pseudomonas fluorescens* at 2g/ *Glomus mosseae* at 25g along with press mud at 3 kg/plant. These treatments also enhanced the plant height, pseudo stem girth, number of leaves, leaf area and bunch weight (Jonathan and Cannayane, 2002).

Citrus: Management of the citrus nematode, *Tylenchulus semipenetrans* based on the application of neem cake/castor cake each at 10 g per plant along with an antagonistic fungus *Paecilomyces lilacinus* at 4 g (grown on paddy seeds) per plant was found to be extremely effective in reducing the citrus nematode population with a consequent increase in the growth of acid lime seedlings (Parvatha Reddy *et al.*, 1991a).

Trichoderma harzianum in combination with neem/karanj/castor oil cake was effective in increasing the growth of acid lime seedlings and reducing the citrus nematode population both in soil and roots. The parasitization of citrus nematode females with *T. harzianum* increased in the presence of the oil cakes (Parvatha Reddy *et al.*, 1996a). *Verticillium lecanii* in combination with the oil cakes was effective in increasing the growth of acid lime and in reducing the citrus nematode population both in soil and roots. Between the two parasitic fungi tested in combination with oil cakes,

V. lecanii performed better than *Pochonia chlamydosporia*. The parasitization of nematode eggs was also maximum when *V. lecanii* was used in combination with the oil cakes (Parvatha Reddy *et al.*, 1996b). This strategy forms the basis for the effective management of citrus nematode under field conditions with minimum amount of botanicals and bio-agents.

(ii) Vegetable Crops

Okra: Application of 5 per cent inoculum of *Arthrobotrys conoides* to the pot soil amended with FYM, effectively reduced the larval penetration of *M. incognita* and root galling was reduced by 30-40 per cent in okra (Srivastava and Swarup, 1986).

Neem cake suspension (5%) mixed with the spores of *P. lilacinus* is also effective for the treatment of okra seeds for the management of *M. incognita* under field conditions.

Tomato: Incorporation of neem cake/castor cake/neem leaf/castor leaf/pongamia leaf in nursery beds at 400 g/m^2 and application of spore suspension of *P. lilacinus* or *P. chlamydosporia* in tomato nursery beds amended with the above plant products facilitated the effective management of root-knot nematode, *M. incognita* and reniform nematode, *R. reniformis*. Further, root-dip treatment of seedlings in 5% aqueous suspension of neem leaf/ cake mixed with the spores of *P. lilacinus* or *P. chlamydosporia* for 20-30 minutes before transplanting was effective in the management of root-knot and reniform nematodes in the main field (Rao and Parvatha Reddy, 1992a, b; 1993; 1994; 1996; Rao *et al.*, 1994; 1997b,c).

Combined application of *P. penetrans* + *T. viride* + neem/castor cake at 1/3 doses was significantly superior compared to their individual applications in terms of increased plant growth and reduced root galling, egg mass production and final populations of *M. incognita* infecting tomato. Similarly, total parasitization by bioagents was higher in plants treated with these bioagents in terms of number of juveniles encumbered and adult females infected with *P. penetrans* and egg masses parasitized by *T. viride*, compared to their individual applications (Rangaswamy *et al.*, 1999).

Transplanting of tomato seedlings raised in nursery beds treated with neem cake and *P. penetrans* in pits incorporated with *P. lilacinus* gave effective management of *M. incognita* and increased tomato fruit yield (Parvatha Reddy *et al.*, 1997).

Soil amendment with crab shell meal and addition of *Catenaria anguillulae* reduced *M. incognita* population and significantly reduced egg mass production in tomato (Roy, 1984).

The integration of castor leaves with *P. lilacinus* increased the efficiency of nematode control (*M. javanica*) in tomato. The residual effect on the second crop of tomato revealed the reduction in root-knot index (29.6 to 356.8%) and J2 population in soil (26.4 to 57.9%) (Zaki, 1998).

Integration of neem cake with *Trichoderma harzianum* gave effective control of *M. incognita* infecting tomato and increased root colonization with biocontrol agent (Rao *et al.*, 1997c).

Brinjal: Root dip treatment of eggplant seedlings in *P. lilacinus* grown on neem cake extract resulted in significant increase in plant growth parameters and drastic reduction in root-knot index and final nematode population of *M. incognita*. Further, significant increase in root colonization, propagule density in soil and egg mass infection by *P. lilacinus* revealed the complimentary effect of these two components (*P. lilacinus* and neem cake extract) required for their effective use in sustainable management of *M. incognita* in eggplant (Rao *et al.*, 1998d, 1994).

(iii) Ornamental Crops

Crossandra: Combination of *V. lecanii* and *P. lilacinus* with neem leaf extract resulted in increased flower yield of crossandra. Root galling was least when neem leaf extract was integrated with *P. lilacinus* (Nagesh and Parvatha Reddy, 1995).

Tuberose: Treatment of tuberose bulbs with neem and calotropis leaf extracts mixed with *P. lilacinus* spores significantly reduced *M. incognita* infection and multiplication besides stimulating the plant growth (Nagesh *et al.*, 1997).

Carnation: Three antagonistic fungi viz., *T. harzianum*, *V. lecanii* and *P. lilacinus* at 2 x 10^4 spores/ml in combination with neem cake reduced *M. incognita* population in both soil and roots of carnation (Nagesh and Parvatha Reddy, 1996a).

Chrysanthemum: Field application of two formulations of *P. lilacinus* (talc and pesta granules) in combination with neem cake reduced root gall index, nematode populations and enhanced flower yield of chrysanthemum by 23 to 28% and increased the *P. lilacinus* spore viability in the rhizosphere for longer time (Nagesh *et al.*, 2001).

Identification of unique properties of the botanical extracts to support the growth of the promising bio-agents, has not only helped in evolving integrated methods by rational combination of bio-agents and botanicals, but also in standardizing the protocol for mass production of these bio-agents.

Table 11.1. Integrated nematode management in horticultural crops using bioagents and botanicals.

Crop	Nematode	INM Components	Reference
Citrus	*T. semipenetrans*	*P. lilacinus* + Neem/Castor cake	Parvatha Reddy *et al.*, 1991a
		T. harzianum + Neem/ Karanj/ Castor cake	Parvatha Reddy *et al.*, 1996a
		V. lecanii + oil cakes	Parvatha Reddy *et al.*, 1996b
		P. fluorescens + Neem cake	Parvatha Reddy *et al.*, 2000
Banana	*R. similis*	*P. penetrans* + Neem cake	Channabasappa *et al.*, 1995

Grapevine	*M. incognita*	*Phosphobacterium* + *Azospirillum* + Neem cake	Sivakumar & Vadivelu, 1999
Tomato	*M. incognita*	*P. chlamydosporia* + Neem leaves	Parvatha Reddy *et al.*, 1999
		T. harzianum + Neem cake	Rao *et al.*, 1997c
		T. viride + *P. penetrans* + Neem/Castor cake	Rangaswamy *et al.*, 1999, 2000
		P. lilacinus + Neem leaf extract	Parvatha Reddy *et al.*, 1998
		P. lilacinus + Castor leaf extract	Rao *et al.*, 1999
		P. lilacinus + Wellgro (neem product)	Rao & Parvatha Reddy, 1993a
		P. chlamydosporia + Neem cake	Rao *et al.*, 1998b
		P. lilacinus + *P. penetrans* + Neem cake	Parvatha Reddy *et al.*, 1997
		Oil cake based formulation of *T. harzianum*	Rao *et al.*, 1998c
		Catenaria anguillulae + Crab shell meal (Chitin)	Roy, 1984
Brinjal	*M. incognita*	*P. chlamydosporia* + Castor cake	Rao & Parvatha Reddy, 1993b
		Neem cake extract formulation of *P. lilacinus*	Rao *et al.*, 1998d
		P. lilacinus + Neem leaf suspension	Rao *et al.*, 1997b
		P. lilacinus + Neem cake extract	Rao *et al.*, 1998d
		P. lilacinus + Neem kernel	Sharma & Kamra, 2001
		Castor cake formulation of *T. harzianum*	Rao *et al.*, 1998d
Okra	*M. incognita*	*P. lilacinus* + Castor cake suspension	Rao *et al.*, 1997a
		Arthrobotrys conoides + FYM	Srivastava & Swarup, 1986
Tuberose	*M. incognita*	*P. lilacinus* + Neem leaf extract	Nagesh *et al.*, 1996

		P. lilacinus + Neem cake	Nagesh *et al.*, 1998
		P. lilacinus + *T. viride* / *T. harzianum* + Neem cake	Nagesh *et al.*, 1997
Crossandra	*M. incognita*	*V. lecanii* + Neem leaf extract	Nagesh & Parvatha Reddy, 1995
Chrysanthemum	*M. incognita*	*P. lilacinus* + Neem cake	Nagesh *et al.*, 2001
Carnation & Gerbera	*M. incognita*	*P. lilacinus* / *T. harzianum* + Neem cake	Nagesh & Parvatha Reddy, 1996a
Betel vine	*M. incognita*	*P. lilacinus* + Neem cake	Anon, 2003

11.2.2. Arbuscular Mycorrhizal Fungi and Botanicals

(i) Fruit Crops

Banana: Integration of karanj cake with *G. mosseae* gave maximum banana fruit yield, while neem cake + *G. mosseae* was most effective in reducing *Radopholus similis* population both in soil and roots (Parvatha Reddy *et al.*, 1997).

Citrus: Inoculation of endomycorrhiza, *Glomus fasciculatum* in the soils amended with neem cake was found to be effective in the management of *T. semipenetrans* on acid lime. This strategy can help in combating the menace of citrus nematode at nursery stage and also provide highly mycorrhizal seedlings of acid lime for transplanting in the main field for the management of *T. semipenetrans* under field conditions (Parvatha Reddy *et al.*, 1995).

(ii) Vegetable Crops

Tomato: Inoculation of *G. mosseae* or *G. fasciculatum* in the nursery beds amended with neem cake/castor cake/neem leaf/calotropis leaf reduced the infestation of root-knot and reniform nematodes to the maximum extent on tomato. Amendment of botanicals in the nursery beds increased the multiplication of endomycorrhizae and colonization of tomato root which in turn could protect the crop from these nematodes in the main field resulting in increased yields (Rao *et al.*, 1995; 1996; 1998).

Brinjal: Inoculation of *G. mosseae*/*G. fasciculatum* in the brinjal nursery beds and subsequent application of 5% aqueous extracts of neem cake/castor cake/neem leaf/calotropis leaf in the nursery beds resulted in the effective management of root-knot and reniform nematodes and yielded healthy brinjal seedlings which would withstand the attack of these nematodes after transplanting in the main field (Rao *et al.*, 1998).

Cowpea: Integration of chopped leaves of *Prosopis fuliflora* with *G. fasciculatum* to cowpea increased the spore production and root colonization of *G. fasciculatum* which resulted in reduced *M. incognita* population (Shanthi and Rajeswari, 1998).

(iii) Ornamental Crops: The root colonization of mycorrhizal fungi increased significantly in the presence of neem cake which in turn improved their efficacy in reducing *M. incognita* population in crossandra roots (Nagesh and Parvatha Reddy, 1996b).

Table 11.2. Integrated nematode management in horticultural crops using endomycorrhizae and botanicals.

Crop	Nematode	INM Components	Reference
Citrus	*T. semipenetrans*	*G. fasciculatum* + Neem cake	Parvatha Reddy *et al.*, 1995
Banana	*R. similis*	*G. mosseae* + Neem/Karanj cake	Parvatha Reddy *et al.*, 1997, 2002
Tomato	*M. incognita*	*G. mosseae* + Neem cake	Rao *et al.*, 1995 Parvatha Reddy *et al.*, 1998
		G. fasciculatum + Calotropis leaves	Rao *et al.*, 1996
		G. fasciculatum + Castor cake	Rao *et al.*, 1994, 1997
Brinjal	*M. incognita*	*G. fasciculatum* + Castor cake	Rao *et al.*, 1998
Crossandra	*M. incognita*	*G. mosseae* + Neem/Karanj/Castor cake	Nagesh & Parvatha Reddy, 1996b
		G. fasciculatum / *G. mosseae* + Neem cake	Nagesh *et al.*, 1998

11.2.3. Bio-agents and Arbuscular Mycorrhizal Fungi

(i) Vegetable Crops: Interestingly, integration of bio-agent (*P. lilacinus*) and arbuscular mycorrhizal fungi (*G. mosseae*/*G. fasciculatum*) have culminated into the successful management of *M. incognita* infecting tomato. This phenomenon facilitated standardization of a strategy wherein inoculation of mycorrhizae and bio-agent in the nursery beds protected the seedlings of tomato from the attack of *M. incognita*. Further, these mycorrhizal seedlings (colonized either with *G. mosseae* or *G. fasciculatum*) can be given a root-dip treatment with spore suspension of *P. lilacinus* for 5-10 minutes for the effective management of these nematodes in the main field after transplanting (Rao *et al.*, 1993). Integration of *Glomus deserticola* with *P. chlamydosporia* gave effective management of *M. incognita* on tomato seedlings and increased the egg parasitization by the bioagent (Rao *et al.*, 1997).

Table 11.3. Integrated nematode management in horticultural crops using bioagents and arbuscular mycorrhizal fungi.

Crop	Nematode	INM Components	Reference
Tomato	*M. incognita*	*P. lilacinus* + *G. mosseae*	Rao *et al.*, 1998
		P. chlamydosporia + *G. deserticola*	Rao *et al.*, 1997
		P. lilacinus + *G. fasciculatum*	Rao *et al.*, 1993
		P. penetrans + *G. mosseae*	Rao *et al.*, 2000
Brinjal	*M. incognita*	*P. lilacinus* + *G. mosseae*	Rao & Parvatha Reddy, 2001; Rao *et al.*, 1998
		P. chlamydosporia + *G. fasciculatum*	Rao *et al.*, 2003

11.2.4. Bio-agents, Arbuscular Mycorrhizal Fungi and Botanicals

(i) Vegetable Crops: After investigating the interactive effects of useful components of management (botanicals, bio-agents and endomycorrhizae), strategies to integrate these components for the management of root-knot and reniform nematodes infecting tomato and brinjal have been developed. These strategies include inoculation of *G. fasciculatum* in the castor cake amended nursery beds followed by the root-dip treatment of mycorrhizal seedlings of brinjal in spore suspension of *P. lilacinus* for the management of *R. reniformis* (Rao *et al.*, 1993). Similarly, inoculation of *G. mosseae* in neem leaf/neem cake amended nursery beds followed by the root-dip treatment of mycorrhizal seedlings of tomato in spore suspension of *P. lilacinus* was effective for the management of *M. incognita* under field conditions (Rao *et al.*, 1995a).

(ii) Ornametal Crops: Integration of a bioagent (*V. lecanii*), endomycorrhiza (*G. mosseae*) with botanicals improved the growth of crossandra and reduced the population of *M. incognita*. *G. mosseae* reduced the requirement of phosphatic fertilizer and favoured the antagonistic potential of *V. lecanii* against *M. incognita* (Nagesh and Parvatha Reddy, 1995).

Table 11.4. Integrated nematode management in horticultural crops using bioagents, arbuscular mycorrhizal fungi and botanicals

Crop	Nematode	INM Components	Reference
Tomato	*M. incognita*	*P. lilacinus* + *G. mosseae* + Neem leaf/ cake	Rao *et al.*, 1995a
Brinjal	*M. incognita*	*T. harzianum* + *G. fasciculatum* + Neem cake	Parvatha Reddy *et al.*, 1998
Okra	*R. reniformis*	*P. lilacinus* + *G. fasciculatum* + Castor cake	Rao *et al.*, 1993
	M. incognita	*P. fluorescens* + *G. fasciculatum* + FYM	Verma, 2001
Crossandra	*M. incognita*	*V. lecanii* + *G. mosseae* + Botanicals	Parvatha Reddy & Nagesh, 1995

11.2.5. Bio-agents and Chemicals

(i) Fruit Crops: An integrated management of the citrus nematode, *T. semipenetrans* infecting acid lime based on application of the parasitic fungus, *P. lilacinus* at 4 g per plant (grown on paddy seeds) along with carbofuran at 1 kg a.i./ha has been developed. It is found to substantially reduce the nematode population both in soil and roots and to increase plant growth parameters (Parvatha Reddy *et al.*, 1996c).

(ii) Vegetable Crops

Tomato: Integration of a bio-agent, *P. lilacinus* with carbofuran at 1 kg a.i./ha was found effective in the management of reniform nematode, *R. reniformis* infecting tomato (Parvatha Reddy and Khan, 1988).

Maheshwari *et al.* (1987) reported that application of *P. penetrans* in combination with carbofuran, aldicarb, miral, sebufos and phorate significantly improved plant growth of tomato by greatly reducing galling due to *M. javanica*.

Integration of *P. chlamydosporia* with carbofuran resulted in maximum plant growth parameters, minimum root-knot index and soil nematode populations. This combination also recorded maximum number of tomato fruits/plant and yield/plant. Maximum parasitization of root galls was recorded when *P. chlamydosporia* was integrated with carbofuran (Gopinath *et al.*, 2002).

Brinjal: Application of *Pseudomonas fluorescens* at 10g/ m^2 + carbofuran at 0.15g a.i./m^2 effectively controlled initial root-knot nematode (*M. incognita*) population in the nursery which in turn protected the brinjal crop in the main field also and increased the yield by 17 to 22% (Sheela *et al.*, 2002).

Integration of a bio-agent, *P. lilacinus* with carbofuran at 1 kg a.i./ha was found effective in the management of reniform nematode, *R. reniformis* infecting brinjal (Parvatha Reddy and Khan, 1989).

Cowpea: A significant increase in cowpea plant growth was observed when carbofuran was combined with *P. penetrans* over the corresponding treatments of carbofuran alone. The percentage reduction in cyst production of *H. cajani* was from 72.1 to 97.2 in treatments with carbofuran along with *P. penetrans* (1 x 10^7 spores/pot) compared to 44.9 to 71.4% in treatments with carbofuran alone. The study indicated the compatibility of *P. penetrans* with carbofuran, with no adverse effect on the growth of *P. penetrans* (Singh and Dhawan, 1998).

(iii) Ornamental Crops: Integrated management of root-knot nematode infecting crossandra was obtained by rational combination of bio-control agents *P. chlamydosporia/T. harzianum* with aldicarb at 0.5 kg a.i./ha (Parvatha Reddy and Nagesh, 1995).

(iv) Medicinal Crops: *P. lilacinus* at 4 g per plant + carbofuran at 0.75 kg a.i. per ha significantly reduced *M. incognita* population in soil and root-knot index and increased plant growth parameters of betel vine (Nakat *et al.*, 1995).

Table 11.5. Integrated nematode management in horticultural crops using bioagents and chemicals.

Crop	Nematode	INM Components	Reference
Citrus	*T. semipenetrans*	*P. lilacinus* + Carbofuran	Parvatha Reddy *et al.*, 1996c
Tomato	*M. incognita*	*P. chlamydosporia* + Carbofuran	Gopinath *et al.*, 2002
	M. javanica	*P. penetrans* + Carbofuran/ Aldicarb/ Phorate	Maheshwari *et al.*, 1987
	R. reniformis	*P. lilacinus* + Carbofuran	Parvatha Reddy & Khan, 1988
Crossandra	*M. incognita*	*P. chlamydosporia* + Aldicarb *T. harzianum* + Aldicarb	Parvatha Reddy & Nagesh, 1995
Carnation & Gerbera	*M. incognita*	*P. lilacinus*/ *T. harzianum* + Fenamiphos	Nagesh & Parvatha Reddy, 1996a
Betel vine	*M. incognita*	*P. lilacinus* + Carbofuran	Nakat *et al.*, 1995

11.2.6. Arbuscular Mycorrhizal Fungi and Chemicals

(i) Ornamental Crops: Integration of *G. mosseae*/*G. fasciculatum* with aldicarb at 0.5 kg a.i./ha gave effective management of root-knot nematodes infecting tuberose (Parvatha Reddy and Nagesh, 1995).

Table 11.6. Integrated nematode management in horticultural crops using endomycorrhiza and chemicals.

Crop	Nematode	INM Components	Reference
Tuberose	*M. incognita*	*G. fasciculatum* / *G. mosseae* + Aldicarb	Parvatha Reddy & Nagesh, 1995

11.2.7. Two or more Bio-agents

(i) Fruit Crops

Grapevine: The combined application of bioagents *P. lilacinus* and *P. penetrans* was most effective in increasing the fruit yield of grapevine besides checking *M. incognita* with cost: benefit ratio of 1:3.3 (Sivakumar and Vadivelu, 1999).

Papaya: Nursery bed treatment with *T. harzianum* and *P. lilacinus* each at 5 or 10 g/kg soil resulted in production of highly vigorous papaya seedlings whose roots were colonized with both the bioagents. There was significant reduction in root galling in the combination treatments (2.9 to 3.2) compared to control (8.9) (Rao and Naik, 2003).

(ii) Vegetable Crops

Tomato: In nursery beds, *P. chlamydosopria* in combination with *P. penetrans* was effective in increasing plant growth parameters of tomato seedlings. The seedlings were properly colonized with *P. chlamydosporia* and *M. incognita* females in seedlings were also adequately infected with *P. penetrans*. The above tomato seedlings when transplanted in the main field had significantly lower root-knot index, number of eggs per egg mass, nematode population in roots and soil. The above treatment also increased root colonization and egg parasitization with *P. chlamydosporia,* infection of *M. incognita* females with *P. penetrans* and increased tomato fruit yield. This method of integrated nematode management resulted in significant reduction in the amount of inocula of the bioagents required for the treatment of soil only in the nursery beds (Rao *et al.*, 1998a).

Integration of two bio-agents, *P. lilacinus* and *P. chlamydosopria* resulted in combined and complimentary effects for the successful management of *M. incognita* infecting tomato (Rao and Parvatha Reddy, 1992c).

Maheshwari and Mani (1988) observed that population of *M. incognita* and *M. javanica* were suppressed effectively and yields of tomato were greater when *P. penetrans* and *P. lilacinus* were applied together.

Brinjal: The nursery seedling stand index was good with seed treatment with *P. fluorescens* (50 g/kg) + soil application (10 g/m^2 seed bed) followed by soil application of neem based *T. harzianum* + *P. fluorescens* and neem based *P. fluorescens*, where the crop stand index was 4.6 and 4.0, respectively.

Chilli: Root dip treatment with *Pseudomonas aeruginosa* with or without *T. harzianum*, *T. koningii* and *T. hamatum* significantly controlled infection of roots by *F. solani*, *R. solani* and *M. javanica* on *Capsicum annum* (chilli). Combined use of *T. harzianum* with *P. aeruginosa* caused the greatest reduction in gall formation by *M. javanica*. Infection of *M. phaseolina* and *R. solani* was comparatively greater in *M. javanica* infested soil than in non-infested soil, whereas *F. solani* infection was greater in non-infested soil. *P. aeruginosa* mixed with *Trichoderma* spp. also significantly increased plant growth (Siddiqui *et al.*, 1999).

Capsicum: Integration of *P. fluorescens* and *P. chlamydosporia* in the nursery bed has proved significantly effective in reducing the root-galling index, number of nematodes in the roots and soil and increasing the yield of the capsicum crop (Naik, 2004).

In capsicum, the seedling stand was good where the combination of neem based *P. fluorescens* and *T. harzianum* was used (4.4), followed by seed treatment with *P. fluorescens* + soil application of *T. harzianum* (Naik, 2004).

(iii) Ornamental Crops: Combined application of *P. lilacinus* and *V. lecanii* enhanced plant growth and flower yield of crossandra besides reducing root galling due to *M. incognita* (Nagesh and Parvatha Reddy, 1995).

(iv) Spice Crops: Black pepper vines combinedly inoculated with *P. lilacinus* and *P. penetrans* had put out 23-112 per cent increase in plant growth over control and were very effective in the management of root-knot nematodes (Sosamma and Koshy, 1995).

Table 11.7. Integrated nematode management in horticultural crops using two bioagents

Crop	Nematode	INM Components	Reference
Citrus	*T. semipenetrans*	*P. lilacinus* + *P. penetrans*	Parvatha Reddy & Nagesh, 2000
Grapevine	*M. incognita*	*P. lilacinus* + *P. penetrans*	Sivakumar & Vadivelu, 1999
Papaya	*M. incognita*	*T. harzianum* + *P. lilacinus*	Rao & Naik, 2003
Tomato	*M. incognita*	*P. lilacinus* + *P. chlamydosporia*	Rao & Parvatha Reddy, 1992c
		P. chlamydosporia + *P. penetrans*	Rao *et al.*, 1998a
	M. javanica & *M. incognita*	*P. lilacinus* + *P. penetrans*	Maheshwari & Mani, 1988
Capsicum	*M. incognita*	*P. chlamydosporia* + *P. fluorescens*	Naik, 2004
Crossandra	*M. incognita*	*P. lilacinus* + *P. penetrans*	Nagesh *et al.*, 1995
		P. lilacinus + *V. lecanii*	Nagesh & Parvatha Reddy, 1995
Black pepper	*M. incognita*	*P. lilacinus* + *P. penetrans*	Sosamma & Koshy, 1995

11.2.8. Bioagents and Physical Methods

(i) Spice Crops: Soil solarization of nursery beds and application of bioagents such as *P. lilacinus* or *Trichoderma* spp. improved growth of cardamom seedlings by suppressing root-knot nematode population (Eapen and Venugopal, 1995).

Table 11.8. Integrated nematode management in horticultural crops using bioagents and physical methods.

Crop	Nematode	INM Components	Reference
Cardamom	Root-knot	Soil solarization + *P. penetrans*	Eapen & Venugopal, 1995

11.2.9. Cultural and Chemical Methods

(i) Vegetable Crops

Tomato: Integration of 3 summer ploughings at 10 days interval during May – June with spot application of aldicarb at 1.0 kg a.i./ha while transplanting healthy tomato seedlings of 8-weeks-old gave effective control of *M. javanica* (85.5 to 93.7% reduction in nematode population) and gave maximum fruit yield (45% increase over control) of tomato (Jain and Bhatti, 1989).

Treatment combination of deep ploughing (20 cm deep) during June (atmospheric temperature between 42–47°C) (92.3% reduction in *M. javanica* population) + aldicarb treated nursery (0.4 g a.i./m²) + aldicarb application at 1 kg a.i./ha at transplanting of tomato seedlings gave maximum yield (7.8 kg/plot) with minimum root-knot index (1.3) and final population in comparison to 3.8 kg/plot fruit yield in check (Jain and Gupta, 1990).

Brinjal: Nursery treatment with carbofuran at 0.3g a.i./m² along with main field treatment with ploughing + exposing the field and ploughing + covering with polythene sheets for 15 days improved the plant growth and yield of brinjal by 20-32% (Sheela *et al.*, 2002).

Okra: Integration of deep summer ploughing (20 cm) in June (76 to 79% reduction in *M. javanica* population) with application of aldicarb at 1 kg a.i./ha while sowing okra seeds gave significantly higher yield and lower root-knot index than control (Jain and Gupta, 1990).

Table 11.9. Integrated nematode management in horticultural crops using cultural and chemical methods.

Crop	Nematode	INM Components	Reference
Tomato	*M. javanica*	Deep summer ploughing + Aldicarb	Jain & Bhatti, 1989 Jain & Gupta, 1990
		Soil solarization + Carbofuran	Dwivedi *et al.*, 2001
Okra	*M. javanica*	Deep summer ploughing + Aldicarb	Jain & Gupta, 1990

11.2.10. Botanicals and Chemical Methods

(i) Vegetable Crops

Tomato: Integration of chopped castor leaves (40 to 60 g/kg soil) 1 or 2 weeks before transplanting with application of aldicarb or carbofuran each at 2 kg a.i./ha at transplanting, significantly reduced number of galls due to *M. javanica* and enhanced the growth of tomato (Ravi Dutt and Bhatti, 1986).

Integration of soil application of castor leaves with inorganic fertilizer (75 kg N/ha) enhanced plant growth of tomato and reduced *M. javanica* infestation (95% and 78% reduction in root galls and egg mass production, respectively) (Zaki and Bhatti, 1989).

Application of neem cake in the nursery at 100 g/ m² followed by carbofuran at 1 kg a.i./ha in the main field significantly reduced the soil population of *R. reniformis* and enhanced the fruit yield of tomato by 67% (Anitha and Subramanian, 1998).

Okra: Application of aldicarb at 1 kg a.i./ha + neem cake at 0.5 ton/ha followed by carbofuran at 1 kg a.i./ha + neem cake at 0.5 ton/ha proved most effective in reducing the *R. reniformis* population and in increasing the growth of okra plants (Krishna Rao *et al.*, 1987).

Application of subabool (*Leucaena leucophila*) leaves at 40 g/kg soil which are allowed to decompose for four weeks before sowing + carbofuran at 1 kg a.i./ha while sowing of okra seeds resulted in minimum galling (35.4 galls/plant) in comparison to control (64 galls/plant) and better plant growth parameters in okra (Paruthi *et al.*, 1987).

(ii) Ornamental Crops: Application of phorate at 4 g a.i. per plant during May and September months and incorporation of 20 kg FYM per plant increased flower yield of jasmine by 50% and reduced the root-knot nematode population by 70% (Sundarababu, 1992).

(iii) Plantation Crops: Integration of neem oil cake application at 1 kg/plant with phorate at 15 g a.i./plant gave effective control of *R. similis* in arecanut. In arecanut + banana + black pepper cropping system, integration of phorate at 15 g a.i./plant with neem oil cake at 500 g/plant was found most effective in reducing the nematode population (Sudha and Sundararaju, 1998).

Table 11.10. Integrated nematode management in horticultural crops using botanicals and chemicals.

Crop	Nematode	INM Components	Reference
Tomato	*M. javanica*	Castor leaves + Aldicarb/ Carbofuran	Ravi Dutt & Bhatti, 1986
		Castor leaves + 75 Kg N/ha	Zaki & Bhatti, 1989
	R. reniformis	Neem cake + Carbofuran	Anitha & Subramanian 1998
Okra	*M. incognita*	Neem cake + Carbofuran	Parvatha Reddy & Khan 1991
	R. reniformis	Achook (Azadirachtin) + Carbofuran	Singh *et al.*, 1998
		Neem cake + Aldicarb/ Carbofuran	Krishna Rao *et al.*, 1987
		Subabool leaves + Carbofuran	Paruthi *et al.*, 1987
Jasmine	*M. incognita*	FYM + Phorate	Sundarababu, 1992
Arecanut	*R. similis*	Neem cake + Phorate	Sudha & Sundararaju, 1998

11.2.11. Cultural and Physical Methods

(i) Fruit Crops: Integration of paring of banana suckers and hot water treatment was most effective in reducing *R. similis* population and improving the plant growth (Ravichandra and Krishnappa, 1985).

Table 11.11. Integrated nematode management in horticultural crops using physical and cultural methods.

Crop	Nematode	INM Components	Reference
Banana	*R. similis*	Paring of suckers + hot water treatment	Ravichandra & Krishnappa, 1985

11.2.12. Bioagents, Botanicals and Physical Methods

(i) Vegetable Crops: Better control of *Globodera rostochiensis* by *P. penetrans* was possible with soil solarization combined with application of FYM or karanj cake which resulted in greater kill of nematodes (Sitaramaiah and Naidu, 2003).

Table 11.12. Integrated nematode management in horticultural crops using bioagents, botanicals and physical methods.

Crop	Nematode	INM Components	Reference
Potato	*G. rostochiensis*	Soil solarization + FYM/ Karanj cake + *P. penetrans*	Sitaramaiah and Naidu, 2003

11.2.13. Using Bioagents, Botanicals and Chemicals

Table 11.13. Integrated nematode management in horticultural crops using bioagents, botanicals and chemicals.

Crop	Nematode	INM Components	Reference
Banana	*R. similis*	*T. viride* + Neem cake + Carbofuran	Ravi *et al.*, 2000
Carnation & Gerbera	*M. incognita*	*P. lilacinus* + *P. chlamydosporia* + Neem cake + Dazomet	Nagesh & Parvatha Reddy, 2001

11.2.14. Using Physical, Cultural, Chemicals and Botanicals

Table 11.14. Integrated nematode management in horticultural crops using physical, cultural, chemical methods and botanicals.

Crop	Nematode	INM Components	Reference
Banana	*M. incognita*	Paring of suckers + hot water treatment +Carbofuran + Neem cake	Mohanty & Mohapatra, 2001

11.2.15. Using Bioagents, Botanicals and Cultural Methods

Table 11.15. Integrated nematode management in horticultural crops using bioagents, botanicals and cultural methods.

Crop	Nematode	INM Components	Reference
Tomato	*M. incognita*	*T. viride* + *P. penetrans* + Neem cake + Crop rotation with marigold	Parvatha Reddy *et al.*, 2000

11.2.16. Using two Botanicals

Table 11.16. Integrated nematode management in horticultural crops using two Botanicals

Crop	Nematode	INM Components	Reference
Henbane	*M. incognita*	Neem cake + *Murraya konengii* / *Adathoda vasica* / *Mentha arvensis*	Pandey, 1997

11.2.17. Using Arbuscular Mycorrhizal Fungi and Cultural Methods

Table 11.17. Integrated nematode management in horticultural crops using endomycorrhiza and cultural methods.

Crop	Nematode	INM Components	Reference
Brinjal	*M. incognita*	*G. mosseae* + Crop rotation with pearl millet	Jothi & Sundarababu, 2001

In the above manner, INM strategies developed are eco-friendly and aim at efficient utilization of promising bio-agents, botanicals, chemicals, physical methods and arbuscular mycorrhizal fungi for obtaining higher yields in horticultural crops by suppressing parasitic nematodes.

11.3. Low Input Sustainable INM Strategy for Low and Medium Value Crops

11.3.1. Summer

- Two or three deep summer ploughings with a soil turning harrow at fortnightly intervals, preferably with a light irrigation between two ploughings.
- Soil solarization of nursery beds or pit soil using clear thin polyethylene mulch for 3-6 weeks before sowing.
- Rabbing of nursery beds (burning of crop residues over the soil surface), preferably after turning of soil, before sowing.
- Application of FYM/compost enriched with *Trichoderma harzianum* / *Paecilomyces lilacinus* / *Pochonia chlamydosporia* to nursery beds.
- Growing of non-host green manure/antagonistic crops/trap crops and incorporating in the soil at the time of flowering.
- Ploughing back non-commercial crop residues into the soil before monsoon.

11.3.2. Monsoon

- Growing of non-host green manure/antagonistic crops/trap crops and incorporating in the soil at the time of flowering.
- Application of organic amendments (FYM/compost/oil cakes/crop residues) to the soil while preparing the land.

11.3.3. *Kharif*

- Growing of non-host or nematode-resistant crop cultivars.
- Application of FYM/compost enriched with *Trichoderma harzianum* / *Paecilomyces lilacinus* / *Pochonia chlamydosporia* to nursery beds.
- Uprooting and burning of roots of host crops and weeds after harvest.

11.3.4. *Rabi*

- Delaying sowing/planting to mid-November (when soil temperature falls to 15-18°C).
- Restricting of growing susceptible or tolerant crops to *rabi* season with a suitable period of rotation.
- Removal and burning of non-commercial but disease-free crop residues.

Part - II

PRACTICES FOR INTEGRATED NEMATODE MANAGEMENT IN HORTICULTURAL CROPS

Chapter 12

FRUIT CROPS

12.1. BANANA

12.1.1. The Burrowing Nematode, *Radopholus similis*

(i) Physical methods: To avoid introducing inoculum into a new plantation, sets may be disinfected by paring and heat therapy. Corms of suckers should be trimmed carefully to remove the nematode affected necrotic tissue. Deep paring of infected banana suckers used for planting helps to reduce nematode population. The pared sets can be disinfected by dipping them in a hot water bath at 55°C for 25 minutes or 43°C for 30 minutes without causing significant losses. Sun drying the rhizomes before planting has been shown to be effective in the control of nematodes, which is being commonly practiced in Kerala.

(ii) Cultural Methods: Nematode-free suckers should be selected for planting. The land should be ploughed thoroughly during summer and left fallow or planted with cover crops/green manure crops (sunnhemp) not affected by nematodes for at least 6 months. Sugarcane has been shown to be a good alternate crop.

The number of nematodes in soil may be decreased either by fallowing, crop rotation or intercropping. The banana biotype of *R. similis* has a narrow host range and limited survival period in soil in the absence of host. Therefore, destruction of the infested area either mechanically with a bulldozer or by injection of a herbicide into the pseudo stem and replanting the area with nematode-free sets seems a practical approach to rehabilitate the infested plantations. After destroying an infested plantation, planting of cover crops like *Panicum maximum* var. *trichoglume* and *Phaseolus atropurpureus* did not support *R. similis* population. Planting of sugarcane or pangola grass (*Digitaria decumbens*) following the destruction of bananas eliminated *R. similis* after 10 weeks (Loos, 1961).

Fallowing for a period of 3 months after banana effectively suppressed the burrowing nematode population. Flood fallowing for about 5 months destroyed not only the *Fusarium* but also the burrowing nematodes (Rajendran *et al.*, 1979).

Rotation of banana with paddy, sugarcane, green gram, cotton, pangola grass, sorghum, tobacco, cassava or turmeric suppressed the nematode population and increased the yield (Rajendran *et al.*, 1979). Crop rotation with rice in wet lands and with vegetables, cotton and cereals in garden lands helps to reduce the nematode population. Rice or green gram after banana suppressed the population of *R. similis, P. coffeae* and *H. multicinctus*.

The nematode population both in soil and roots were considerably reduced when banana was intercropped with nematode antagonistic plants such as marigold. *Crotalaria juncea* was most effective in reduction of soil nematodes and *Coriandrum sativum* was more effective in reduction of root nematodes. *C. juncea* gave maximum bunch weight (16.7 kg) recording 51% more than control (11.03 kg) followed by *Tagetes errecta* (16.2 kg), *Sesamum indica* (15.87 kg), *Acorus calamus* (15.67 kg) and carbofuran (15.23 kg) (Charles and Venkitesan, 1993).

Intercropping with sunnhemp, coriander, marigold, radish or lucerne reduced nematode population and increased yield (Vadivelu *et al.*, 1987). Intercropping sunnhemp in banana field and ploughing the green manure 45 days after sowing was found effective in reducing *R. similis, P. coffeae* and *H. multicinctus* population by 38.4%, followed by marigold and cowpea which recorded 29.0 and 22.3% reduction, respectively (Shanthi *et al.*, 2001). Banana intercropped with *Crotalaria* reduced *R. similis* population in root and increased plant growth and yield (Charles *et al.*, 1985). Lakshmana Murthy (1983) reported that when banana was intercropped with betel vine, ginger, papaya and marigold, *R. similis* was not encountered. Subramaniam and Selvaraj (1988) reported that intercropping with *Tagetes, Crotalaria* and radish significantly reduced the *R. similis* population.

Mulching with black polythene at 20 per cent moisture depletion recorded highest yield with low population of *R. similis* (Vadivelu *et al.*, 1987). Covering soil with black polyethylene, sugarcane trash and banana trash are reported to reduce *R. similis* and *Pratylenchus* sp. in banana roots (Bhattacharya and Rao, 1984).

Botanicals: Application of neem cake at 800 g per plant (two split applications) or neem coated urea at 110 g per plant reduced the population of the burrowing nematode and also increased the yield of banana (Vadivelu *et al.*, 1987). Neem formulations like Econeem and Nimbicidine showed maximum efficacy in reducing the nematode population with increased plant growth and yield.

Application of press mud obtained from sugarcane waste treated plants has shown significant reduction in nematode populations (Sundararaju *et al.*, 2004).

(iii) Biological Methods: Soil application of *Pseudomonas fluorescens* at 20 g/plant (2.5 x 10^8 cfu/g) was effective in increasing the plant growth parameters (pseudo stem height and girth, number of leaves, root length and weight) and in reducing the nematode population in soil (62.34, 67.12 and

58.87% of *R.similis, P. coffeae* and *H. multicinctus*, respectively) and roots (60.35, 62.00 and 56.54% of *R.similis, P. coffeae* and *H. multicinctus*, respectively). Rhizosphere treatment with *P. fluorescens* at 20 g/250 ml water along with soil treatment with 100 g powder/rhizome gave 56.5% reduction in nematode population. During the second month the nematode reduction was 63.8% (Nirmal Johnson and Devarajan, 2004). Application of fungal antagonists like *Paecilomyces lilacinus, Trichoderma viride, Pochonia chlamydosporia* and bacterial antagonists like *Pasteuria penetrans* are effective in reducing the nematode population in soil and roots of banana.

In an interaction study of arbuscular mycorrhizal fungus (AMF), *Glomus fasciculatum* and *R. similis*, the multiplication of nematodes in the root was suppressed and soil population was significantly lower. The banana plants inoculated with mycorrhiza also recorded higher N, P, K, Ca, Mg, sugars, phenols and amino acids in their root system suggesting AMF as a potential biocontrol agent in the management of *R. similis* (Umesh *et al.*, 1988). Inoculation of *G. mosseae* during raising of banana nursery was found effective in reducing the nematode populations of *R. similis* and *M. incognita* besides enhancing plant growth and bunch weight under field conditions. There was 35.7% increase in bunch weight, 30% increase in number of hands and 39.6% increase in number of fingers in the mycorrhizae colonized plants (Sosamma *et al.*, 1998).

Soil application of biofertilizer, *Azospirillum* at 20 g/plant had significant effect in enhancing the plant growth and reduced the population of *R. similis, P. coffeae* and *H. multicinctus* in soil to an extent of 50.1, 45.5 and 48.9%, respectively over control (Shanthi *et al.*, 2004).

(iv) Host Resistance: Banana cvs./hybrids viz., Gold Finger, Kadali, Octoman, Then Kunnan, Kunnan, Dulmal Bhog, Malbhug, Tongat, Pedalimoongil, Pisang Lilin, Pisang Jari Buaya, Poovan, Anil Vazhai, Velatothan, Dudhsagar, Chinachampa, Chinia Kurlan, Benid, Ayiramkapoovan, Pisang Seribu, Vennettukunnan, Veneetu Mannan, Palayankodan, Beecha, Brilechinia, Local Dwarf, Manick Champa, Singhlal, Pey Kunnan, Elakkibale, Paka, Annaikomban, Bodles Altafort and Hybrid-74 (Matti x Pisang Lilin) were found tolerant/resistant to *R. similis* (Parvatha Reddy *et al.*, 1989; Ravichandra and Krishnappa, 1985; Charles *et al.*, 1983; Vadivelu *et al.*, 1987; Sundararaju *et al.*, 2004). Pisang Jari Buaya from Malaysia, is known to be resistant to *R. similis* (Pinochet and Rowe, 1979).

(v) Integrated Methods: Integration of neem cake at 400 g per plant with *P. penetrans* at 100 g soil (300 spores/g)/*T. harzianum* / *T. viride* at 250 g / plant while planting was found effective in reducing nematode population both in soil and roots of banana by more than 50 per cent and increased plant growth parameters. The treatment should be repeated 4 months after planting. The root-knot and the burrowing nematode on banana were effectively managed by integration of neem cake at 500 g and FYM enriched with *T. harzianum* at 2 kg/plant. The above treatment increased the fruit yield to 45 kg/plant and bunches came to harvest 65 to 75 days earlier.

Integration of neem cake (at 200 g/plant) with *Glomus mosseae* (100 g/plant with 25-30 chlamydospores/g) was most effective in reducing the burrowing nematode population both in soil and roots, while karanj cake with *G. mosseae* gave maximum increase in plant growth parameters and fruit yield of banana. Mycorrhizal root colonization and number of chlamydospores of *G. mosseae* were maximum in neem cake amended soil (Parvatha Reddy *et al.*, 2002).

Soil application of 2 kg FYM with *P. fluorescens* (with 1 x 10^9 spores/g) and *P. chlamydosporia* (with 1 x 10^6 spores/g) per plant at the time of planting and at an interval of 4 months significantly reduced the burrowing nematode by 64% compared to control.

The following INM measures are suggested for the effective management of the burrowing nematode:

- Mulching with black polythene at 20 per cent moisture can reduce the populations of nematodes.
- Intercropping with sunnhemp, coriander, marigold or lucerne can reduce nematode population in the soil.
- Fallowing for a period of 3 months after banana can suppress the burrowing nematode population.
- Flood fallowing for about 5 months destroy not only the Panama wilt fungus - *Fusarium* but also the nematodes.
- Use of resistant/tolerant cultivars.
- Use of nematode-free certified planting material.
- Increased use of green manure and other organic amendments.
- Crop rotation with paddy, sugarcane and coriander.
- Application of neem cake at 800 grams per plant (two split applications - while planting and 1 month after planting) can reduce the population of the burrowing nematode and also increase the yield of banana.
- Paring and pralinage with carbofuran at 5 grams per sucker reduce the incidence of nematode. For doing this, chop the rhizome portion thoroughly. Prepare mud slurry and dip this chopped portion of rhizome of the sucker in the mud slurry and sprinkle 5 grams of carbofuran on the rhizome.
- Dipping of banana suckers in 0.1 per cent aldicarb solution increase plant height, bunch weight, fingers per bunch, induce early flowering and effectively reduce the burrowing nematode population.
- Application of bio-control agents like *Paecilomyces lilacinus* and *Trichoderma harzianum* to the plants along with neem cake and farm yard manure (FYM) is beneficial. One kg of each of these bio-agents can be used to enrich 1 ton of farm yard manure. Farm yard manure added with these bio-agents should be left under shade for a period of 15 days and in between it is advisable to mix FYM thoroughly at an interval of 3 days to enrich FYM with the bio-agents. This enriched

FYM can be added to the pit at 1 kg per pit at the time of planting and thrice after planting with an interval of 2 months at the same dosage.

12.1.2. The Spiral Nematode, *Helicotylenchus multicinctus*

(i) Cultural Methods: Rice or green gram after banana suppressed the population of *H. multicinctus*. Intercropping with sunnhemp in banana field was found effective in reducing *H. multicinctus* population by 38.4%, followed by marigold and cowpea which recorded a 29.0 and 22.3% reduction, respectively.

(ii) Biological Methods: Soil application of *Pseudomonas fluorescens* at 20 g/plant (2.5 x 10^8 cfu/g) was effective in increasing the plant growth parameters (pseudo stem height and girth, number of leaves, root length and weight) and in reducing the spiral nematode population in soil (58.87%) and roots (56.54%).

(iii) Host Resistance: Banana cvs./hybrids viz., Neyvannan, Rasthali, Wather, Peyan, Kullan, Krishna Vazhai, Karpurvalli, Sirumalai, CO-1, Kunnan, Robusta, Gros Michel, Vennettu Kunnan, Hybrid-94, 100, 106 and 109 (all Matti x Tongat) and Hybrid-74 (Matti x Pisang Lilin) were found tolerant/resistant to *H. multicinctus* (Vadivelu *et al.*, 1987).

(iv) Integrated Methods: Significant reduction in population of *Helicotylenchus multicinctus* was observed in banana plants treated with *Pseudomonas fluorescens* at 2g/ *Glomus mosseae* at 25g along with press mud at 3 kg/plant. These treatments also enhanced the plant height, pseudo stem girth, number of leaves, leaf area and bunch weight (Jonathan and Cannayane, 2002).

12.1.3. The Lesion Nematode, *Pratylenchus coffeae*

(i) Cultural Methods: A significant reduction in *Pratylenchus* sp. population (85%) was observed in the banana field where *Tagetes erecta* was grown as intercrop. The yield of the plants significantly increased (12 kg/plant) when intercropped with *Tagetes* spp. compared with the untreated control (7 kg/plant) (Sundararaju *et al.*, 2002). Intercropping with sunnhemp in banana field was found effective in reducing *P. coffeae* population by 38.4%, followed by marigold and cowpea which recorded a 29.0 and 22.3% reduction, respectively. Rice or green gram after banana suppressed the population of *P. coffeae*.

Botanicals: A significant reduction in *P. coffeae* population and yield increase was recorded in plants that have received 50% N applied through neem cake (Sundararaju and Kumar, 2003). Among neem formulations evaluated for the management of *P. coffeae* and *M. incognita,* Econeem and Nimbicidine showed maximum efficacy in reducing the nematode population with increased plant growth and yield (15 kg/plant) (Sundararaju and Cannayane, 2002). A significant reduction in *P. coffeae* population and increased yield in plants treated with press mud (15 t/ha) was recorded.

Application of press mud is economical and eco-friendly (Sundararaju *et al.*, 2002). Application of distillery sludge at 2.5 kg + vermicompost at 1 kg + neem cake at 1 kg + poultry manure at 2.5 kg/plant at 3, 5, and 7 months after planting significantly reduced *P. coffeae* population. It was on par with distillery sludge at 2.5 kg + neem cake at 1 kg treatment. Neem cake at 200 g/plant recorded the maximum increase in plant growth parameters with significant reduction in lesion nematode (*P. coffeae*) population in banana cv. Nendran.

(ii) Biological Methods: Application of *Trichoderma viride* at 20 g/plant at the time of planting and repeated 3 months after planting is effective in controlling *P. coffeae* as well as reducing the incidence of Panama wilt in banana. Soil application of *Pseudomonas fluorescens* at 20 g/plant (2.5 x 10^8 cfu/g) was effective in increasing the plant growth parameters (pseudo stem height and girth, number of leaves, root length and weight) and in reducing the lesion nematode population in soil (67.12 %) and roots (62.00%).

Application of *Paecilomyces lilacinus* at 10 g along with FYM at 500 g or neem cake at 250 g/plant effectively suppressed the lesion nematode population by 60 and 63%, respectively (Poornima *et al.*, 2004).

(iii) Host Resistance: Banana cvs./hybrids viz., Kunnan, Pey Kunnan, Then Kunnan, Nattu Poovan, Pidi Monthan, Chirapunji, Singhlal, Sakkarachayam, Malai Kala, Manik Champa, Kartobium Tham, Vennettu-kunnan and Hybrid-74 (Matti x Pisang Lilin), H-21, H-55, H-59, H-65, H-84, H-109, H-110 were found tolerant/resistant to *P. coffeae* (Vadivelu *et al.*, 1987; Sundararaju *et al.*, 2004).

12.1.4. The Cyst Nematode, *Heterodera oryzicola*

(i) Physical Methods: The nematode can be controlled by treatment of suckers (by hot water or chemicals) before planting.

(ii) Botanicals: Application of distillery sludge at 2.5 kg + vermicompost 1 kg + neem cake at 1 kg + poultry manure at 2.5 kg/plant at 3, 5, and 7 months after planting significantly reduced *H. oryzicola* population. It was on par with distillery sludge at 2.5 kg + neem cake at 1 kg treatment (Sundararaju *et al.*, 2002).

12.1.5. Root-knot Nematodes, *Meloidogyne* spp.

(i) Physical Methods: The suckers can be pared to a depth of 1 cm and treated in hot water at 55°C for 20 min (Inomoto and Monteiro, 1989).

(ii) Cultural Methods: Suckers should be selected from nematode-free mother plants. In wet lands, banana can be rotated with rice. Marigold can be grown in basins and incorporated into the soil around the plants.

Botanicals: Neem cake at 2 t/ha, FYM at 25 t/ha and press mud at 15 t/ha can be applied to encourage predacious nematodes and antagonistic fungi which in turn kill the nematodes. Application of distillery sludge at 2.5 kg + vermicompost 1 kg + neem cake at 1 kg + poultry manure at 2.5 kg/plant at 3, 5, and 7 months after planting significantly reduced *M. incognita* population. It was on par with distillery sludge at 2.5 kg + neem cake at 1 kg treatment.

(iii) Biological Methods: Prior colonization of banana suckers with *Glomus mosseae* was found to reduce the root and soil population of *M. incognita*.

Soil application of *Pasteuria penetrans* at 1 x 10^8 spores/g root powder mixed with 200 cc soil/plant and evenly distributed 45 days after planting recorded lowest final nematode population (473) at harvest compared to control (620). The yield was higher in plants treated with *P. penetrans* (3.5 kg/plant) compared to check (2.1 kg/plant) (Karuna *et al.*, 2004).

(iv) Host Resistance: Banana cvs. Patkapuri, Pisang Jari Buaya, Poovan, Pedali Moongil, Kunnan, Pisang Seribu, Veneetu Mannan, Elakkibale, Palayankodan, Singhlal, Karpuravalli, Manik Champa, Boddida Bukisa, Sappumala Anamulu, Sabari, Hoobale, Mendi and Kothia were found resistant to *M. incognita* in Orissa (Ray and Parija, 1987). In Tamil Nadu, cvs./hybrids such as Chakia, CO 1, Pisang Lilin, H 74, H 94 and H 110 were found resistant to root-knot nematodes (Jonathan, 1994; Sundararaju *et al.*, 2004).

(v) Integrated Methods: Soil application of 2 kg FYM with *P. fluorescens* (with 1 x 10^9 spores/g) and *P. chlamydosporia* (with 1 x 10^6 spores/g) per plant at the time of planting and at an interval of 4 months significantly reduced the root-knot nematode by 76% compared to control.

12.1.6. Burrowing Nematode, *Radopholus similis* and Wilt, *Fusarium oxysporum* f. sp. *cubense* Disease Complex

(i) Integrated Methods: Soil application of neem cake + *T. viride* + carbendazim was found effective in reducing the disease complex and in increasing the banana fruit yield (15.147 t/ha). This treatment also gave minimum lesion index as compared to control (4.0) with cost benefit ratio of 1:2.72 (Ravi *et al.*, 2001).

12.2. CITRUS

12.2.1. The Citrus Nematode, *Tylenchulus semipenetrans*

(i) Physical Methods: The nematode can be controlled by dipping the bare roots of citrus seedlings in hot water at 45°C for 25 min or 46.7°C for 10 min (Bindra *et al.*, 1967). This treatment leaves no adverse effect on seedlings. During summer, exposing the soil to hot sun can control the nematodes substantially (Sehgal *et al.*, 1990).

(ii) Cultural Methods: Populations of the citrus nematode in soil and roots can be reduced by removing the old feeder roots before the growth flush followed by application of FYM.

The citrus nematode is easily transported to new orchards through infested seedlings and budded plants which are usually carried 'balled' with infested soil. Therefore, nurseries should never be established on or near old citrus orchards and nursery soil should be sterilized before planting. Care should also be taken not to spread infection through tools, machinery and irrigation water used in infested groves.

Interculture of acid lime with marigold and mustard markedly reduced the rate of multiplication of *T. semipenetrans* to 10.03 and 8.59, respectively (Mani, 1988). Interculture of onion, garlic or marigold in the citrus orchards not only reduces the nematode population but also provides additional income.

Botanicals: Singh and Sitaramaiah (1973) reported that castor cake was highly effective in reducing 50% of the citrus nematode population. Mobin and Khan (1969) have also observed that neem, mahua, groundnut and mustard oil cakes were effective in reducing the nematode population in citrus rhizosphere. Application of neem cake at 20 kg per tree basin at an interval of 4 months can effectively reduce the nematode population of *T. semipenetrans* and *Meloidogyne javanica* (Mani and Murty, 1986).

Irrigation of citrus orchards with sewage water reduces the citrus nematode population to a very low level.

A number of plants possess nematicidal properties. Different plant parts and their extracts have been tested against *T. semipenetrans* by different workers. Aqueous and methanol garlic extracts were toxic to *T. semipenetrans in vitro* (Nath *et al.*, 1982). The extracts of leaves, stem and buds of *Datura stramonium, Ipomea carnea, Tagetes patula* and *Lowsonia alba* caused 50-100% mortality at 4 mg/ml or 1:5 dilution after 48 hrs (Kumari *et al.*, 1986). Similarly, water extract of *Calotropis procera* and *Nerium oleander* and essential oil of *Cymbopogon* grasses have been found toxic to *T. semipenetrans* (Sangwan *et al.*, 1985; Verma *et al.*, 1989). Castor (*Ricinus communis*), Aak (*Calotropis procera*) and Datura (*Datura strumonium*) chopped leaves applied at 20 and 40 g/kg soil in 9 sq. m. area around tree trunks can result in reduction of nematode population and increase in yield (Baghel, 1995).

(iii) Biological Methods: Several nematophagous fungi such as *Arthrobotrys cladodes, Dactylaria* sp., *Monacrosporium gephyrophagum, Laginidium* sp. etc., can trap the *T. semipenetrans* and could be exploited for nematode control (Gowda *et al.*, 1982). Endoparasitic fungus, *Paecilomyces lilacinus* is a potential biocontrol agent for controlling citrus nematode (Mani *et al.*, 1989). Arbuscular mycorrhizal fungus, *Glomus mosseae* when added to the soil with *T. semipenetrans* in *Citrus jambhiri,* limited the nematode development and partially neutralized the adverse effect of the nematode (Baghel *et al.*, 1990).

Since the parasitic fungus, *Paecilomyces lilacinus* has many attributes of a successful biological control agent, experiments were conducted in nematode infested citrus grove to determine its efficacy in controlling *T. semipenetrans.* It was observed that fruit diameter of the orange cv. 'Valencia' was significantly affected by application of 3 nematicides and *P. lilacinus*. The fruit diameter from fungus treated tree was greater than that from trees treated with temik, vydate and mocap. Similarly, the number of nematodes in roots and soil around citrus trees inoculated with *P. lilacinus* was significantly lower than that in the non-treated or the nematicide treated trees (Jatala, 1986). Application of *P. lilacinus* in the form of spore suspension checked the multiplication of *T. semipenetrans* to 71.7 to 97.8% (Mani and Murthy, 1989).

Application of *Pseudomonas fluorescens* at 20 g/tree three times in a year at 15 cm depth and 50 cm away from tree trunk helps in reducing the nematode population.

In citrus, a monthly rate of 1.1 kg a.i. per ha of avermectins (*Streptomyces avermitilis*) for 7 months gave maximum increase in yield and reduction of *T. semipenetrans* population (Garabedian and Van Gundy, 1983).

(iv) Host Resistance: Citrus rootstocks which are resistant or highly tolerant to the citrus nematode would offer practical and effective means of control. Trifoliate orange (*Poncirus trifoliata*), Troyer Citrange and Swingle citrumello and its hybrids (Citrumello) were highly resistant to the citrus nematode (Parvatha Reddy and Singh, 1978; Chhabra and Bindra, 1974; Parvatha Reddy and Agarwal, 1987; Mani and Reddy, 1986). The citrus rootstock hybrids evolved at the Indian Institute of Horticultural Research, Bangalore, by crossing Rangpur lime (*Citrus limonia*) with *P. trifoliata* viz., CRH-3, CRH-5 and CRH-41 were highly resistant to the citrus nematode (Parvatha Reddy *et al.*, 1987).

Inter-generic hybrids were evolved, involving the commercial rootstocks Rough lemon, Rangpur lime, Cleopatra mandarin and Orlando tangelo, as seed parents and *Poncirus trifoliata* (the citrus nematode resistant compatible donor) as pollen parent to overcome the citrus nematode problem. Twenty-five promising hybrids and the 5 parents were screened against *T. semipenetrans* by artificial inoculation under glasshouse conditions. Nine hybrids: CRH-57, 47, 10, 35, 17, 5, 41, 3 and 12 were resistant while 2 hybrids, CRH-8 and 24 were susceptible (Prasad *et al.*,1998, 1999). The remaining 14 hybrids were moderately resistant. Among the parents, *Poncirus trifoliata* was resistant while all the others were highly susceptible.

Resistance, based on the number of larvae and adult females on the feeder roots, was evaluated in 57 selections belonging to 4 genera (*Citrus, Fortunella, Poncirus* and *Severinia*) and 24 species/hybrids. All *Severinia disticha* and *P. trifoliata* strains and hybrids except 2 were rated resistant. Pummelo (*C. maxima*) x *P. trifoliata* and Smooth Flat Seville (*C. aurantium*) x Swingle citrumelo (*C. paradisi* x *P. trifoliate*) exhibited a moderately resistant reaction. It is suggested that Swingle citrumello and *P. trifoliata* strains should be used as citrus rootstock parents. Rangpur lime *(C. limonia)*, grapefruit *(C. paradisi)* and *C. pectinfera* were moderately resistant. Of 13 Rangpur lime (*Citrus limonia*) and rough lemon (*C. jambhiri*) strains tested, none were found to be resistant to *T. semipenetrans*. Rough lemon (Poona), Rough lemon (Madhya Pradesh), Rough lemon 14, Rough lemon 8779 and

Rough lemon 8782 were moderately resistant, while all the others were susceptible.

(v) Integrated Methods

Using Bioagents and Botanicals: Management of the citrus nematode based on the application of neem cake and castor cake both at 10 kg per plant along with a nematophagous fungus, *Paecilomyces lilacinus* at 250 g (grown on paddy seeds) per plant 3 times in a year at 15 cm depth and 50 cm away from the trunk was found to be extremely effective in reducing the citrus nematode population with a consequent increase in the growth of acid lime trees. The above treatment also gave highest parasitization of egg masses and eggs of *T. semipenetrans* and increased spore density of *P. lilacinus* in soil (Parvatha Reddy *et al.*, 1991a, 1993).

Trichoderma harzianum in combination with neem oilcake was effective in increasing the growth of acid lime trees and reducing the citrus nematode population both in soil and roots. The parasitization of citrus nematode females with *T. harzianum* increased in the presence of oil cakes (Parvatha Reddy *et al.*, 1996b).

Incorporation *Verticillium lecanii* with neem cake facilitated the effective management of *T. semipenetrans* on acid lime (Parvatha Reddy *et al.*, 1996a).

Integration of neem cake with *Pseudomonas fluorescens* gave maximum reduction in citrus nematode population both in soil and roots and increased plant growth of acid lime (Parvatha Reddy *et al.*, 2000).

Using Arbuscular Mycorrhizal Fungi and Botanicals: Inoculation of endomycorrhiza, *Glomus fasciculatum* in the soils amended with neem cake was found effective for the management of *T. semipenetrans* on acid lime. This strategy can help in combating the menace of citrus nematode at nursery stage and also provide highly mycorrhizal seedlings of acid lime for transplanting in the main field for the management of *T. semipenetrans* under field conditions (Parvatha Reddy *et al.*, 1995).

Using Two Bioagents: Application of bacterial bioagent, *Pasteuria penetrans* (at 2 x 10^9 spores/plant) and fungal bioagent, *P. lilacinus* (at 50 g/plant with 4 x 10^7 spores/g) was effective in reducing the *T. semipenetrans* population and in increasing the parasitization of eggs by both the bioagents (Parvatha Reddy and Nagesh, 2000).

Using Nematicides and Botanicals: Neem cake at 1 kg/plant combined with carbofuran at 2 kg/ha reduced 33.0% nematode population and increased the yield by 33.8% (Baghel, 1995).

Using Bioagents and Chemicals: Integration of *P. lilacinus* (at 4 g/plant) with carbofuran (at 30 mg a.i./plant) was found effective in increasing the plant growth parameters of acid lime and in reducing the citrus nematode population both in soil and roots (Parvatha Reddy *et al.*, 1996c).

12.2.2. The Lesion Nematode, *Pratylenchus coffeae*

(i) Host Resistance: Out of 125 selections of citrus, most of the selections were considered highly susceptible supporting population more than 1000 per gram root. Only four selections of *Microcitrus australis* x *M. australasica* hybrid and Rubiboux 70-A5, Trifoliate orange (*Poncirus trifoliata*) had nematode populations of less than 100/g root. They were considered to show some resistance to *P. coffeae* because of low numbers.

12.2.3. Root-knot Nematodes, *Meloidogyne javanica, M. indica*

(i) Physical Methods: Soil solarization is one of the effective ways of suppressing nematode populations and can be employed mostly during hot weather days. The nursery area should be ploughed well and leveled after breaking the clods. The prepared field may be covered with polythene sheet of 100 gauge thickness for about 7-10 days before sowing.

(ii) Cultural Methods: At least two to three deep ploughings should be given during bright sunny days at an interval of 10-15 days.

It is essential to select root-knot nematode-free planting material to prevent the introduction of nematode inoculum into orchards. It is preferable to avoid growing vegetables as intercrops or as rotational crops in citrus orchards and nurseries. Cereals and millets can be grown in nursery area according to grower's choice.

Growing sunnhemp in tree basin reduced the population of root-knot nematodes. *Crotalaria* roots exert a toxic effect on *Meloidogyne* population. *Meloidogyne* larvae freely entered roots of *Crotalaria* but failed to survive. The possibility of growing *Crotalaria* in tree basin for the control of root-knot nematodes infecting citrus needs further investigation.

Once the life of the citrus orchard comes to an end, it is highly preferable to grow cereals and millets for at least 2-3 seasons before initiating new orchard in the same field. This will help in reducing the populations of *Meloidogyne* species and *T. semipenetrans* which are common in citrus orchards.

Botanicals: Field experiments carried out in lime orchards indicated that application of neem cake at 20 kg per tree basin reduced the population of *M. javanica* by 25 per cent three months after application. Repeated applications once in four months significantly reduced the root-knot nematode population. FYM or compost at 20-25 t/ha or oil cakes, particularly neem or karanj cakes at 2-3 t/ha may be incorporated in soil before sowing or transplanting.

(iii) Host Resistance: It was observed that *C. reticulata*, Palestine and Rangpur limes, Rough lemon, Sour orange, Sexton, Thentriton tangelo and Volkamer lemon were not infected by *M. javanica*. Sour orange, Lime, Rough lemon, Grape fruit, Troyer citrange, Cleopatra mandarin and Trifoliate orange were highly resistant to *M. incognita* (Race 1 and 3) as well as *M. javanica*.

The response of *Citrus* spp. and related rootstocks to a population of *Meloidogyne javanica* was evaluated by some researchers in a screen-house experiment. Palestine and Rangpur lime, rough lemon, sour orange, Sexton and Thentriton tangelos, and Volkamer lemon were not infected by *M. javanica*. Galls and tip swellings were observed on the roots of *Poncirus trifoliata* and Troyer citrange. There was no evidence of nematode development Symptoms induced by the nematode were stelar division, syncytia formation in the vascular tissues and necrotic cells.

(iv) Biological Methods: A fungal parasite, *Paecilomyces lilacinus* consistently and effectively controlled the population of *Meloidogyne* spp.

12.2.4. The Burrowing Nematode, *Radopholus citrophilus*

(i) Cultural Methods: Barriers on a small scale are being used to confine *R. citrophilus* within limited areas of infestation (Poucher *et al.*, 1967). These man-made barriers surrounding the infested area consist of a strip of land 3 to 8 meter wide, free of citrus trees, and treated with a nematicide at 6 months intervals to keep the soil in the barrier zone free of roots as well as burrowing nematodes. The surface of the soil is also treated with herbicides to keep the barrier free of all plant growth. A biological barrier composed of a band of resistant citrus trees or a non-host of *R. citrophilus* has been suggested to block the migration of nematodes from infested areas (DuCharme and Suit, 1956; Ford, 1967).

Attempts are being made to eradicate *R. citrophilus* from infested groves (Lakeland fine silt) in Florida by the "pull and treat" method (Poucher *et al.*, 1967). This procedure involves destruction of all trees in an area overlapping the infested site by at least 16 meters in all directions and removal of as many roots as possible from the soil. The cleaned soil is treated with DD at 650 liters/ha, and the area is then kept free of all vegetation for 2 years. After the period of fallow, the site is replanted preferably with trees grafted on a tolerant rootstock.

(ii) Host Resistance: Citrus rootstocks such as Milam, Ridge Pineapple, Carrizo Citrange and Algerian Navel were reported to be resistant to the burrowing nematode (Ford and Feder, 1969).

12.3. PAPAYA

12.3.1. Root-knot Nematodes, *Meloidogyne* spp.

(i) Physical Methods: Hot water treatment of *M. incognita* infected papaya seedling roots at 50°C for 10 min. gave satisfactory control provided the main field was not infested.

(ii) Host Resistance: The papaya cv. Pusa 22-3 was resistant, while Thailand, Waimanalo, Mukund Farm, Pusa 1-15 and Pusa 1-45 were moderately resistant (Parvatha Reddy *et al.*, 1988). Naik (1986) reported that cvs. CO 2 and CO 3 were found to inhibit the development and reproduction of *M. incognita*. Khan *et al.* (1992) found that cv. CO 2 was resistant, while Peradeneya, Selection No.7 and Washington were moderately resistant against *M. incognita* Race 1.

(iii) Integrated Methods: Nursery bed treatment with *T. harzianum* and *P. lilacinus* each at 5 or 10 g/kg soil resulted in production of highly vigorous papaya seedlings whose roots were colonized with both the bioagents. There was significant reduction in root galling in the combination treatments (2.9 to 3.2) compared to control (8.9) (Rao and Naik, 2003).

Simultaneous application of *Paecilomyces lilacinus* and *Trichoderma harzianum* both at 1 g per plant gave maximum increase in plant growth parameters and highest reduction in reproduction factor and root galling in papaya (Khan, 1991).

12.3.2. The Reniform Nematode, *Rotylenchulus reniformis*

(i) Cultural Methods: Neem cake at 2.5 t/ha satisfactorily controlled *M. reniformis* and increased yield by 260% (Nayak, 1986).

(ii) Host Resistance: Papaya cultivars CO-1, CO-2, CO-3, CO-4, CO-5, Thailand and Giant were tolerant/resistant to *R. reniformis*. CO-2 and CO-3 were found to inhibit the development and reproduction of *M. incognita* (Nayak, 1986). Patel *et al.* (1989) reported that papaya cvs. Solo and Washington were resistant, while Coorg Honey Dew was moderately resistant. Papaya cv. Mammoth was found resistant, while CO1, Peradeneya, Sel.7 and Washington were moderately resistant (Khan and Khan, 1998).

12.3.3. Root-knot, *Meloidogne incognita* and Wilt, *Fusarium solani* Disease Complex

(i) Integrated Methods: Khan *et al.* (1997) reported that application of both the bioagents (*P. lilacinus and T. harzianum*) limited the damage caused by the disease complex and gave a 35% increase in plant growth compared to individual applications.

15.4. PINEAPPLE

15.4.1. The root-knot Nematode, *Meloidogyne incognita, M. javanica*

(i) Cultural Methods: Yield increases have been observed when planting was done in a field, which was previously had pigeon pea, owing to decrease in infestation by *M. incognita*. Planting of pineapple after pangola grass reduced the population of *M. incognita, Criconemoides* spp., *Helicotylenchus* spp. and *R. reniformis* and increased the fruit yield in the planted crop (23.5 MT/acre as against 15.1 MT/acre in control) and first ratoon crop (18.4 MT/acre as against 9.3 MT/acre in control) (Ayala *et al.*, 1967).

(ii) Host Resistance: *Ananas ananasoides* and three of its hybrids are highly resistant to attacks by *M. incognita*.

12.4.2. The Reniform Nematode, *Rotylenchulus reniformis*

(i) Cultural Methods: Rotation of pineapple with sugarcane or pangola grass is effective in controlling *R. reniformis*.

Irrigated fallow would reduce reniform nematode population as nematode prefers low moisture regimes in pineapple crop.

(ii) Host Resistance: Pineapple varieties/species such as Venezolana, *Ananas ananasoides* and several of its hybrids are highly resistant to *R. reniformis* and do not allow the initial infestation to be maintained.

12.5. GRAPEVINE

12.5.1. The Root-knot Nematode, *Meloidogyne* spp.

(i) Physical Methods: Hot water treatment of lightly root-knot infected rooted cuttings at 50°C for 10 min or at 48°C for 30 min before planting has been used to disinfect planting material.

(ii) Cultural Methods: Use of nematode-free planting material is one of the most important cultural practices adopted to avoid nematode infection. *Tagetes patula* used as intercrop, reduced 39-44% nematode population and significantly increased vine yield. Under field conditions, 27.0 to 38.7% soil population and 30.1 to 35.7% root population of nematodes on grapevine was reduced when intercropped with marigold, asparagus or sunnhemp. Marigold was found most effective (Bhagel and Gupta, 1986). Intercropping with marigold, asparagus or sunnhemp gave effective control and resulted in increased fruit yield. Cultivation of onion or garlic in the basin area brings additional remuneration besides reducing the nematode population.

(iii) Biological Methods: Sundarababu *et al.* (1999) observed that commercial formulation of *Pseudomonas fluorescens* reduced root-knot nematode population effectively and enhanced yield in grapevine. Soil application of talc formulation of *P. fluorescens* containing 15x10^8 cfu/g at 4 g/vine around root-knot infested grapevine at 15 cm depth in the basin at the time of pruning (during July) significantly reduced root galling due to *M. incognita* (39%), number of egg masses (250%) and increased fruit yield (166%) (Shanthi *et al.*, 1998). The above treatment also showed significant reduction in root galling (46%), number of egg masses (205%) and increased the yield (212%). Natural suppression of *M. incognita* by *Pasteuria penetrans* has been reported on grapevine in Tamil Nadu (Mani, 1996). Endomycorrhizal fungus, *Glomus mosseae* adversely affected *M. javanica* on grapevine cv. Perlette. *Paecilomyces lilacinus* when applied at 20 g/sq.m./vine reduces *M. incognita* infestation by 33% on grapevine.

(iv) Host-Resistance: Use of nematode resistant root-stocks has an added advantage over other methods of control. The cost factor is reduced to minimum. In grapevine, Dogridge (*Vitis champini)*, Salt creek (*V. champini*), 1613 (*V. solanis* x Othello), Harmony (1613 x Dogridge), St. George (*V. rupestris*), A x G1 (Aroman x Ganzin1), are well known nematode resistant root-stocks. Lake Emerald and Tompa are recently recognized as resistant root-stocks (Mortensten, 1985). Banquabad, Cardinal, Early Muscat, Joazbeli, Loose Perlette and Reisling (all commercial varieties) have been recognised as resistant to *M. javanica* (Anon, 1989). Black Champa, Dogridge, 1613, Salt Creek, Cardinal, Banquabad were recognized as resistant to *M. incognita*. Deshmukh *et al.* (2004) reported that grape varieties Black Champa, Dogridge and Degrasset resistant, while Pandhri Sahebi and Champanel moderately resistant to *M. incognita*.

(v) Integrated Methods: Application of neem cake (200 g/plant) + carbofuran (10 g/plant) reduced the nematode populations in the soil and roots. Combination of garlic and carbofuran at 0.1 g a.i./sq. m. reduced nematode population (44.2 – 47.6%) and increased yield (88.3 – 105.6%) over control with cost – benefit ratio of 1:4 to 1:5 (Anon, 1993).

The combined application of bioagents *P. lilacinus* and *P. penetrans* was most effective in increasing the fruit yield of grapevine besides checking *M. incognita* with cost: benefit ratio of 1:3.3 (Sivakumar and Vadivelu, 1998).

Application of neem cake at 200 g and *Paecilomyces lilacinus* at 50 g/vine helps in improving vine growth and yield. Further, a technology was developed to enrich farm yard manure (FYM) by applying nematode bio-control agents such as *Trichoderma harzianum* and *Paecilomyces lilacinus* . To achieve this, one kilogram of neem based formulation of this bio-agent was applied to 1 ton of farm yard manure. One hundred kgs of neem cake was applied to the farm yard manure to enhance the rate of enrichment by bio-agents. The farm yard manure was kept moist for 15 days under shade with thorough mixing at 5 day interval. This enriched FYM was applied to the field at the rate of 2 kgs per plant for 5-6 times at an interval of 2 months. This integrated method reduced the nematode problems significantly and improved the yield levels of grapes.

Combined application of neem cake and nematicide at reduced doses, along with intercropping with onion and garlic is highly profitable.

12.5.2. The Reniform Nematode, *Rotylenchulus reniformis*

(i) Host Resistance: Grapevine cvs. Dakshi, Joazbeli and Mukchilani are reported to be resistant to the reniform nematode.

12.5.3. The Dagger Nematode, *Xiphinema index*

(i) Cultural Methods: In France, to save grapevines from attack by *Xiphinema index* and subsequently infection by grapevine fan leaf virus, a 7-year rotation with cereal and alfalfa crop has been recommended (Lamberti, 1981).

12.5.4. Root-knot, *Meloidogyne incognita* and Wilt, *Fusarium moniliformae* Disease Complex

(i) Biological Methods: Soil application of *Pseudomonas fluorescens* at 100 g and FYM at 20 kg/vine gave effective management of disease complex and improved the plant stand by reducing the final soil nematode population (56.9%), root gall index (1.8) and % disease incidence (15.67%). This treatment also increased the number and weight of fruit bunches (17.83 and 155.40%, respectively) and fruit quality (more TSS – 13.53 Brix, TSS:acid ratio – 14.87, lower acidity – 0.91%) (Senthil Kumar and Rajendran, 2004).

12.6. MULBERRY

12.6.1. Root-knot Nematodes, *Meloidogyne* spp.

(i) Cultural Methods: Intercropping of mulberry with marigold at wider spacing of 30 to 45 cm controlled *M. incognita*, increased leaf moisture and leaf yield of mulberry (Govindaiah *et al.*, 1990).

(ii) Integrated Methods: Combined inoculation of AMF and *Pseudomonas fluorescens* was found effective against root-knot nematodes on mulberry (Kumutha, 2001).

12.7. PEACH, Prunus persica

12.7.1. Root-knot Nematodes, *Meloidogyne* spp.

(i) Physical Methods: Hot water treatment of peach seedlings at 50-52°C for 5-10 min eliminates the nematode infection.

(ii) Cultural Methods: Nematode-free planting stock is essential for transplanting in field.

(iii) Biological Methods: In simultaneous inoculations of *Glomus mosseae* and *M. javanica* on peach cv. Floridasun, the fungus negated the growth suppressive effect of nematode and reduced the multiplication of nematodes.

(iv) Host Resistance: Nemagaurd, Nemared and Okinawa are the most popular root-stocks resistant to *M. incognita, M. javanica* and *M. arenaria* (Sharpe *et al.*, 1983). Yunnan, Shalil and Bokhara have been found resistant to *M. incognita* but susceptible to *M. javanica*. Peach cvs. Chinaflat, Bidbillos Early, Japanese Double and Elberta showed no gall formation in screening tests against *M. incognita* (Verma, 1987).

12.7.2. The Ring Nematode, *Macroposthonia xenoplax*

(i) Physical Methods: A 37-100% reduction in population of *M. xenoplax* and other nematodes has been achieved by soil solarization in South Africa (Barberecheck and Broembsen, 1986).

(ii) Biological Methods: *Hirsutella rhossiliensis* reduced nematode population by 50%. Rate of nematode decay was much faster for parasitized second or third-stage juveniles (Jaffee *et al.*, 1988).

(iii) Host Resistance: Lorell rootstock of peach has been recorded as most tolerant to *M. xenoplax* (Rom *et al.*, 1985).

12.7.3. The Lesion Nematode, *Pratylenchus vulnus, P. penetrans*

(i) Physical Methods: Hot water treatment of peach seedlings at 52°C for 15 min eliminates *P. penetrans* infection.

(ii) Biological Methods: Pinochet *et al.* (1995) found that precolonization of peach roots with *Glomus mosseae* was effective in reducing the *P. vulnus* population.

12.8. CHERRY

12.8.1. Root-knot Nematodes, *Meloidogyne* spp.

(i) Physical Methods: Hot water treatment of cherry seedlings at 50-52°C for 5-10 min eliminates the nematode infection.

(ii) Host Resistance: Cherry rootstocks – Mazzard, Stockton Morello, English Morello and Montmorency were entirely free of root-knot galls.

12.8.2. The Lesion Nematode, *Pratylenchus vulnus, P. penetrans*

(i) Host Resistance: Cherry replants on Mahaleb rootstock grew well and *P. vulnus* and *P. penetrans* caused no root lesions.

12.9. STRAWBERRY

12.9.1. The Bud and Leaf Nematode, *Aphelenchoides fragariae*

(i) Physical Methods: Hot water treatment of strawberry runners at 50°C for 5 min or 48°C for 15 min or 46°C for 20 min eliminates the nematode infection.

(ii) Cultural Methods: Uprooting and destroying infected plants, and propagation through healthy planting stock are recommended for the control of nematodes.

12.9.2. The Lesion Nematode, *Pratylenchus penetrans*

(i) Physical Methods: Hot water treatment of dormant strawberry plants at 54.4°C for 1 min or 52.8°C for 3 min or 51°C for 7.5 min or 49.4°C for 17.5 min eliminates the nematode infection.

(ii) Integrated Methods: When the soil is heavily infested with *P. penetrans*, intelligent combination of nematicides and crop rotation is desirable:

- Soil application of nematicides reduce nematode population to a very low level.
- An advisable cropping sequence could then be: (a) strawberry which is susceptible but does not cause a large increase of the residual nematode population, (b) potato which is moderately susceptible and

causing moderate nematode reproduction, (c) oats or barley which are not susceptible but causing nematode reproduction and (d) beet which is not susceptible and suppressing nematode reproduction, after which moderately susceptible crops can be grown again successfully.

12.10. APPLE

12.10.1. The Lesion Nematode, *Pratylenchus vulnus*

(i) Physical Methods: Hot water treatment of apple rootstocks at 50°C for 15 min eliminates the nematode infection.

12.11. PASSION FRUIT

12.11.1. The Root-knot Nematode, *Meloidogyne incognita*

(i) Host Resistance: Passion fruit cvs. Yellow and Kaveri were found highly tolerant to *M. incognita*.

12.12. POMEGRANATE

12.12. 1. The root-knot Nematode, *Meloidogyne incognita*

(i) Cultural Methods: Application of neem cake at 3 kg/tree was effective in reducing the nematode population (57.83%) and in increasing the fruit yield (33.64%) over control.

Table 12.1. Biological control of plant parasitic nematodes in fruit crops.

Fruit crop	Nematode	Effective Bioagents	Reference
Banana	*M. incognita*	*P. lilacinus*	Devarajan & Rajendran, 2002
		G. fasciculatum	Prakash Babu, 2000
			Suma *et al.*, 2002
			Vidya & Reddy, 1998
			Umesh *et al.*, 1988
	R. similis	*P. lilacinus*	Devarajan & Rajendran, 2001
		T. viride	Sosamma *et al.*, 1994
		G. fasciculatum	Shanthi, 2003
			Prakash Babu *et al.*, 2002
			Prakash Babu, 2000
			Channabasappa *et al.*, 1995
		P. penetrans	Tabassum *et al.*, 2002
	M. incognita & *R. similis*	*G. mosseae*	Vidya & Reddy, 1998
			Parvatha Reddy *et al.*, 1997

Citrus	*T. semipenetrans*	*P. lilacinus*	Manzoor *et al.*, 2002
		P. fluorescens	Shanthi *et al.*, 1999 Parvatha Reddy *et al.*, 2000
		P. penetrans	Parvatha Reddy & Nagesh, 2000; Mani, 1988a
Grapevine	*M. incognita*	*P. fluorescens*	Shanthi *et al.*, 1999
		B. subtilis	Mani *et al.*, 1999
		P. penetrans	Stirling & White, 1982

Chapter 13

VEGETABLE CROPS

13.1. POTATO

13.1.1. Root-knot Nematodes, *Meloidogyne* spp.

(i) Physical Methods: Incubation of potato tubers at 45°C for 48 hours has been shown to kill about 98.9 per cent *M. incognita* without affecting tuber viability (Nirula and Bassi, 1965). Hot water treatment of potato tubers at 46-47.5°C for 120 min gave effective control of root-knot nematode infection. In small holdings, burning of trash in the field before taking up tuber planting not only helps in reducing the nematode population but also enriches the soil.

(ii) Cultural Methods: Tubers from root-knot infested potato crops should not be used as seed, as even outwardly clean looking seed may be infested. The movement of soil and water from the infested fields should be avoided. The field should be kept free from weeds, since root-knot nematodes have a wide host range and most of the weeds help in the build up of nematodes. Thus, clean cultivation (free from weeds which act as hosts) reduces nematode population to a great extent.

Potatoes grown with white mustard in a pot of infested soil were less heavily attacked by root-knot nematodes than potatoes growing alone. Potato root diffusate was ineffective in the presence of leachings from the roots of mustard seedlings. Mustard oils increased the yield of potatoes by reducing the severity of nematode attack. The active principle involved in mustard is allyl isothiocyanate which is toxic to the nematodes.

In long season, warm climate regions, potatoes may be planted during the winter months and harvested before injury occurs in the spring without visible infection. Frequently potatoes are planted in January-February and harvested in April-May before the second generation of nematodes develops and attacks the tubers. Planting of potato during third or fourth week of March in Shimla Hills would reduce the damage due to *M. incognita* (Prasad *et al.*, 1983). In Shimla Hills, early planting in third or fourth week of March reduces both root and tuber infestation without affecting the yield (Ahuja, 1983a). The yields were maximum which was concomitant with the lowest tuber infestation and lowest larval population in the soil at harvest. Late planting of autumn crop and early planting of spring crop in North Western Plains, while in hills, early planting of summer crop in fourth week of March is ideal. At Jalandhar, early planting of spring crop in first week of January and late planting of autumn crop in second and third week of October reduced the root-knot infection in potato. The interplanting of marigold with potato has resulted in a 7% increase in tuber yield.

Deep ploughing and drying of soil in the summer months facilitates the drying and death of infective larvae therby reducing initial inoculum in the soil. Dry fallow with 2 or 3 deep ploughings during the hot summer months gives excellent control of root-knot nematodes. Following 2 years crop rotational sequence of maize-wheat-potato-wheat coupled with summer fallow after 2 or 3 deep ploughings in North Western Plains reduced root-knot damage significantly. Similarly, in North Western Hills, rotation of potato with French beans, leafy vegetables, maize and wheat helped in minimizing the nematode damage and increasing the productivity of the crop (Raj and Nirula, 1969). Crop rotation with non-host crops like maize, wheat, barley, beans, etc. reduces nematode population. Growing of trap crop like French marigold, *Tagetes patula* in alternate rows with potato reduced the root-knot nematode larval population in soil, root galling and tuber infestation while the yields increased up to 123% over control. Similarly, growing *Cassia* sp. and incorporation of saw dust at 2.6 t/ha also helped in reducing tuber infestations.

Intercropping of onion and maize reduce galling by *M. incognita* on potato roots. Potatoes intercropped with maize have been found to be extremely vigorous with well developed root systems.

Botanicals: The oil cakes (neem, groundnut, mustard and castor) were effective in reducing parasitic nematode population (*Hoplolaimus*, *Tylenchorhynchus, Meloidogyne* and *Helicotylenchus* species) and in increasing yields of potato. Application of eucalyptus leaf waste from oil distillation plants at 2.5 t/ha helps in reducing the nematode population.

(iii) Biological Methods: The bacterium, *Pasteuria penetrans* has offered possibilities of biocontrol. Arbuscular mycorrhizal fungi such as *Glomus fasciculatum* and *G. mosseae* have shown promise in reducing the root-knot nematode infestation.

(iv) Host Resistance: Potato cv. Kufri Dewa was found resistant to root-knot nematodes.

(v) Integrated Methods: Application of neem cake/FYM/compost enriched with *Trichoderma harzianum*/*Paecilomyces lilacinus* will give effective control of root-knot nematodes.

13.1.2. Cyst Nematodes, *Globodera rostochiensis, G. pallida*

(i) Regulatory Methods: In India, there is a Quarantine Act against the cyst nematode of potato (*G. rostochiensis*) in Nilgiris. Infected potatoes are not allowed to be transported from Nilgiris to other parts of India for seed and table purposes in order to prevent the spread of potato cyst nematode from Tamil Nadu to other states and Union Territories.

Potato seed pieces free of the cyst nematode can be produced commercially by seed certification.

(ii) Physical Methods: Fassuliotis and Sparrow (1955) demonstrated that irradiation of potato tubers with x-rays inhibited sprouting and also development of *G. rostochiensis*. Cysts of this nematode exposed to 20,000 r contained only brown and dead eggs, while at 40,000 r the eggs lost their contents completely.

Careful washing of soil adhering to potato tubers can considerably reduce the risk of spreading cyst nematodes. Commercial potato tuber washing apparatus has been developed.

(iii) Cultural Methods: The potato cyst nematode (*G. rostochiensis*) can be eliminated by selecting nematode-free planting material.

Early maturing varieties of potato like Irish Cobbler usually suffers less than do late maturing varieties. These varieties start growth in spring at soil temperature too low for much nematode activity and may be harvested before many of the nematodes can reproduce.

Continuous cropping with potatoes increases the cyst population of *G. rostochiensis*. More than three year rotations with wheat, strawberry, cabbage, cauliflower, peas, maize and beans reduces the nematode population to a safe level. Growing non-host crops like radish, garlic, beet root, French bean, cruciferous vegetables, turnip or green manuring crops bring down the cyst nematode population by more than 50%. A crop rotation pattern (using potato-French bean-peas-potato), wherein potato is grown once after 3-4 other crops, decreased the nematode population by 98 to 99% and increased yields of potato by 90%. In Nilgiris, crop rotation with cabbage and carrot gave effective control of cyst nematodes. Growing of vegetables and resistant potato in a long term rotation trial brought down the cyst propagule population up to 92% and was effectively used for nematode management at Nilgiris.

Botanicals: Incorporation of eucalyptus litter at 10 t/ha at the time of preparation of land reduces the population of the potato cyst nematodes. Application of eucalyptus distillation waste at 2.5 t/ha also helps to reduce the cyst nematode population in soil.

(iv) Biological Methods: Application of *Pseudomonas fluorescens* at 10 g/m^2 gives good control. Application of neem cake at 1 t/ha/FYM at 20 t/ha/compost at 10 t/ha enriched with *Trichoderma harzianum* / *Paecilomyces lilacinus* will give effective control of cyst nematodes.

(v) Host Resistance: In the United States, both USDA and Cornell University scientists have evolved two commercial potato varieties – Peconic and Wauseon – which are resistant to the cyst nematode.

In India, Potato variety Kufri Swarna developed by the Central Potato Research Institute Substation at Ooty is resistant to both the species of cyst nematode and the late blight disease. The variety is also draught resistant and has shorter dormancy which is advantageous to Nilgiris region where potato is taken up mainly as a rainfed crop almost throughout the year and thus covers about 40% of potato area (Anon, 1985).

(vi) Integrated Methods: The combination of disease escape (by planting early maturing varieties) and the use of nematicides gave good control of the potato cyst nematode. Hygiene in the form of seed certification, combined with crop rotation in seed growing areas is effective. Use of a resistant variety followed by a nematicide would kill 99% of the nematode population. The potato cyst nematode problem in Nilgiris is being managed by chemical treatments, crop rotations and utilizing the available sources of resistance (Kufri Swarna). The most effective management combines crop rotation, use of nematicides and resistant varieties to keep the nematode at an economically acceptable level (Jones, 1969).

Table 13.1. Integrated management of the potato cyst nematode.

Sl. No.	Management method (s)	Resulting kill popn. (% initial popn)	Kill (%)	Popn. after growing & harvesting a susc. cv., calculated at 2 assumed nema multpln. rates (% initial popn)	
				30 x	70 x
1	4 years without potato	3	97	90	< 100
2	1 year with resistant Potato	20	80	> 100	> 100
3	Nematicide treatment (Fumigant)	25	75	> 100	> 100
4	1 and 2	0.6	99.4	18	42
5	1 and 3	0.75	99.25	22.5	52.5
6	2 and 3	5	95	> 100	> 100
7	All 3 methods	0.15	99.85	4.5	10.5

Better control of *Globodera rostochiensis* by *P. penetrans* was possible with soil solarization combined with application of FYM or karanj cake which resulted in greater kill of nematodes (Sitaramaiah and Naidu, 2003).

A well known integration of methods which can control *G. rostochiensis* in potato is a 4-year rotation combining soil fumigation (causes 30% reduction in nematode population), cultivation of non-host (50% reduction), cultivation of resistant potato variety (30% reduction), cultivation of non-host (50% reduction) and then cultivation of susceptible variety of potato (Oostenbrink, 1972).

13.1.3. The Potato Rot Nematode, *Ditylenchus destructor*

(i) Cultural Methods: The use of healthy seeds, destruction of infected plant parts in the field, control of weed hosts, late planting when soil moisture is inadequate for optimum invasion and early harvest reduces losses. Two to three years rotation of potato with beans, corn and fallow kept free of weeds gave a very marked reduction in nematode populations. Satisfactory control of *D. destructor* can also be obtained by crop rotation with small grains, vetch and lupine provided potatoes are grown once in 3-4 years.

(ii) Integrated Methods: The control of potato rot nematode was achieved by the combination of disease escape, hygiene, in the form of seed certification and crop rotation as follows (Winslow and Willis, 1972):

- Healthy seed, planted late, harvested early and stored as cool and dry as possible.
- Proper rotation of potatoes with non-host crops and growing of potatoes not more frequently than once in 3 or 4 years.
- Field hygiene in the form of removal of old infested tubers and weed control.

13.1.4. The Lesion Nematode, *Pratylenchus penetrans*

(i) Host Resistance: Potato cv. Russet Burbank is tolerant, while Peconic appears to be less susceptible to the lesion nematode.

13.1.5. Root-knot Nematode, *Meloidogyne* spp. and Wilt, *Verticillium dahliae*

Disease Complex

(i) Cultural Methods: In fields infested with root-knot nematodes and *V. dahliae,* Scholte (1990) found that a five-year rotation of maize-sugar beet-barley-barley-potato had significantly higher potato yields than continuous potato.

13.2. TOMATO

13.2.1. Root-knot Nematodes, *Meloidogyne* spp.

(i) Physical Methods: The most important nematode pests controlled by greenhouse steaming are cyst nematode of potato attacking tomatoes and root-knot nematodes on tomatoes. Soil solarization or burning of paddy husk or saw dust on infested nursery proved to be reasonably effective. Rabbing of root-knot infested tomato nursery with bajra husk at 7 kg/sq.m. a week prior to seeding is recommended for economic and effective management of root-knot disease (ICBR 1:5.12). Soil solarization of nursery beds with 100 gauge LLDPE clear plastic film during summer (in between April 15th and June 15th) for 15 days is recommended. It increased the number of transplantable seedlings by 615% giving net profit of Rs.7,661 from 1000 sq.m. (ICBR 1:5.65) and decreased the root-knot disease and weeds by 66% and 93%, respectively (Table 13.2).

Table 13.2. Effect of soil solarization of nursery beds on root galling and number of transplantable seedlings of tomato in 1000 sq. m.

Treatment	No. of seedlings	Weight of 100 seedlings	Root-knot index	Net profit	Cost:Benefit Ratio
With plastic	2,60,400	36 g	1.37	Rs. 7,661	1:5.65
Without plastic	36,400	26 g	3.97	---	---
% Increase (+)/decrease (-) over control	(+) 615	(+) 38	(-) 66	---	---

(ii) Cultural Methods: Several cultural methods like selection of nematode-free nursery sites, destruction of infested roots after crop harvest, crop rotation with non-host (maize, wheat, sorghum) or antagonistic crops (marigold, mustard, sesame), flooding, fallowing, deep summer ploughing and harrowing either alone or in combination proved to be reasonably effective and economical to check multiplication of root-knot, reniform and lesion nematodes on the transplanted vegetable crops like tomato, brinjal and capsicum. Ploughing not only leads to disturbance and instability in nematode community but also causes their mortality by exposing them to solar heat and desiccation. Three summer ploughings each at 10 days interval during June (average atmospheric temperature ranging between 40-46°C) at Hisar (Haryana) led to 96.5 % reduction in *M. javanica* population, while fallowing itself during the same period registered 44.5% reduction (Jain and Bhatti, 1987). Additional use of plastic sheets for covering soil either in nursery beds or in field further enhance nematode reduction. Such an approach also helps in reducing the intensity of weeds, fungi and bacteria in the soil.

Jain and Bhatti (1987) reported that planting of nematode-free seedlings in an infested field which received three summer ploughings increased the yield of tomato by 55%.

Decrease in *M. incognita* population on tomato occurs when the field was left fallow or following crop rotation with marigold, spinach and bottle gourd (Khan *et al.*, 1975). To control *M. incognita* and *M. javanica*, a tomato crop can be followed by peanut without risk of damage to the peanuts. While the peanuts are growing, the nematodes can not reproduce. Instead many of the larvae in the soil die or become non-infective because of starvation and the attacks of predators, fungi and diseases. If the population is reduced sufficiently, tomatoes can be grown again after peanuts without serious injury (Taylor and Sasser, 1978). Tomato rotated with cotton-wheat-cotton resulted in low gall index due to *M. javanica* (root-knot index 1-2) and gave maximum yield (38% increase over continuous tomato).

Sesamum-tomato sequence produced the least reproduction rate (0.47), least root-knot index (1.33) and highest yield (86 q/ha) of tomato which was at par with niger-tomato sequence that had reproduction rate of 0.56, root-knot index of 1.66 and highest yield of 75.5 q/ha of tomato (Sahoo *et al.*, 2004a).

In cropping sequence studies for the management of *M. javanica* on tomato, nematode population decreased to low levels when carrot, capsicum and [illegible]e grown (Kanwar and Bhatti, 1992). The population of *M. incognita* was reduced considerably under groundnut-mustard-tomato cropping sequence (Sharma *et al.*, 1980).

Intercropping with marigold or mustard with tomato reduced the damage of root-knot and reniform nematodes. The root-knot development on tomato was low when interplanted with *Tagetes erecta*. The population of *Tylenchorhynchus, Helicotylenchus, Hoplolaimus, Rotylenchulus* and *Pratylenchus* was also markedly reduced (Alam *et al.,* 1977; Khan *et al.,* 1971). α-Terthienyl is the active principle in *Tagetes* spp. which is toxic to these nematodes.

Tomato nursery beds previously planted with trap crop (marigold) gave maximum reduction in root-knot nematode population in soil which also increased the germination of tomato seeds and production of more healthy (nematode-free) seedlings (Rangaswamy *et al.*, 1995).

Marigold intercropped with tomato at 1:4 and 1:6 ratio and mustard at 1:2 ratio was found to be effective in reducing root galls, egg masses and nematode population at harvest and gave cost: benefit ratio of 1:8.36, 1:7.88 and 1:3.31, respectively (Rangaswamy *et al.*, 1999).

Interculture of onion with tomato reduced gall index due to *M. javanica* significantly and improved tomato fruit yield. Intercropping with 2 rows of onion with one row of tomato was the most effective treatment in reducing the nematode infection and population and enhancing tomato yield (Ram and Gupta, 2001).

In pots having *Zinnia elegans* with tomato, the population of *M. incognita* and *R. reniformis* declined. The root-knot index on tomato was reduced to 1.08 compared to 3.75 in control. The nematode reproduction factor of *M. incognita* and *R. reniformis* was reduced to 0.88 and 0.80, respectively compared to 3.86 and 2.43 in control (Tiyagi *et al.*, 1986).

Cultivation of tomato with castor results in considerable reduction in number of root-knot galls on tomato (Hackney and Dickerson, 1975).

Inclusion of rice, maize, groundnut and *Stylozanthes* in rotation sequences controlled root-knot nematodes and increased yields of tomato. In crop rotation trials, it was found that a crop of groundnut or strawberry grown prior to tomato in soil heavily infested with root-knot nematodes greatly increased yields of tomato. Rotation of tomato with leek (*Allium porrus*), onion, groundnut and sunnhemp reduces the population of root-knot nematodes.

Increased levels of potash have significantly reduced the number of galls by *M. javanica* in tomato (Gupta and Mukhopadhyaya, 1971).

The growth of tomato increased (35%) and root galling due to *M. incognita* (root- knot index 1 as against 4 in control), nematode population in soil (80% reduction) and egg mass production reduced (egg mass index 1 as against 4 in control) in soil amended with 20% fly ash (Haq *et al.*, 1985). The increase in plant growth may be ascribed to increased nutrient availability (N, K, Ca, Mg, Na, B, SO_4) and the reduction in nematode population might be due to toxic compounds (polycyclic aromatic hydrocarbons, dibenzofuran and dibenzo-p-dioxin mixtures) present in fly ash. Row application of fly ash at 0.6 kg/sq. m. enhanced the yield of tomato by 90.4% (Khan and Ghadipur, 2004).

Red plastic mulch suppresses root-knot nematode damage in tomatoes by diverting resources away from the roots (and nematodes) and into foliage and fruit (Adams, 1997).

Botanicals: Soil application of neem and subabool leaves to nursery beds at 0.5 t/ha gave better seedling growth and reduced root galling in tomato (Jain *et al.*, 1988).

Singh and Sitaramaiah (1966) applied finely divided oil cakes to root-knot infested soil and noted a reduction in disease incidence in tomato. The oil cakes (neem, groundnut, mustard and castor) were effective in reducing parasitic nematode population (*Hoplolaimus, Tylenchorhynchus, Meloidogyne* and *Helicotylenchus* species) and in increasing yields of tomato.

Application of neem cake at 15 g/spot or 100 g/m furrow, three weeks prior to transplanting of tomato and brinjal have been reported to give 36-450% enhanced yield with corresponding reduction in gall index to 2.3 as against 4.3 in control (Anon, 1992).

Best results in respect of root-knot reduction due to *M. javanica* and increase in yield of tomato were obtained by amending the soil with saw dust at 2.5 tons per ha 3 weeks before planting and then applying N through urea at 120 kg per ha (Singh and Sitaramaiah, 1971b). Rice hull ash at 2.5 tons per ha increased the yield of tomato by 133 to 317 per cent and reduced root-knot incidence by 46 to 100 per cent (Sen and Dasgupta, 1981). Application of poultry manure at 2.0, 2.5 and 3.0 t/ha 15 days prior to tomato seeding is recommended for effective management of root-knot disease and production of more transplantable seedlings in tomato nursery giving ICBR of 1:4.54, 1:5.63 and 1:6.68, respectively.

Green manuring with naffatia/besharmi or neem leaves (ICBR 1:2.87) or water hyacinth (ICBR 1:2.527) or calotropis (ICBR 1:2.26) or congress grass (ICBR 1:2.09) each at 3 kg/sq.m. 15 days prior to seeding is recommended for economic and effective management of root-knot disease in tomato nursery. Incorporation of green materials of naffatia and neem leaves each at 2 kg/1.44 sq.m. (ICBR 1:15.7) or congress grass and neem leaves each at 2 kg/1.44 sq.m. (ICBR 1:15.6) or calotropis, naffatia, congress grass and neem leaves each at 1 kg/1.44 sq.m. (ICBR 1:14.7) 15 to 20 days prior to tomato seeding gave effective management of root-knot nematodes and production of more healthy seedlings in tomato nursery during kharif season.

(iii) Biological methods: Application of 'Royal 350' (*Arthrobotrys irregularis* cultured on oat seed medium) at 140 g/m^2 a month before transplantation of tomato resulted in good protection against root-knot nematodes. Among the egg parasites, efficacy of *Paecilomyces lilacinus* (commercially formulated as 'Biocon' in Philippines) has been found to be comparatively higher in suppressing the population of *Meloidogyne* spp. and *R. reniformis* on tomato. Bare root-dip treatment of tomato seedlings in a spore suspension of *P. lilacinus* at 4 x 10^5 cfu/ml reduces *M. incognita* infestation in roots by 50% and increases the growth of seedlings by 20%. Application of organic amendments increases the effectiveness of *P. lilacinus* against *M. incognita*.

Two bacteria *Bacillus licheniformis* and *Pseudomonas mindocina* and 2 fungi *Acrophialophora fusispora* and *Aspergillus flavus* were evaluated for their efficacy as bio-control agents individually and in various combinations against *Meloidogyne incognita* infecting tomato. Individually, *B. licheniformis* gave best control although when all 4 agents were used, nematode multiplication was reduced by 82.06% (Siddiqui and Husain, 1991).

Application of *Bacillus thuringiensis* causes 95% mortality of *M. javanica* juveniles due to β-exotoxin production and suppresses gall formation, egg mass production and nematode population in soil. Field application of a nematicidal *B. thuringiensis* strain to tomato in Puerto Rico reduced galling in roots due to *M. incognita* and soil population of *Rotylenchulus reniformis* and increased yield significantly (Zuckermann *et al.*, 1993).

Root material containing *Pasteuria penetrans* spores at 212-600 mg powder/kg soil, when broadcast in soil, showed reduction in galling on tomato roots and final nematode population. Tomato yield increased by 20% over untreated check, when *P. penetrans* was introduced to the nursery soil at 1x10^5 spores/g soil. The persistence of *P. penetrans* in the field soil was evident since about 50% of the final juvenile population was encumbered with the bacterial spores (Walia and Dalal, 1994). *P. penetrans* is highly resistant to heat, desiccation and has ability to survive for more than 2 years in soil, qualifies as the most potential biocontrol agent against root-knot nematodes infesting vegetable crops.

Application of *Pseudomonas fluorescens* at 10 g/m^2 in nursery beds gave good control of root-knot nematodes (Shanthi and Sivakumar, 1995).

Arbuscular mycorrhizal fungus (AMF), *Glomus fasciculatum* was effective against root-knot nematodes and *G. mosseae* against reniform nematodes on tomato. Organic amendments (oil cakes, calotropis leaves) in combination with AMF enhance the colonization of AMF on tomato roots which further increased plant growth and reduced gall index. Sundaram and Arangarasan (1996) observed that inoculation of *G. fasciculatum* at 10 g/plant was effective in increasing the uptake of nutrients and yield and reducing root galling (10 galls compared to 135.3 in control), followed by *G. mosseae, Gigaspora margarita* and *Acaulospora laevis*. Application of *G. fasciculatum* (Shimoga banana isolate) at 50 g/sq. m. (200 spores/g of soil) gave minimum number of galls per root system (39.2) at harvest and recorded maximum yield (26.47 t/ha) with an increased cost benefit ratio (1:3.34) and decreased nematode population (Kiran Kumar *et al.*, 2004).

Avermectins B_1 and B_{2a} (applied to soil through drip irrigation systems) at rates ranging from 0.093 to0.34 kg a.i./ha applied as a single dose or 0.24 kg a.i./ha applied as three doses each at 0.08 kg on tomatoes against *M.*

incognita were as effective as oxamyl and aldicarb at 3.36 kg a.i./ha (Garabedian and Van Gundy, 1983). The aqueous solution of avermectins (250 ml of 0.001%/m^2) significantly reduced root galls of tomato seedlings raised in root-knot infested nursery beds which yielded robust and healthy seedlings with no root infection (Parvatha Reddy and Nagesh, 2002). In a root-knot infested micro-plot study, avermectins at 0.015 kg/ha effectively controlled *M. incognita* on tomato and increased yield by 11 and 8% compared to the untreated and carbofuran treated plots, respectively (Parvatha Reddy and Nagesh, 2002). Seedling bare root-dip in avermectin 100% gave highest reduction in root-knot index (1.3), soil nematode population (132.8), number of females/g root (12.8), egg masses/g root (5.4) and eggs/egg mass (125.2) in tomato (Jayakumar *et al.*, 2005).

(iv) Host Resistance

Tomato cvs. Nematox, SL-120, NTR-1, SL-12, Patriot, VFN-8, VFN Bush, Piersol, Radiant, Nemared, Ronita, Anahu, Bresch, Helani, Campbell-25, Punuui, Arka Vardan, Pelican, Hawaii-7746, Hawaii-7747 and Hisar Lalit have been reported to be resistant to root-knot nematodes.

In a field with resistant tomato (Hisar Lalit) grown for a year, the yield was 27% higher than susceptible variety (HS-101) (Kanwar and Bhatti, 1990).

(v) Integrated Methods

Bio-agents and Botanicals: In nursery, integration of *Pasteuria penetrans* (at 28 x 10^4 spores/m^2), *P. lilacinus* (at 10 g/ m^2 with 19 x 10^9 spores/g) and neem cake (at 0.5 kg/ m^2) gave maximum increase in plant growth and number of seedlings/bed. Parasitization of *M. incognita* females was highest when neem cake was integrated with *P. penetrans*, while parasitization of eggs was highest when neem cake was integrated with *P. lilacinus*. In field, planting of tomato seedlings (raised in nursery beds amended with neem cake + *P. penetrans*) in pits incorporated with *P. lilacinus* (at 0.5 g/plant) gave least root galling and nematode multiplication rate and increased fruit weight and yield of tomato (Parvatha Reddy *et al.*, 1997). *P. lilacinus* in combination with castor leaves reduced the nematode population up to 89% and increased plant growth and yield in tomato (Zaki and Bhatti, 1991a).

Soil drenching with 5% Wellgro solution (at 200 ml/seed pan) along with *P. lilacinus* (at 2 x 10^4 spores/ml) gave maximum increase in seedling weight and highest reduction in root galling. Roots of tomato seedlings raised in the above treatment dipped in 5% Wellgro solution mixed with *P. lilacinus* spores (at 2 x 10^4 spores /ml) for 20 min. when transplanted in pots gave maximum increase in plant growth, root colonization and parasitization of egg masses of *M. incognita* by the bioagent and highest reduction in root galling and final nematode population both in soil and roots (Rao and Parvatha Reddy, 1993).

Bare root-dip treatment of tomato seedlings in 10% castor leaf extract mixed with *P. lilacinus* spores (at 1.5 x 10^8 spores /ml) for 20 min. significantly increased the plant growth and reduced root galling and final nematode population. The above treatment also gave significant increase in parasitization of eggs and egg masses and propagule density of *P. lilacinus* in roots (Rao *et al.*, 1999). Similarly, root-dip treatment of tomato seedlings in 5 and 10% neem leaf extract mixed with *P. lilacinus* spores (at 6 x 10^4 spores /ml) for 30 min. gave significant reduction in root galling and final nematode population. Significant increases in root colonization, parasitization of eggs and egg masses and propagule density of the bioagent in soil were also noticed in the above treatment (Parvatha Reddy *et al.*, 1998).

Application of *Pochonia chlamydosporia* (100 ml/seed pan containing 4 x 10^5 spores/ml) with neem leaves (150 g/seed pan) increased seedling weight and colonization of roots with the bioagent. Tomato seedlings raised in the above treatment transplanted in pots resulted in maximum increase in plant growth, least root galling, nematode population both in soil and roots and highest parasitization of eggs, egg masses and propagule density of *P. chlamydosporia* in roots (Parvatha Reddy *et al.*, 1999).

Incorporation of neem cake (20 g/pot) along with *P. chlamydosporia* (10 ml/pot containing 4 x 10^5 spores/ml) gave maximum increase in plant growth and significant reduction in root galling and nematode population both in soil and roots. The above treatment also gave highest parasitization of eggs and egg masses of *M. incognita* and maximum propagule density of the bioagent in soil and roots (Rao *et al.*, 1998).

Integration of neem cake (40 g/plant) with *T. harzianum* (at 5 g/plant with 4 x 10^8 spores/g) was effective in increasing the plant growth and reducing root galling and final population of *M. incognita* infecting tomato. The above treatment also gave maximum reduction in number of eggs /egg mass and increased root colonization, spore density in soil and parasitization of adult females with *T. harzianum* (Rao *et al.*, 1997c).

Combined application of *P. penetrans* + *T. viride* + neem/castor cake at 1/3 doses was significantly superior compared to their individual applications in terms of increased plant growth and reduced root galling, egg mass production and final populations of *M. incognita*. Similarly, total parasitization by bioagents was higher in plants treated with these bioagents in terms of number of juveniles encumbered and adult females infected with *P. penetrans* and egg masses parasitized by *T. viride*, compared to their individual applications (Rangaswamy *et al.*, 1999).

Nursery bed treatment with *P. fluorescens* (with 1 x 10^9 spores/g) and *P. chlamydosporia* (with 4 x 10^6 spores/g) each at 20 g/m^2 and field application of 5 tonnes of enriched FYM with the above bioagents each at 5 kg, significantly reduced root-knot nematodes in tomato by 76% over untreated check. The yield increase was up to 21.7% with a cost: benefit ratioof 1: 4.9.

Arbuscular Mycorrhizal Fungi and Botanicals: In nursery, integration of neem cake (at 500 g/ m^2) with *Glomus mosseae* (at 250 g/ m^2 containing 16 chlamydospores/g) significantly reduced *M. incognita* population in soil, root galling, egg mass production, fecundity and produced vigorous tomato seedlings with increased root colonization with *G. mosseae*. The seedlings raised in the above treatment when planted in the main field significantly reduced root galling and increased fruit yield, root colonization with *G. mosseae* and chlamydospore population in soil (Parvatha Reddy *et al.*, 1998).

Application of calotropis leaves at 400 g/ m^2 along with *G. fasciculatum* (250 g/ m^2 containing 25-30 chlamydospores/g) in nursery beds gave significant reduction in root galling and fecundity of *M. incognita*. The above treatment also gave maximum increase in plant growth and root colonization with *G. fasciculatum* (Rao *et al.*, 1996).

Bio-agents and Arbuscular Mycorrhizal Fungi: Interestingly, integration of bio-agent (*P. lilacinus*) and endomycorrhizae (*G. mosseae*/*G. fasciculatum*) have culminated in the successful management of *M. incognita* infecting tomato. This phenomenon facilitated standardization of a strategy wherein inoculation of mycorrhizae and bio-agent in the nursery beds protected the seedlings of tomato from the attack of *M. incognita*. Further, these mycorrhizal seedlings (colonized either with *G. mosseae* or *G. fasciculatum*) can be given a root-dip treatment with spore suspension of *P. lilacinus* for 5-10 minutes for the effective management of these nematodes in the main field after transplanting (Rao *et al.*, 1993).

Integration of *G. deserticola* with *P. chlamydosporia* gave effective management of *M. incognita* in tomato, gave increased seedling weight, root colonization with *G. deserticola* and *P. chlamydosporia* and parasitization of eggs of *M. incognita* with *P. chlamydosporia* and reduced root galling, egg mass production and fecundity of root –knot nematodes (Rao *et al.*, 1997).

Using Bio-agents, Arbuscular Mycorrhizal Fungi and Botanicals: Strategies to integrate botanicals, bio-agents and endomycorrhizae components for the management of root-knot and reniform nematodes infecting tomato have been developed. These strategies include inoculation of *G. mosseae* in neem leaf/neem cake amended nursery beds followed by the root-dip treatment of mycorrhizal seedlings of tomato in spore suspension of *P. lilacinus* was effective for the management of *M. incognita* under field conditions (Rao *et al.*, 1995).

In nursery, integration of neem cake (at 500 g/m^2) with *G. fasciculatum* (200 g/m^2 containing 15 spores/g) gave maximum increase in plant growth and highest root colonization with *G. fasciculatum* and least root galling. Integration of neem cake (at 500 g/m^2) with *T. harzianum* (at 100 g/ m^2 with 4 x 10^8 spores/g) gave maximum parasitization of eggs with *T. harzianum*. In field, planting of tomato seedlings raised in nursery beds treated with neem cake + *G. fasciculatum* in pits incorporated with *T. harzianum* (at 0.5 g/ plant with 4 x 10^8 spores/g) was effective in increasing tomato fruit yield, root colonization with *G. fasciculatum* and egg parasitization with *T. harzianum* and significant reduction in root galling and multiplication rate of *M. incognita* (Parvatha Reddy *et al.*, 1998).

Using Bio-agents and Chemicals: Integration of a bio-agent, *P. lilacinus* with carbofuran at 1 kg a.i./ha was found effective in the management of reniform nematode, *R. reniformis* infecting tomato (Parvatha Reddy and Khan, 1988).

Maheshwari *et al.* (1987) reported that application of *P. penetrans* in combination with carbofuran, aldicarb, miral, sebufos and phorate significantly improved plant growth of tomato by greatly reducing galling due to *M. javanica*.

Using Two Bio-agents: Integration of two bio-agents, *P. lilacinus* and *P. chlamydosopria* resulted in combined and complimentary effects for the successful management of *M. incognita* infecting tomato (Rao and Parvatha Reddy, 1992c).

In nursery, integration of *P. chlamydosopria* (200 g/ m^2 containing 16 chlamydospores/g) with *P. penetrans* at 1 g mixed with 100 g of sand/m^2 (containing 10^5 spores/10 mg) was effective in increasing the seedling weight and root colonization with *P. chlamydosopria* and adult females of *M. incognita* adequately infected with *P. penetrans*. The tomato seedlings raised in the above treatment when transplanted in the field gave maximum reduction in root galling, fecundity, nematode population both in soil and roots and increased root colonization with *P. chlamydosopria*, infection of *M. incognita* females with *P. penetrans* and tomato fruit yield (Rao *et al.*, 1998).

Maheshwari and Mani (1988) also observed that population of *M. incognita* and *M. javanica* were suppressed effectively and yields of tomato were greater when *P. penetrans* and *P. lilacinus* were applied together.

Combined soil application of *P. lilacinus* and *Aspergillus niger* at the time of transplanting tomato is very effective in reducing root-knot nematodes.

The combination treatment with *T. harzianum* + *T. viride* each at 50 g (4 x 10^8 cfu/g) considerably increased the plant growth, yield and reduced the root galls and soil nematode population (Hassan and Sobita Devi, 2004).

Using Cultural and Chemical Methods: Combinations of deep ploughing (up to 20 cm) and nursery bed treatment with aldicarb at 0.4 g per m^2 and main field treatment with aldicarb/carbofuran at 1 kg a.i./ha proved effective in the control of root-knot nematodes in tomato which also registered maximum yield (Jain and Bhatti, 1985).

In tomato, application of aldicarb and carbofuran each at 1 kg a.i. per ha in combination with neem cake and urea each at 10 kg N per ha, at transplanting, produced maximum yield with lowest gall index (2.5) and nematode population, 90 days after planting (Routaray and Sahoo, 1985).

Using Botanicals and Physical Methods: Integration of soil solarization of nursery beds (with LLDPE transparent film of 25 μ for 15 days) with soil incorporation of Calotropis leaves at 4 kg/1.44 m^2 resulted in increased plant height (29.9 cm compared to 20 cm in control), fresh shoot weight (176.49 g compared to 100 g in control), transplantable seedlings (546/ m^2 compared to 297 in control) with minimum root-knot index (1.83 compared to 3.85 in control) (Patel *et al.*, 2006). Soil solarization with clear LLDPE film (25μ) for 15 days in hot summer in combination with poultry manure at 2 t/ha proved effective in the management of nematodes and higher production of transplants in tomato nursery with cost benefit ratio of 1:3.10.

Botanicals and Chemical Methods: Application of carbofuran at 1 kg a.i./ha in the nursery beds followed by neem cake at 400 kg /ha in the main field increased yield and reduced the gall index (Singh and Gill, 1998) (Table 13.3).

Table 13.3. Integrated management of root-knot nematodes in tomato with carbofuran and neem cake.

Treatment/Dose	% Reduction in root galling	% increase in yield
Carbofuran (1 kg a.i./ha) + Neem cake at 400 kg/ha	77.00	61.50
Carbofuran (1 kg a.i./ha) + Urea (23.8 kg/ha) + Neem cake at 200 kg/ha	67.90	49.20
Phenamiphos (1 kg a.i./ha) + Neem cake at 400 kg/ha	74.90	39.70
Phenamiphos (1 kg a.i./ha) + Urea (23.8 kg/ha) + Neem cake at 200 kg/ha	67.90	31.43

Integration of chopped castor leaves (40 to 60 g/kg soil) 1 or 2 weeks before transplanting with application of aldicarb or carbofuran each at 2 kg a.i./ha at transplanting, significantly reduced number of galls due to *M. javanica* and enhanced the growth of tomato (Ravi Dutt and Bhatti, 1986).

Integration of soil application of castor leaves with inorganic fertilizer (75 kg N/ha) enhanced plant growth of tomato and reduced *M. javanica* infestation (95% and 78% reduction in root galls and egg mass production, respectively) (Zaki and Bhatti, 1989).

13.2.2. The Reniform Nematode, *Rotylenchulus reniformis*

(i) Cultural Methods: Soil amendment with fresh glyricidia leaves, fresh sawdust and chicken manure controlled the root-knot and reniform nematode population on tomato (Castillo, 1985). The yield increases and the nematode suppressive effect of fresh chicken manure surpassed even phenamiphos (Table 13.4).

Table 13.4. Effect of soil amendment with organic materials on *M. incognita* and *R. reniformis* population and tomato fruit yield.

Treatment	No. of nematodes/300 ml soil & 1 g roots			Yield (t/ha)
	No. of weeks after treatment			
	2	6	11	
Phenamiphos at 10 kg a.i./ha	96	366	465	31.1
Rice straw at 10 t/ha	215	500	1921	29.1
Glyricidia leaves at 10 t/ha	201	533	1885	27.3

Saw dust at 10 t/ha	342	859	958	25.3
Chicken manure at 10 t/ha	17	169	1078	32.6
Control	552	816	2515	22.2

(i) Host Resistance: Balasubramanian and Ramakrishnan (1983) observed that Kalyanpur Selection-I, Kalyanpur-III, CA-121 exhibited resistant reaction.

(ii) Integrated Methods: Nursery bed treatment with *P. fluorescens* (with 1 x 10^9 spores/g) and *P. chlamydosporia* (with 4 x 10^6 spores/g) each at 20 g/m^2 and field application of 5 tonnes of enriched FYM with the above bioagents each at 5 kg, significantly reduced reniform nematodes in tomato by 72% over untreated check. The yield increase was up to 21.7% with a cost: benefit ratio of 1: 4.9.

Application of neem cake in the nursery at 100 g/ m^2 followed by carbofuran at 1 kg a.i./ha in the main field significantly reduced the soil population of *R. reniformis* and enhanced the fruit yield of tomato by 67% (Anitha and Subramanian, 1998).

13.2.3. The Stubby Root Nematode, *Trichodorus christiei*

(i) Cultural Methods: *Asparagus officinalis* would not support populations of *Trichodorus chiristiei* for more than 40 to 50 days. Tomato, normally a good host of this nematode, supported only a low population when asparagus was growing in the same pots. A glycoside (asparagusic acid) is the active principle involved which is toxic to *T. christiei* and several other nematode species.

13.2.4. Root-knot Nematode, *Meloidogyne incognita* and Wilt, *Fusarium oxysporum* f. sp. *lycopersici* Disease Complex

(i) Integrated Methods: Deep ploughing and exposing soil to hot sun in summer, removal and burning of crop debris, soil application of *Trichoderma viride* and *Paecilomyces lilacinus*, use of wilt resistant varieties like Utkal Pallavi, Utkal Deepti, Utkal Kumari, Utkal Urbasi, etc. help in controlling the disease complex. Application of *T. harzianum* at 50 kg/ha (2 x 10^8 cfu/g) was effective in controlling the disease complex and improving the plant growth parameters and fruit weight (Haseeb *et al.*, 2006).

Integration of mustard cake with *Glomus etunicatum* was effective in reducing the damage caused by *M. incognita* and *F. o.* f. sp. *lycopersici* on tomato (Bhagawati *et al.*, 2000).

Tomato roots that received *P. lilacinus* along with *T. harzianum* and neem cake were free from root-knot nematodes (*M. incognita*) and did not wilt due to *F.o.* f. sp. *lycopersici* till harvest. The roots were also free from *Fusarium* infection. The above treatment also reduced the % of wilt (10% compared to 90% in control), root-knot index (1.8 compared to 4.4 in control) and increased root colonization with bioagents (74% compared to 0% in control), parasitization of egg masses (56% compared to 0% in control) and eggs (44% compared to 0% in control) (Nagesh *et al.*, 2006).

13.3. BRINJAL

13.3.1. Root-knot Nematodes, *Meloidogyne* spp.

(i) Physical Methods: Burning of paddy husk or saw dust on infested nursery proved to be reasonably effective.

(ii) Cultural Methods: Several cultural methods like selection of nematode-free nursery sites, destruction of infested roots after crop harvest, crop rotation with non-host (maize, wheat, sorghum) or antagonistic crops (marigold, mustard, sesame), flooding, fallowing, deep summer ploughing and harrowing either alone or in combination proved to be reasonably effective and economical to check multiplication of root-knot, reniform and lesion nematodes on brinjal. Ploughing not only leads to disturbance and instability in nematode community but also causes their mortality by exposing them to solar heat and desiccation. Generally, 2 to 3 summer ploughings each at 10 days interval during April – May (40-46°C) have been recorded to reduce 96.5 % *M. javanica* population, while fallowing itself during the same period registered 44.5% reduction. Additional use of plastic sheets for covering soil either in nursery beds or in field further enhance nematode reduction. Such an approach also helps in reducing the intensity of weeds, fungi and bacteria in the soil.

Nursery bed treatment with neem cake at 200 g/sq.m. gave maximum reduction in nematode population giving an yield of 20.75 q/ha (Sheela and Nisha, 2004).

The maximum reduction in *M. javanica* population and increase in fruit yield was recorded in brinjal by soil ploughing + covering with polythene sheet, followed by soil ploughing + exposure to sun (Anon, 1989).

Decrease in *M. incognita* population on brinjal occurs when the field was left fallow or following crop rotation with marigold, spinach and bottle gourd (Khan *et al.*, 1975). Rotation of brinjal with sweet potato (cv. Sree Bhadra) reduced the root-knot nematode population by 47% and increased the fruit yield by 22%. Since the sweet potato cv. Sree Bhadra is a high yielding one, the farmer will get a good crop within 3 months. Thus it becomes an additional income generating practice (Sheela *et al.*, 2002). Crop rotation with sorghum, wheat and chilli reduced the root-knot nematode population. The population of *M. incognita* larvae decrease in combination of brinjal-wheat, brinjal-chilli-wheat and red gourd-brinjal-mustard. The root-knot larvae were completely eliminated in marigold-marigold-fallow and chilli-cauliflower-cauliflower. Garlic following brinjal also brought about considerable reduction in larvae of root-knot nematode. Intercropping with marigold, onion and garlic is also recommended. The population levels of *M. incognita* on highly infested land dropped to zero after 18 months of continuous cultivation of *Panicum maximum*. Growing of brinjal after *Panicum maximum* was completely free from root-knot nematodes (Netscher, 1983).

Intercropping brinjal with marigold, garlic or resistant tomato (SL-12) improved plant growth and reduced number of galls and final soil population of nematode (Jain *et al.*, 1990). Marigold (cv. Calcutta Yellow)-brinjal cropping sequence has been recommended to the farming community for management of *M. incognita* affecting brinjal.

Incorporation of neem leaves at 500 g/sq.m. or 250 ml of 5% neem leaf extract/ 5% neem cake extract/sq.m. in nursery beds of brinjal gave effective control of root-knot nematodes.

Row application of fly ash at 0.6 kg/sq. m. enhanced the fruit yield by 27.7% and inhibited the reproduction of *M. incognita* (Khan and Ghadipur, 2004).

(iii) Biological Methods: Application of 'Royal 350' (*Arthrobotrys irregularis* cultured on oat seed medium) at 140 g/m^2 a month before transplantation of brinjal resulted in good protection against root-knot nematodes. Among the egg parasites, efficacy of *Paecilomyces lilacinus* (commercially formulated as 'Biocon' in Philippines) has been found to be comparatively higher in suppressing the population of *Meloidogyne* spp. and *R. reniformis* on brinjal. Soil application of *P. lilacinus* at 20 g/sq. m. to brinjal nursery reduced infection of *M. incognita* by 50% and improved growth of seedlings by 30%.

Application of *Pasteuria penetrans* in nursery soil of brinjal resulted in 31 and 36% suppression in root galling when seedlings were transplanted to sterilized and nematode-infested soils, respectively. The results were also indicative of easy dissemination of *P. penetrans* to the main field through infected seedlings (Walia *et al.*, 1992).

Nursery bed treatment with *Bacillus macerans* at 25 g/sq.m. + soil drench (2% solution) ten days after sowing gave maximum reduction in nematode population and maximum yield of 38.25 q/ha (Sheela and Nisha, 2004).

Application of *P. chlamydosporia* at 50 g/m^2 in the nursery beds significantly increased the plant growth parameters (42 and 50% increase in height and weight of seedlings, respectively over control) and root colonization by the bioagent on transplants. When the above seedlings were transplanted in field, there was reduction in root galling index (5.2 compared to 7.8 in control), number of egg masses in 5 g root (20 compared to 49 in control), number of juveniles in 5 g root (62 compared to 162 in control) and number of juveniles in 100 cc of soil (67 compared to 152 in control) and increased plant growth and yield in eggplant at harvest (Naik, 2004).

Avermectins effectively reduced root-knot nematode infection in brinjal under field conditions (Parvatha Reddy and Nagesh, 2002).

(iv) Integrated Methods

Using Bio-agents and Botanicals: Integration of *P. chlamydosporia* (at 100 ml/seed pan containing 1.2 x 10^4 spores/ml) in castor cake (at 40 g/seed pan) amended soil was effective in increasing the seedling weight, colonization of roots with the bioagent. Brinjal seedlings raised in the above treatment transplanted in pots gave maximum increase in plant growth, root colonization and parasitization of eggs of *M. incognita* by *P. chlamydosporia*. The above treatment also gave least root galling and final nematode population both in soil and roots (Rao and Parvatha Reddy, 1993b).

Borkakaty (1993) observed that inoculation of *P. lilacinus* at 4 g/kg of soil in combination with mustard oil cake at 0.5 and 1.0 t/ha increased plant growth with corresponding decrease in number of galls, egg masses and eggs/egg mass of *M. incognita* on brinjal.

Application of 10% neem cake extract (at 20 ml/pot) mixed with spores of *P. lilacinus* (at 1 x 10^6 spores/ml) was effective in increasing plant growth and reducing root galling and final nematode population both in soil and roots. The above treatment also gave maximum parasitization of egg masses of *M. incognita* and spore density of *P. lilacinus* in soil (Rao and Parvatha Reddy, 1994).

Significant reduction in root galling and final nematode population of *M. incognita* were observed in brinjal seedlings which were given bare root-dip treatment in 10% neem leaf suspension mixed with spores of *P. lilacinus* (at 4 x 10^5 spores/ml) for 30 min. Significant increases were also observed in root colonization, parasitization of eggs and spore density of *P. lilacinus* in soil (Rao *et al.*, 1997b).

Root-dip treatment of brinjal seedlings in neem cake extract based formulation of *P. lilacinus* (at 5 x 10^6 spores/ml) for 20 min. and planted in pots gave significant increase in plant growth, root colonization, propagule density in soil and parasitization of eggs of *M. incognita* by *P. lilacinus* and drastic reduction in root galling, fecundity and final nematode population in soil and roots (Rao *et al.*, 1998).

Application of castor cake extract based formulation of *T. harzianum* (at 500 ml/m^2 containing 9.9 x 10^3 spores/ml) to nursery beds of brinjal was effective in producing vigorous seedlings (with maximum seedling weight) with least root galling. The above treatment also increased root colonization and parasitization of *M. incognita* females by *T. harzianum* (Rao *et al.*, 1998d).

Using Arbuscular Mycorrhizal Fungi and Botanicals: Brinjal seedlings raised in nursery treated with *G. fasciculatum* (at 250 g/m^2 containing 16 chlamydospores/g) planted in pots amended with castor cake (applied at 10 g/kg soil 15 days before transplanting) resulted in maximum plant growth, root colonization with *G. fasciculatum* and chlamydospore density in soil and least root galling, fecundity and final nematode population in soil and roots (Rao *et al.*, 1998c).

Integration of *G. fasciculatum* and neem cake at 0.5 t/ha was found to be effective in increasing plant growth parameters and yield (30 t/ha compared to 17 t/ha in control) and in reducing the root-knot index (2.6 compared to 4.8 in control) and final nematode population (173.3/250 ml soil compared to 510.8/250 ml soil in control) (Borah and Phukan, 2004).

Using Bio-agents and Arbuscular Mycorrhizal Fungi: Brinjal seedlings raised in seed pans treated with *G. mosseae* (at 100 g/seed pan containing 28-32 chlamydospores/g) dipped in *P. lilacinus* spore suspension (containing 4 x 10^5 spores/ml) for 5 min. transplanted in pots gave maximum increase in plant growth, root colonization, propagule density in soil of both *G. mosseae* and *P. lilacinus* and parasitization of eggs of *M. incognita*. The above treatment also gave least root galling, fecundity and final nematode population in soil and roots (Rao *et al.*, 1998).

Using Bio-agents, Arbuscular Mycorrhizal Fungi and Botanicals: Strategies to integrate botanicals, bio-agents and endomycorrhizae components for the management of root-knot and reniform nematodes infecting brinjal have been developed. These strategies include inoculation of *G. fasciculatum* in the castor cake amended nursery beds followed by the root-dip treatment of mycorrhizal seedlings of brinjal in spore suspension of *P. lilacinus* for the management of *R. reniformis* (Rao *et al.*, 1993b).

In nursery, soil application of neem cake at 400 g/m^2 along with *G. mosseae* (at 500 g/ m^2 containing 26-32 chlamydospores/g) and *P. lilacinus* (at 2 liters/ m^2 containing 6 x 10^5 spores/ml) resulted in production of healthy and vigorous brinjal seedlings colonized with *G. mosseae* and *P. lilacinus* and with least root galling. In field, transplanting of brinjal seedlings raised in the above treatment gave maximum reduction in root galling, egg mass production, fecundity and final nematode population in soil and roots. The above treatment also gave highest fruit yield, root colonization with *G. mosseae* and *P. lilacinus* and egg parasitization with *P. lilacinus* (Rao and Parvatha Reddy, 2001).

Using Two Bioagents: Combined soil application of *P. lilacinus* and *Aspergillus niger* at the time of transplanting brinjal is very effective in reducing root-knot nematodes.

In egg-plant, the nursery seedling stand index was good with seed treatment with *P. fluorescens* (50 g/kg) + soil application (10 g/m^2 seed bed) followed by soil application of neem based *T. harzianum* + *P. fluorescens* and neem based *P. fluorescens*, where the crop stand index was 4.6 and 4.0, respectively.

Using Bio-agents and Chemicals: Integration of a bio-agent, *P. lilacinus* with carbofuran at 1 kg a.i./ha was found effective in the management of reniform nematode, *R. reniformis* infecting brinjal (Parvatha Reddy and Khan, 1989).

Split application of *T. harzianum* at 50 kg/ha (10^8 cfu/g) (before transplanting and 45 days after transplanting) + carbofuran at 16.5 kg/ha was effective in increasing the brinjal fruit yield (Haseeb *et al.*, 2004).

Using Cultural Methods and Chemicals: The planting of marigold combined with application of carbofuran at 1 kg a.i./ha controls *M. javanica* infestation on brinjal (Singh, 1991).

Treatment of nursery beds with carbofuran at 3 kg a.i./ha along with summer ploughing and covering with polythene sheets gave better yield and controlled root-knot nematode in brinjal (Anon, 1989c) (Table 13.5).

Table 13.5. Integrated management of root-knot nematodes by summer ploughing/ fallowing and nursery bed treatment in brinjal.

Treatment	Reduction in nematode population (%)		Yield in kg/sq.m.	
	Hisar	Vellayani	Hisar	Vellayani
Ploughing + Fallowing for 15 days	87.00	66.20	28.98	---
Ploughing + Covering with polythene sheet for 15 days	94.90	77.00	18.84	78.00
No ploughing + No covering	77.60	---	---	---

Application of aldicarb at 1.0 kg a.i./ha in the nursery beds along with neem cake at 400 kg/ha increased yield and reduced the root galling (Singh and Gill, 1998) (Table 13.6).

Table 13.6. Integrated management of root-knot nematodes in brinjal with neem cake and nematicides.

Treatment	% Reduction in gall index	% Increase in yield
Aldicarb (1.0 kg a.i./ha) + Neem cake (400 kg/ha)	92.90	58.70
Carbofuran (1.0 kg a.i./ha) + Neem cake (400 kg/ha)	68.10	68.10

Nursery treatment with carbofuran at 0.3g a.i./m^2 along with main field treatment with ploughing + exposing the field and ploughing + covering with polythene sheets for 15 days improved the plant growth and yield of brinjal by 20-32% (Sheela *et al.*, 2002).

Using Cultural and Physical Methods: Integrating summer ploughing with soil solarization with polythene mulching effectively reduced *M. javanica* population in brinjal (Jain and Gupta, 1991). Soil solarization of nursery beds (using 100 gauge LDPE clear film for 15 days) and application of neem cake at 200 g/sq.m. proved most effective in reducing root-knot nematode infestation and increasing yield in brinjal. Solarization of nursery beds for 15 days in summer and application of poultry manure at 200 kg/ha gave maximum yield.

13.3.2. Root-knot Nematode, *Meloidogyne incognita* and Bacterial Wilt, *Ralstonia solanacearum* Disease Complex

(i) Integrated Methods

Combined application of *Trichoderma harzianum* and *Pochonia chlamydosporia* gave effective management of the disease complex and gave least root gall index (RGI), least wilt index (WI) and maximum fruit yield (1.8 RGI, 22.18% WI and 2.024 kg/3 sq.m. yield) followed by *Pseudomonas fluorescens* + *P. chlamydosporia* (1.81 RGI, 25.13 WI and 1.782 kg/3 sq.m. yield) and *Trichoderma harzianum* + *P. fluorescens* (1.93 RGI, 27.28 WI and 1.680 kg/3 sq.m. yield) (Naik, 2004).

13.4. CHILLI

13.4.1. Root-knot Nematodes, *Meloidogyne* spp.

(i) Cultural Methods: Decrease in *M. incognita* population on chillies occurs when the field was left fallow or following crop rotation with marigold, spinach and bottle gourd (Khan *et al.*, 1975). Growing of chilli along with marigold gave highest degree of reduction in root-knot nematode population on chilli roots, followed by onion, garlic and asparagus (Trivedi and Tiagi, 1984).

Row application of fly ash at 0.6 kg/sq.m. enhanced the fruit yield by 21.3% and inhibited the reproduction of *M. incognita* (Khan and Ghadipur, 2004).

Botanicals: Soil application of decaffeinated tea leaves effectively reduced root galling on chilli plant (Trivedi and Tiagi, 1984).

(ii) Biological Methods: Pandey and Trivedi (1992) reported a significant reduction in number of galls, eggs/egg mass, hatching and final soil population of *M. incognita* infecting chillies in the presence of *Paecilomyces lilacinus*.

Avermectins effectively reduced root-knot nematode infection in chilli under field conditions (Parvatha Reddy and Nagesh, 2002).

(iii) Host Resistance: Chilli cvs. Pusa Jwala, Wonder Hot, Teja, Utkal Abha, Utkal Rashmi were found resistant to root-knot nematodes (Patnaik *et al.*, 2004).

13.4.2. Root-knot Nematode, *Meloidogyne javanica* and *Fusarium oxysporum* Disease Complex

(i) Integrated Methods: *Pseudomonas aeruginosa* and *Paecilomyces lilacinus* when used together significantly reduced infection of the disease complex on chilli (Perveen *et al.*, 1998).

13.5. CAPSICUM

13.5.1. Root-knot Nematodes, *Meloidogyne* spp.

(i) Physical Methods: Burning of paddy husk or saw dust on infested nursery proved to be reasonably effective.

(ii) Cultural Methods: Several cultural methods like selection of nematode-free nursery sites, destruction of infested roots after crop harvest, crop rotation with non-host (maize, wheat, sorghum) or antagonistic crops (marigold, mustard, sesame) or intercropping with marigold, onion and garlic, flooding, fallowing, deep summer ploughing and harrowing either alone or in combination proved to be reasonably effective and economical to check multiplication of root-knot, reniform and lesion nematodes on capsicum. Ploughing not only leads to disturbance and instability in nematode community but also causes their mortality by exposing them to solar heat and desiccation. Generally, 2 to 3 summer ploughings each at 10 days interval during April – May (40-46°C) have been recorded to reduce 96% *M. javanica* population while fallowing itself during the same period registered 45.5% reduction. Additional use of plastic sheets for covering soil either in nursery beds or in field further enhance nematode reduction. Such an approach also helps in reducing the intensity of weeds, fungi and bacteria in the soil.

(iii) Biological Methods: Application of 'Royal 350' (*Arthrobotrys irregularis* cultured on oat seed medium) at 140 g/m^2 a month before transplantation of capsicum resulted in good protection against root-knot nematodes. Among the egg parasites, efficacy of *Paecilomyces lilacinus* (commercially formulated as 'Biocon' in Philippines) has been found to be comparatively higher in suppressing the population of *Meloidogyne* spp. and *R. reniformis* on capsicum.

Field application of a nematicidal *B. thuringiensis* strain to capsicum in Puerto Rico reduced galling in roots due to *M. incognita* and soil population of *Rotylenchulus reniformis* and increased yield significantly (Zuckermann *et al.*, 1993).

Bio-agents *P. chlamydosporia* and *Pseudomonas fluorescens* were evaluated under field conditions for their efficacy against root-knot nematode *M. incognita* infecting capsicum. Treatment of the nursery bed with the formulation of *P. chlamydosporia* at 50g/m^2 was significantly effective in reducing the galling index, number of nematodes in roots and soil, increasing the percent parasitization of eggs by bioagent and also yields. Seed treatment with *P. fluorescens* alone and the nursery bed treatment with *P. chlamydosporia* alone were effective. The main purpose of these studies was to produce capsicum seedlings that are colonized by *P. chlamydosporia* before transplanting so that they could carry the bio-agent to the field. During this process the field soil would be enriched with the propagules of eco-friendly component of management (here it is a bio-agent) in two or three seasons (Naik, 2004).

(iv) Host Resistance: Capsicum cvs. Mississipi-68, Santanka, Anaheim Chile and Italian Pickling are reported to be resistant to root-knot nematodes (Hare, 1951).

(v) Integrated Methods: Integration of *P. fluorescens* and *P. chlamydosporia* in the nursery bed has proved significantly effective in reducing the root-galling index (*M. incognita*), number of nematodes in the roots and soil and increasing the yield of the capsicum crop under field conditions (Naik, 2004).

In capsicum, the seedling stand was good where the combination of neem based *P. fluorescens* and *T. harzianum* was used (4.4), followed by seed treatment with *P. fluorescens* + soil application of *T. harzianum* (Naik, 2004).

13.5.2. Root-knot Nematode, *Meloidogyne incognita* and Bacterial Wilt, *Ralstonia solanacearum* Disease Complex

(i) Integrated Methods: Combined application of neem based formulations of *Pseudomonas fluorescens* and *Pochonia chlamydosporia*/ *Trichoderma harzianum* at 25 g/m^2 in nursery beds and root dip treatment with either of these bioagents before transplanting resulted in significant reduction in disease complex in capsicum to the tune of 70% and increased the crop yield by 37% (Rao *et al.*, 2002) (Table 13.7).

Table 13.7. Effect of integration of neem-based bioagents on the growth of transplants and management of disease complex and yield of capsicum.

Treatment	Seedling weight (g)	Root-knot index (1-10)	Disease index (1-9)	Yield in kg/ 4 sq.m.
Seed treat. with *P. fluorescens*	421	5.6	6.4	4.3
Seed treat. with neem-based *P. fluorescens*	428	5.2	6.7	4.7
Nursery treat. with *P. fluorescens*	435	4.6	5.4	4.8
Nursery treat. with *P. chlamydosporia*	364	4.4	7.3	3.2
Nursery treat. with *T. harzianum*	374	4.8	7.0	3.4
Nursery treat. with *P. fluorescens* + *P. chlamydosporia*	463	4.1	5.2	4.0
Nursery treat. with *P. fluorescens* + *T. harzianum*	493	3.8	3.5	5.1
Control	340	8.7	8.2	2.6
CD at 5%	27.20	0.49	0.38	0.25

Naik (2004) reported that combination of *T. harzianum* along with *P. fluorescens* increased the yield (3.320 kg/3m^2 plot) followed by combination of *T. harzianum* and *P. chlamydosporia* (2.960 kg/3m^2). The least gall index was present where the combination of *T. harzianum* and *P. fluorescens* was used (1.5) followed by *T. harzianum* + *P. chlamydosporia* (1.7) and *P. fluorescens* + *P. chlamydosporia* (1.7). But among combination treatment of bio-agents, all the treatments were on par (*P. fluorescens* + *T. harzianum*, *P. fluorescens* + *P. chlamydosporia*, *T. harzianum* + *P. chlamydosporia*). Results related to the percent bacterial wilt disease (*Ralstonia solanacearum*) incidence also showed similar trend in which *T. harzianum* + *P. fluorescens* performed very well and the percent incidence was 16.90 followed by *T. harzianum* + *P. chlamydosporia* (17.20).

13.6. CUCUMBER

13.6.1. Root-knot Nematodes, *Meloidogyne* spp.

(i) Physical Methods: The most important nematode pests controlled by greenhouse steaming are root-knot nematodes attacking cucumbers .

(ii) Integrated Methods: Combined inoculation of AMF and *Pseudomonas fluorescens* was found effective on root-knot nematodes on cucumber (Jakobsen, 1999).

13.7. LETTUCE

13.7.1. Root-knot Nematodes, *Meloidogyne* spp.

(i) Physical Methods: The most important nematode pests controlled by greenhouse steaming are root-knot nematodes on lettuces.

(ii) Cultural Methods: Planting of 4 and 5 seedlings of *Tagetes erecta* in a pot around a lettuce plant reduced root galling by 56 and 72% and nematode population by 73 and 94%, respectively (Pervez *et al.*, 1988).

13.7.2. The Reniform Nematode, *Rotylenchulus reniformis*

(i) Physical Methods: Under field conditions, soil solarization significantly reduced soil population of the reniform nematode up to 15 cm deep and increased the yield of lettuce.

13.8. CABBAGE AND CAULIFLOWER

13.8.1. The Stunt Nematode, *Tylenchorhynchus brassicae*

(i) Cultural Methods: There was sudden decline in population of *T. brassicae* when cabbage and cauliflower were rotated with wheat (Siddiqi *et al.*,1973). Intercropping of cabbage and cauliflower with neem reduced the multiplication rate of *T. brassicae* and improved the plant health (Siddiqi and Saxena, 1987b). The mustard (rabi), radish (summer), sesame (kharif) sequence considerably reduced the population of *Tylenchorhynchus* spp. (Haque and Gaur, 1985).

Botanicals: Amendment of soil with oil cakes of neem, groundnut, mustard and castor was effective in reducing the population of *T. brassicae* around the roots of cabbage and cauliflower (Siddiqi *et al.*, 1976).

13.8.2. Root-knot Nematode, *Meloidogyne incognita* and Club Root, *Plasmodiophora brassicae* Disease Complex

(i) Integrated Methods: PGPR strains (*Pseudomonas fluorescens, Bacillus subtilis*) combined with fungal biocontrol agents (*Trichoderma viride, T. harzianum*) were found to be effective in reducing the nematode-fungal disease complex in cabbage (Loganathan *et al.*, 2001). The bioformulation mixture of *P. fluorescens, T. viride* and chitin effectively reduced the disease complex both in cabbage and cauliflower both under greenhouse and field conditions (Samiyappan, 2003) (Table 13.8).

Table 13.8. Efficacy of bioformulation mixtures against club root disease – root-knot nematode complex in cabbage under greenhouse conditions.

Treatment	Club root index	Nematode incidence	
		Population	Root-knot index
Trichoderma viride	25.99 (30.65)	129	2.66
Pseudomonas fluorescens	28.20 (32.07)	112	2.33
T. viride + *P. fluorescens*	25.33 (30.22)	114	2.33
T. viride + Chitin	25.44 (30.29)	108	2.33
P. fluorescens + Chitin	25.66 (30.43)	111	2.00
T. viride + *P. fluorescens* + Chitin	22.22 (28.12)	108	2.00
Chitin alone	31.70 (34.26)	139	3.00
Carbendazim	19.90 (26.49)	264	4.66
Carbofuran	39.90 (39.17)	106	1.66
Cabendazim + Carbofuran	15.00 (22.79)	103	1.66
Plasmodiophora brassicae alone	48.90 (44.37)	0.033	0.133
Meloidogyne incognita alone	0.03 (0.60)	280	5.00

13.9. RADISH

13.9.1. Stunt Nematodes, *Tylenchorhynchus* spp.

(i) Cultural Methods: The mustard (rabi), radish (summer), sesame (kharif) rotation sequence considerably reduced the population of *Tylenchorhynchus* spp. (Haque and Gaur, 1985).

13.10. OKRA

13.10.1. Root-knot Nematodes, *Meloidogyne* spp.

(i) Cultural Methods: Decrease in *M. incognita* population in okra, occurs when the field was left fallow (Khan *et al.*, 1975). The maximum reduction in *M. javanica* population and increase in fruit yield was recorded in okra by soil ploughing + covering with polythene sheet, followed by soil ploughing + exposure to sun (Anon, 1989b).

Decrease in root-knot nematode population on okra occurs following marigold, spinach and bottle gourd (Khan *et al.*,1975). Out of 15 different cropping sequences studied, tomato-onion-resistant tomato cv. Hisar Lalit-okra was found to be the best in terms of root-knot nematode reduction and economic returns (Kanwar, 1990). Crop rotations having okra-garlic-cluster bean and okra-coriander-resistant tomato cv. Hisar Lalit effectively managed the root-knot nematode population (Anon, 1993b).

Atwal and Mangar (1969) showed that root exudates from sesame (*Sesamum orientale*) have nematicidal properties against *M. incognita*. When okra was grown in *M. incognita* infested soil it was only slightly attacked and there were fewer nematodes compared with when okra was grown in the absence of sesame. Intercropping of okra with sesame decreased root-knot nematode penetration in okra and also delayed the life cycle of *M. incognita* (Tanda and Atwal, 1988).

Rotation of okra with sweet potato (cv. Sree Bhadra) reduced the root-knot nematode population by 21% and increased the fruit yield by 20.57% (Sheela *et al.*, 2002). Crop rotation with cabbage, marigold and wheat significantly reduced the root-knot nematode population. Crop rotation with cereals and marigold helps in reducing the nematode population in soil. In cropping sequence involving cabbage-marigold-wheat-kochia, the population of root-knot nematode larvae reduced to such an insignificant level that growing okra following the above sequence did not bring about increase in number of larvae (Alam *et al.*, 1977).

Botanicals: Singh and Sitaramaiah (1966) applied finely divided oil cakes to root-knot infested soil and noted a reduction in disease incidence in okra. Soil application of neem cake at 2 t/ha recorded lowest root-knot index (1.6), followed by fresh neem leaves at 5 t/ha (1.6) and silk worm litter at 2.5 t/ha (1.7). Significant highest yield was also obtained with neem oil cake (6.2 t/ha) followed by neem leaves (6.1 t/ha) (Sahoo *et al.*, 2004b).

Best results in respect of root-knot reduction due to *M. javanica* and increase in yield of okra were obtained by amending the soil with saw dust at 2.5 tons per ha 3 weeks before planting and then applying N through urea at 120 kg per ha (Singh and Sitaramaiah, 1971b).

Spot application of groundnut cake and sunflower cake each at 0.25 t/ha, saw dust and coal ash each at 5 t/ha significantly reduced the root-knot nematode population (Bhosle *et al.*, 2006).

Application of press mud at 15 t/ha, a week prior to sowing is recommended for effective management of root-knot nematodes in kharif season with cost benefit ratio of 1:3.1. Application of fresh neem or karanj cake at 1 t/ha at the time of final land preparation is recommended.

Bala and Sukul (1987) observed that Eugenol, an active nematicidal principle from *Ocimum sanctum* showed systemic effect in reducing root-knot infection in okra. Ahuja and Mukhopadhayaya (1984) observed that application of neem leaves and seeds of Persian lilac maximally decreased the root-knot index due to *M. incognita* and significantly increased the yield of okra fruits.

(ii) Biological Methods: Root-knot nematodes in okra can be managed by the application of *Bacillus macerans* at 1.2 x 10^8 cells/sq.m. before sowing.

P. lilacinus reduced root-knot nematode egg masses by 91% after 60 days and by 96% after 90 days, while aldicarb reduced egg masses by 90 and

91% after 60 and 90 days after inoculation, respectively in okra. In general, *P. lilacinus* and aldicarb treatments greatly suppressed the numbers of nematode galls and egg masses on okra plants. However, it was evident that *P. lilacinus* was more effective in reducing nematode egg masses than aldicarb (Ibrahim and Rezk, 1988).

The soil incorporation of *P. lilacinus* (in rice hulls and bran soil mix) was effective in reducing the gall index (2.7 as against 5.0 in control), nematode population in soil (77.3% over control) and in increasing the number of okra fruits (121.7 as against 86.0 in control) and fruit weight (3.26 kg as against 2.11 kg in control) (Davide and Zorilla, 1986).

(iii) Host Resistance: Okra cv. Parbhani Kranthi was found moderately resistant to *M. incognita*.

(iv) Integrated Methods

Using Bio-agents and Botanicals: Integration of *Paecilomyces lilacinus* with neem cake gave effective control of *M. incognita* on okra.

Application of 5 per cent inoculum of *Arthrobotrys conoides* to the pot soil amended with FYM, effectively reduced the larval penetration of *M. incognita* and root galling was reduced by 30-40 per cent in okra (Srivastava and Swarup, 1986).

Soaking of okra seeds in 10% castor cake suspension mixed with spores of *P. lilacinus* (1.5 x 10^6 spores/ml) for 30 min. and sowing in soil drenched with 10% castor cake suspension at 20 litres/ 6 sq.m.) was effective in reducing root galling, final nematode population of *M. incognita* and increasing the fruit yield, root colonization, propagule density in soil and parasitization of eggs by *P. lilacinus* (Rao *et al.*, 1997a).

Using Botanicals and Chemicals: An integrated management of *M. incognita* infecting okra using neem or karanj oil cake at 0.5 t/ha along with carbofuran at 1 kg a.i./ha gave maximum reduction in root galling with consequent increase in okra fruit yield (Parvatha Reddy and Khan, 1991).

Application of aldicarb at 1 kg a.i./ha + neem cake at 0.5 ton/ha followed by carbofuran at 1 kg a.i./ha + neem cake at 0.5 ton/ha proved most effective in reducing the *R. reniformis* population and in increasing the growth of okra plants (Krishna Rao *et al.*, 1987).

Application of subabool (*Leucaena leucophila)* leaves at 40 g/kg soil which were allowed to decompose for four weeks before sowing + carbofuran at 1 kg a.i./ha while sowing of okra seeds resulted in minimum galling (35.4 galls/plant) in comparison to control (64 galls/plant) and better plant growth parameters in okra (Paruthi *et al.*, 1987).

Using Bio-agents, Chemicals and Botanicals: The combined treatment with *P. lilacinus* at 4 g/kg soil + carbosulfan 25 EC at 0.2% + poultry manure at 2.5 t/ha + FYM at 2.5 t/ha gave maximum increase in plant growth parameters and yield (9.2 t/ha compared to 2.0 t/ha in control), and decrease in number of galls (83.5 compared to 183.6 in control), egg masses per root system (23.6 compared to 70.5 in control) and final nematode population in soil (200 compared to 585 in control) (Das and Sinha, 2005).

Using Cultural and Chemical Methods: Deep ploughing (20 cm) followed by fallowing for one month, fallowing for one month after weeding, integration of aldicarb application at 0.8 kg a.i. per ha at sowing after either of the cultural practices or deep ploughing (20 cm) together with carbofuran or aldicarb seed treatment resulted in the control of the reniform nematode (*Rotylenchulus reniformis*) and better yield of okra (Lakshmanan and Sivakumar, 1981).

Using Cultural and Physical Methods: Integration of soil solarization for 15 days in summer and application of neem cake at 200 kg/ha is effective in the management of root-knot nematodes and in getting higher yields.

Using Physical and Chemical Methods: Soil solarization with single layer of polyethylene mulch for 20 days during June and application of carbofuran at 0.5 kg a.i./ha gave least root galling (42/plant compared to 245/plant in control) (Sharma *et al.*, 2005).

13.10.2. The Reniform Nematode, *Rotylenchulus reniformis*

(i) Cultural Methods: Application of potash in combination with phosphorus or nitrogen or potash alone checks the reniform nematode multiplication on okra to a great extent (Sivakumar and Meerazainuddin, 1974).

(ii) Biological Methods: Seed treatment with avermectin 100% recorded maximum fruit yield and highest reduction in *R. reniformis* population. Avermectin treated plants also recorded significantly increased total phenol content and enzymatic activities such as peroxidase and IAA oxidase (Jayakumar *et al.*, 2004).

13.10.3. Root-knot Nematode, *Meloidogyne incognita* and Wilt, *Fusarium oxysporum* f. sp. *vasinfectum* Disease Complex

(i) Biological Methods: Among botanicals and biocontrol agents evaluated, *Trichoderma harzianum* was the most effective treatment in improving plant growth followed by neem seed powder, neem cake and *Aspergillus niger*. Highest suppression of root colonization of the fungus (*F. o.* f. sp. *vasinfectum*) was achieved in plants treated with neem seed powder (7.5), neem cake (12.5), *A. niger* (15.0) and *T. harzianum* (17.5). The greatest suppression of root-knot development was achieved by *T. harzianum* followed by neem seed powder, neem cake and *A. niger* (Abuzar and Haseeb, 2006).

13.10.4. Root-knot Nematode, *Meloidogyne incognita* and Root Rot, *Rhizoctonia solani* Disease Complex

Biological Methods: *Paecilomyces lilacinus* can be effectively used against *Rhizoctonia solani* – *M. incognita* complex of okra (Shahzad and Ghaffar, 1989).

13.11. CARROT

13.11.1. Root-knot Nematodes, *Meloidogyne* spp.

(i) Cultural Methods: The population of *M. incognita* larvae decreased in combination of spinach-marigold-carrot and spinach-radish-carrot (Hasan and Jain, 1998).

Botanicals: The oil cakes (neem, groundnut, mustard and castor) were effective in reducing parasitic nematode population (*Hoplolaimus, Tylenchorhynchus, Meloidogyne* and *Helicotylenchus* species) and in increasing yields of carrot .

ii) Biological Methods: Application of *P. lilacinus* (containing 1 x 10^6 spores at 10 g/ 2.5 sq. m.) recorded least gall index (1.6 as against 5.0 in control), reduced nematode population by 45.23%, reduced forking by 92.4% and increased the yield by 147.6% (Sivakumar, 1998).

(iii) Integrated Methods: A soil fumigant at the beginning of two consecutive carrot crops, followed by two year onion production and at last a cover crop sudax in the course of 5 year rotation can control *M. hapla* problem (Bird, 1981).

Seed treatment with *T. harzianum* at 10 g (with 1 x 10^6 cfu/g)/kg and *P. fluorescens* at 10 g (with 1 x 10^9 cfu/g)/kg and subsequent field application of 5 tonnes of enriched FYM with *T. harzianum* (with 1 x 10^6 cfu/g) and *P. fluorescens* (with 1 x 10^9 cfu/g) per ha significantly reduced root-knot and reniform nematodes in carrot roots by 79 and 75%, respectively. The yield increase was to the tune of 29.8% with a cost: benefit ratio of 1:13.6.

13.12. TURNIP

13.12.1. Root-knot Nematodes, *Meloidogyne* spp.

(i) Cultural Methods: The oil cakes (neem, groundnut, mustard and castor) were effective in reducing parasitic nematode population (*Hoplolaimus, Tylenchorhynchus, Meloidogyne* and *Helicotylenchus* species) and in increasing yields of turnip.

13.13. FRENCH BEAN

13.13.1. Root-knot Nematode, *Meloidogyne* spp.

(i) Cultural Methods: Intercropping of French bean with finger millet, chilli and groundnut helps in the management of *M. incognita* Race 2 (Ramappa, 1983). Growing of chilli or groundnut or finger millet in rotation with French bean significantly reduced the root-knot nematode population (Ramappa, 1988).

Botanicals: Spot application of neem cake at 16.6 g/spot 15 days before sowing was found effective in reducing root galls (64.4 compared to 122.9 in control), number of egg masses (16.0 compared to 44.4 in control) and final nematode population (253 compared to 437 in control) and in increasing plant growth and yield (3.375 t/ha compared to 0.715 t/ha in control) (Ahmed and Choudhury, 2004).

13.13.2. The Reniform Nematode, *Rotylenchulus reniformis*

(i) Cultural Methods: Neem and karanj oil cakes at 2 tons per ha were most effective in reducing *Rotylenchulus reniformis* infecting French bean.

13.13.3. Root-knot Nematode, *Meloidogyne incognita* and Root Rot, *Macrophomina phaseolina* Disease Complex

(i) Cultural Methods: The beneficial effects of neem cake were observed not only against *M. incognita* and the fungus *Macrophomina phaseolina* individually, but also when both the pathogens formed a disease complex on French bean.

13.14. COWPEA

13.14.1. The Root-knot Nematode, *Meloidogyne incognita*

(i) Cultural Methods: The root-knot and reniform nematodes associated with cowpea can be effectively managed by the application of neem or eupatorium leaves at 15 t/ha two weeks before sowing.

(ii) Biological methods: Cowpea seed treatment with *Paecilomyces lilacinus* spores was reported to be effective in reducing the root-knot infestation level (Midha, 1985). Application of certain organic materials, viz. subabool leaves, neem cake or saw dust prior to the application of *P. lilacinus* not only enhanced the growth and multiplication of the fungus resulting in high percentage of egg mass infection but also effectively reduced the soil incidence in cowpea (Hasan, 1988). Application of *P. lilacinus* at 15 kg/ha along with certain organic materials effectively reduced the root-knot nematode incidence in cowpea (Hasan and Jain, 1992).

(iii) Integrated Methods: Hasan and Jain (1992) reported that soil application of *P. lilacinus* cultured on sorghum seeds together with certain organic matter effectively reduced the incidence of *M. incognita* and increased the crop yield of cowpea.

13.14.2. The Reniform Nematode, *Rotylenchulus reniformis*

(i) Physical Methods: Under field conditions, soil solarization significantly reduced soil population of the reniform nematode up to 15 cm deep and increased the yield of cowpea.

(ii) Cultural Methods: Soil application of neem seed kernel at 2.5 t/ha was most effective in reducing the number of females/plant (46.2 compared to 80.7 in check), number of egg masses/plant (26.0 compared to 45.5 in check), number of eggs and larvae/egg mass (46.2 compared to 80.7 in check) and final soil population/200 cc soil (143.5 compared to 831.0 in check) of the reniform nematode (Ram and Baheti, 2004).

(iii) Host Resistance: Cowpea cvs. V-16 and Pusa Phalguni are reported to be resistant to *R. reniformis* (Thaker and Patel, 1984).

13.14.3. The Pigeon pea Cyst Nematode, *Heterodera cajani*

(i) Host Resistance: Cowpea cv. CO 6 was found resistant to the pigeon pea cyst nematode.

13.15. PEA

13.15.1. Root-knot Nematodes, *Meloidogyne* spp.

(i) Cultural Methods: Intercropping with marigold or mustard with pea reduced the damage of root-knot and reniform nematodes.

13.16. CLUSTER BEAN

13.16.1. Root-knot Nematode, *Meloidogyne javanica* and Wilt, *Fusarium solani* Disease Complex

(i) Integrated Methods: *Pseudomonas aeruginosa* and *Paecilomyces lilacinus* when used together significantly reduced infection of the disease complex on cluster bean (Perveen *et al.*, 1998).

13.17. ONION AND GARLIC

13.17.1. Root-knot Nematodes, *Meloidogyne* spp.

(i) Physical Methods: Hot water treatment of onion bulbs at 43.5°C for 120 min eliminates the nematode infection.

(ii) Cultural Methods: The lowest populations of plant parasitic nematodes were recorded in onion-cowpea-fallow cropping sequence (Vetrivelkalai and Subramanian, 2006).

(iii) Integrated Methods: Root dip treatment of onion seedlings in *P. lilacinus* + neem cake suspension for 1 hr before transplanting proved effective.

13.18. PUMPKIN & GOURDS

13.18.1. Root-knot Nematodes, *Meloidogyne* spp.

(i) Physical Methods: Pulverizing the planting pit soil and exposing to sunlight by repeated raking and heaping 4" to 6" dry trash and burning it before application of compost helps in reducing the nematode population.

(ii) Host Resistance: Pumpkin cvs. Jaipuri and Dasna were found resistant to root-knot nematodes.

(iii) Integrated Methods: Addition of 250 g fresh neem cake or karanj cake enriched with *Trichoderma harzianum*/*Paecilomyces lilacinus* per planting pit effectively controlled the nematodes.

13.18.2. Root-knot Nematode, *Meloidogyne javanica* and Root rot, *Macrophomina phaseolina*, *Fusarium oxysporum* Disease Complex

(i) Integrated Methods: *Pseudomonas aeruginosa* and *Paecilomyces lilacinus* when used together significantly reduced infection of the disease complex on pumpkin (Perveen *et al.*, 1998).

13.19. POINTED GOURD

13.19.1. The Root-knot Nematode, *Meloidogyne incognita*

(i) Cultural Methods: Planting of marigold, *Tagetes patula* at 3 to 5 plants/pit will manage attack of root-knot nematodes on pointed gourd.

(ii) Integrated Methods: Integration of *Paecilomyces lilacinus* (50 g) + *Trichoderma harzianum* (100 g) + neem cake (250 g) + marigold (3 plants/pit) increased plant growth parameters and yield (6.6 kg/plant compared to 1.4 kg/plant in control), and reduced root galling (14.3/5 g roots compared to 75.0/ g roots in control), number of egg masses (40/5 g roots compared to 150/ g roots in control) and final nematode population (66.6/100 g soil compared to 86660/ 100 g soil in control) (Verma *et al.*, 2005).

13.20. BOTTLE GOURD

13.20.1. Root-knot Nematodes, *Meloidogyne* spp.

(i) Cultural Methods: Application of press mud at 10 t/ha (ICBR 1:18.95) or mustard cake at 3.333 t/ha (ICBR 1:16.10) or neem cake at 4.286 t/ha one week in advance of sowing gave effective management of root-knot with increased yield of bottle gourd (Dahiya *et al.*, 1998).

13.21. RIDGE GOURD

13.21.1. Root-knot Nematodes, *Meloidogyne* spp.

(i) Host Resistance: Ridge gourd cvs. Panipati and Meerut Special were found resistant to root-knot nematodes.

13.22. WATERMELON

13.22.1. Root-knot Nematodes, *Meloidogyne* spp.

(i) Host Resistance: Watermelon cv. Shehjanpuri was found resistant to root-knot nematodes.

13.23. MUSHROOMS

13.23.1. Nematodes, *Ditylenchus myceliophagus, Aphelenchoides composticola, A. agarici, A. myceliophagus, A. sacchari*

(i) Physical Methods: Use of heat has been the most successful method of nematode control in mushroom cultivation during composting and casing process as well as at the end of cropping. The air as well as bed temperature should be maintained at 60°C for at least 2 hours during peak heating of compost. Thorough pasteurization of casing soil with steam for at least 4 hours gave good control of nematodes. Treatment of used trays and handling tools in boiling water for 1-2 minutes or in disinfectants like formalin is recommended.

An effective method of casing pasteurization, based on the use of solar heat, was developed wherein the thinly spread casing is exposed to the sun under a polythene cover (150 gauge) for 5-6 hours (Grewal and Grewal, 1988).

Wood and Goodey (1957) found that r-rays in between 48,000 and 96,000 r inactivated *Ditylenchus myceliophagus* in mushroom compost.

(ii) Cultural Methods: Incorporation of neem/pongamia/coconut cake at 2kg/100 kg of compost during first or second turning is recommended. Dry leaf powder of neem can be mixed with compost at 2-5% concentration. This has been found useful not only to check nematode multiplication in mushroom growing trays but also helps in proper mycelial growth and improved sporophore yields (Khanna and Sharma, 1994). Addition of oil cakes like neem, karanj, coconut, castor and groundnut have been reported to control nematodes and increase button production (Rao *et al.*, 1992). Further, incorporation of fresh leaves of Pedilanthus, eucalyptus and pongamia in the compost during first or second turning was observed to reduce the population of mushroom nematodes and to increase the yield of button mushrooms. The incorporation of dried leaves of neem, castor and *Cannabis sativa* resulted in the suppression of *A. composticola* below economic injury levels and increased mushroom yields by 6.5 to 19% (Grewal, 1989).

(iii) Bioloical Methods: Nematode trapping fungus, *Arthrobotrys robusta* commercially available as 'Royal 300' (strain Antipolis) in France has been recommended for the management of mushroom nematodes. *Arthrobotrys oligospora* and *A. superba* have also been found to reduce nematode populations and increase the yields. In India, *Arthrobotrys irregularis* and *Candelatrella musiformis* isolated from spent compost have shown promising results in checking nematode population (Khanna and Sharma, 1990). *A. irregularis* was also found to be highly effective against *Aphelenchoides composticola*. Further studies revealed that *A. irregularis* did not compete with *Agaricus bisporus* mushroom and checked the nematode population significantly and increased the yield.

Grewal and Sohi (1988) mass produced *Arthrobotrys conoides* on wheat grain and, after incorporating it into compost at spawning, reduced a mixed nematode population of *A. composticola* and *Rhabditis* sp. by 42.5% in 25 days.

The bacterium, *Bacillus thuringiensis* is found to suppress the population of *Ditylenchus myceliophagus* (Cayrol, 1974).

(iv) Host Resistance: *Agaricus edulis* has been reported to be resistant to *D. myceliophagus*. Thapa *et al.* (1987) reported that *Pleurotus sajor caju* and *Stropharia rugoso annulata* were found resistant to both *A. sacchari* and *D. myceliophagus*. K 30 strain of *A. bitorquis* showed resistance to *A. sacchari* only. Resistance in *Pleurotus sajor caju* against *A. sacchari, A. agarici* and *A. composticola* has been reported by several workers.

(v) Integrated Methods

- Cropping should be done in purposely built mushroom houses with proper ventilation.
- Cleanliness should be maintained inside and surroundings of mushroom farm so that flies, mites and other contaminants are avoided by spraying 2% formalin solution.
- Composting platform must be cemented to prevent the direct contact of compost with soil.
- No person or worker should be allowed to enter into the farm without proper disinfestation of their hands and feet.
- Wooden trays and handling tools should be disinfected before use by dipping them in boiling water or in 5% formalin for 2 minutes before use.
- Compost should be steamed at 60°C for minimum of 2 hours before spawning.
- Casing material should be well sterilized before use. Formalin at 5% concentration can be used for this purpose.
- Irrigation water should be nematode-free.
- Proper cook-out of mushroom houses with steam at 70°C for 5-6 hours or at 80°C for 30-60 minutes is essential after the crop is over to avoid carrying of infestation to subsequent crops.
- The dry neem leaf powder may be mixed with compost at 2-5% concentration which not only check nematode multiplication in mushroom growing trays but also helps in proper mycelial growth and improved sporophore yield.

Table 13.9. Biological control of plant parasitic nematodes in vegetable crops.

Vegetable crop	Nematode	Effective bioagent/s	Reference
Tomato	*M. incognita*	*P. lilacinus*	Ekanayake & Jayasundara, 1994; Cannayane & Sivakumar, 2001; Sharma *et al.*, 1998; Goutam *et al.*, 1995
			Sharma & Trivedi, 1989

		T. viride	Rekha Arya & Saxena, 1998 Xiujuan *et al.*, 2000
		T. harzianum & *T. viride*	Nagesh *et al.*, 2001
		T. virens	Susan *et al.*, 2002b
		P. chlamydosporia	Gopinath *et al.*, 2002 Rao *et al.*, 1998b
		G. fasciculatum	Mishra & Shukla, 1997 Hemavathi, 2002 Kantharaju *et al.*, 2002 Bagyaraj *et al.*, 1979 Suresh & Bagyaraj, 1984 Nagesh *et al.*, 1999
		G. mosseae	Sundarababu & Suguna, 1998
		P. fluorescens	Verma *et al.*, 1999 Shanthi & Sivakumar, 1995
		P. penetrans	Rameshchand & Gill, 2002 Somasekhar & Gill, 1991 Walia & Dalal, 1994 Parvatha Reddy *et al.*, 1997
		P. stutzeri *B. thuringiensis*	Khan & Tarannum, 1999
		A. chroococcum	Chahal & Chahal, 2003
		S. avermitilis	Parvatha Reddy & Nagesh, 2002
	M. javanica	*P. lilacinus*	Zaki, 1998 Khan & Esfahmi, 1990 Zaki & Bhatti, 1991a
		G. fasciculatum	Sundarababu *et al.*, 1993
		P. penetrans	Giannakiou *et al.*, 1999 Maheshwari & Mani, 1988 Rao *et al.*, 1997 Mankau, 1975
	R. reniformis	*P. lilacinus*	Alan & Walters, 1994 Parvatha Reddy & Khan, 1988
		G. fasciculatum	Sitaramaiah & Sikora, 1982
		P. fluorescens	Niknam & Dhawan, 2001b
		B. subtilis	Niknam & Dhawan, 2001a
Brinjal	*M. incognita*	*P. lilacinus*	Pandey & Dwivedi, 2001 Saikia & Das, 2001 Trivedi, 1992 Rao *et al.*, 1998d

		P. chlamydosporia	Naik, 2004
		G. fasciculatum	Sharma & Trivedi, 1989 Jothi & Sundarababu, 2000
		G. mosseae	Borah & Phukan, 2000 Jothi & Sundarababu, 2000 Rao & Parvatha Reddy, 2001
		P. penetrans	Karuna *et al.*, 2001 Singh *et al.*, 1998
		A. chroococcum	Chahal & Chahal, 1988
		S. avermitilis	Parvatha Reddy & Nagesh, 2002
	M. javanica	*P. lilacinus*	Zaki & Maqbool, 1992
		A. chroococcum	Bansal & Verma, 2002
		P. lilacinus	Parvatha Reddy & Khan, 1989
	R. reniformis	*P. lilacinus*	Pandey & Trivedi, 1992
		S. avermitilis	Parvatha Reddy & Nagesh, 2002
Chilli	*M. incognita*	*G. mosseae*	Sundarababu & Suguna, 1998
Bell pepper	*M. incognita*	*T. virens*	Susan *et al.*, 2002a
		P. chlamydosporia	Naik, 2004
Potato	*G. rostochiensis*	*T. harzianum*	Saifullah, 1996
	M. incognita	*P. fluorescens*	Mani *et al.*, 1998
	M. incognita & *G. pallida*	*P. lilacinus*	Jatala *et al.*, 1980
	G. rostochiensis & *Pratylenchus* sp.	*P. lilacinus*	Davide & Zorilla, 1995
	G. rostochiensis & *G. pallida*	*Cylindrocorpon destructus, F. solani, F. oxysporum*	Morgon Jones & Rodriguez Kabana, 1985
Okra	*M. incognita*	*P. lilacinus*	Davide and Zorilla, 1986.
		T. harzianum & *T. viride*	Lal *et al.*, 2002
		G. fasciculatum	Sharma & Trivedi, 1994
		G. mosseae	Sundarababu & Suguna, 1998
		P. fluorescens	Sobita Devi & Dutta, 2002
	M. javanica	*P. lilacinus*	Walia *et al.*, 1991
Carrot	*M. hapla*	*P. lilacinus*	Sivakumar, 1998
	R. reniformis	*P. fluorescens*	Rao & Shylaja, 2004
Pea	*M. incognita*	*B. thuringiensis*	Chahal & Chahal, 2003
Sugar beet	*H. schactii*	*P. fluorescens*	Oostendorp & Sikora, 1989

Chapter 14

ORNAMENTAL CROPS

14.1. ROSE

14.1.1. Lesion Nematodes, *Pratylenchus* spp.

(i) Physical Methods: Hot water treatment of rose seedlings at 50°C for 5 min. or 48°C for 10 min eliminates *P. zeae* infestation.

(ii) Cultural Methods: Intercropping with marigold helps in reducing the nematode population.

(iii) Host Resistance: Ohkawa and Saigusa (1981) found that *Rosa indica* (cv. Major) and *R. multiflora* (cv. 60-5) were resistant to *Pratylenchus penetrans* and *P. vulnus*, respectively.

14.1.2. Root-knot nematodes, *Meloidogyne* spp.

(i) Physical Methods: Hot water treatment of rose roots at 45.5°C for 60 min. eliminates root-knot nematode infestation.

(ii) Host Resistance: Ohkawa and Saigusa (1981) reported that rose cv. Mavetti to be resistant against *M. hapla*.

(iii) Integrated Methods: Pre-treatment of soil with dazomet (at 25 g/m^2) followed by soil amendment with neem cake (at 1 kg/ m^2) along with *Pochonia chlamydosporia* (at 2 x 10^{10} spores/ m^2) recorded reduced root galling and increased flower yield (Nagesh and Janakiram, 2004).

14.2. CARNATION AND GERBERA

14.2.1. The Root-knot Nematode, *Meloidogyne incognita*

(i) Cultural Methods: Application of neem cake at 1 kg per m^2 or carbofuran at 2 kg a.i. per ha gave effective control of *M. incognita*.

(ii) Biological Methods: An experiment carried out for the biological control of root-knot nematode *M. incognita* in carnation (*Dianthus* *caryophyllus*) showed that, though the growth parameters were found to be at par in plants treated with *P. lilacinus* and *P. chlamydosporia*, significantly higher reduction of root galling was achieved with *P. chlamydosporia*. Thus, it was concluded that under polyhouse conditions, *P. lilacinus* and *P. chlamydosporia* were better bio-agents for the control of root-knot nematode compared to *T. harzianum* (Shylaja, 2004).

(iii) Host Resistance: Carnation cvs. Desco, Castelaro, Kappa, Rara, Logupink, Target and Antalia were highly resistant to *M. incognita*.

(iv) Integrated Methods: Integration of neem cake at 0.5 kg per plant with *Trichoderma viride* at 100 g per m^2 was effective in the management of root-knot nematodes and increased flower yields.

Integration of *P. lilacinus*/*Trichoderma harzianum* at 0.5 l per m^2 (aqueous spore suspension containing 2 x 10^4 spores per ml) with neem cake at 0.5 kg per m^2 or fenamiphos at 2 g a.i. per m^2 increased plant growth parameters and flower yield of carnation and gerbera. The above treatments also increased root-knot egg parasitization by the parasitic fungi (Nagesh and Parvatha Reddy, 1996a).

Three antagonistic fungi viz., *T. harzianum*, *V. lecanii* and *P. lilacinus* at 2 x 10^4 spores/ml in combination with neem cake reduced *M. incognita* population in both soil and roots of carnation.

Combined inoculation of AMF and *Pseudomonas fluorescens* was found effective on root-knot nematodes on carnation (Anusuya and Vadivelu, 1999).

Application of *Paecilomyces lilacinus* at 0.5 g/kg of soil along with neem cake at 1.0 t/ha efficiently suppressed the nematode population and checked its build up, enhancing the plant growth parameters resulting in better production with increased flower stalk length and flower diameter in carnation. The plants also came to flower early (Johnson, 2000).

Pre-plant treatment of beds with dazomet (at 40 g/m^2) followed by the application of neem cake (at 1 kg/m^2 15 days later) along with antagonistic fungi, *P. chlamydosporia*/ *P. lilacinus* (at 2 x 10^{12} spores/ m^2) significantly reduced root-knot nematode population (*M. incognita*), mortality of plants and suppressed the nematode infection for 2 years. The antagonistic fungi established better in the beds treated with dazomet. The above treatment also reduced root galling, nematode multiplication rate and plant mortality, and increased spike/stem length, flower yield and root colonization with the bioagents (Nagesh and Parvatha Reddy, 2005).

14.2.2. Root-knot nematode, *Meloidogyne incognita* and Wilt, *Fusarium oxysporum* f. sp. *dianthi* Disease Complex

(i) Integrated Methods: The studies carried out to evaluate combination of bio-agents for the biological control of wilt (*F.o.* f. sp. *dianthi*) and root-knot nematode (*M. incognita*) disease complex in carnation, revealed that a combination of *P. chlamydosporia* + *P. lilacinus* gave the best control of wilt and root knot disease complex (Shylaja, 2004).

14.3. CHRYSANTHEMUM

14.3.1. The Lesion Nematode, *Pratylenchus coffeae*

(i) Integrated Methods: Application of neem cake enriched with *T. harzianum* at 1 t/ha is effective in reducing the nematode population.

14.3.2. Bud and Leaf Nematode, *Aphelenchoides ritzemabosi*

(i) Physical Methods: Hot water treatment of stools at 46°C for 5 min. eliminates nematode infestation.

Sufficient chrysanthemum stools were immersed in hot water at 48°C to bring the temperature down to 46.6°C, no further heating being necessary, and stools were removed after 5.5 min. when the temperature was down to 46.1°C.

(ii) Cultural Methods: Thorough cleaning up and burning of all infected plant material is essential in all control programmes. Plants propagated from top cuttings selected from plants growing under dry conditions offer an opportunity to secure relatively clean stock for foundation plantings.

14.3.3. The Root-knot Nematode, *Meloidogyne incognita*

(i) Biological Methods: Talc and pesta granules of *P. lilacinus* applied at 10 and 15 kg/ha to root-knot infested field reduced root galling (RKI=1.4 to 2.7), nematode multiplication rate (Rf=1.4 to 2.12) and enhanced bioagent propagules in rhizosphere (700-1070), bioagent infectivity (32-52%) and chrysanthemum flower yield by 23 to 28% (Nagesh *et al*., 2003).

14.4. CHINA ASTER

14.5.1. Root-knot Nematodes, *Meloidogyne* spp.

(i) Host Resistance: China aster cv. Shashank was found resistant, while Poornima was moderately resistant to root-knot nematodes (Nagesh *et al*., 1995).

14.5. GLADIOLUS

14.5.1. Root-knot Nematodes, *Meloidogyne* spp.

(i) Physical Methods: Hot water treatment of corms at 58°C for 30 min. eliminates root-knot nematode infection.

(ii) Cultural Methods: Use nematode-free planting material for planting. Crop rotation or intercropping with marigold helps in reducing the nematode population.

(iii) Integrated Methods: Soil application of neem cake enriched with *T. harzianum* at 1 t/ha is also effective.

Application of 5 tonnes of FYM enriched with 5 kgs of *P. fluorescens* (with 1 x 10^9 cfu/g) per hectare significantly reduced *M. incognita* by 61-76%,

R. reniformis by 65-70% in the roots of gladiolus and increased the flower yield by 19-22%.

14.5.2. Root-knot nematode, *Meloidogyne incognita* and Wilt, *Fusarium oxysporum* f. sp. *gladioli* Disease Complex

(i) Biological Methods: Application of *Pseudomonas fluorescens* effectively controlled the disease complex leading to significant improvement in plant growth and flowering (Mustafa and Khan, 2004).

(ii) Integrated Methods: Gladiolus plants treated with *P. lilacinus* + *T. harzianum* + neem cake and *P. lilacinus* + *T. viride* + neem cake combinations not only controlled *M. incognita* infection but also *Fusarium* wilt (*F. oxysporum* f sp. *gladioli*) till the harvest of flower spikes. The corms and cormels obtained from plants treated with these combinations were free from *Fusarium* infection. Fungal colonization of galled roots was maximum in *P. lilacinus* + *T. viride* combination (94%) followed by *P. lilacinus* + *T. harzianum* combination (69%). Similarly, the parasitization of eggs was maximum in *P. lilacinus* + *T. viride* combination (44%) followed by *P. lilacinus* + *T. harzianum* combination (38%), while egg mass parasitization was same in both the combinations (58%) (Nagesh *et al*., 1998).

14.6. TUBEROSE

14.6.1. Root-knot Nematodes, *Meloidogyne* spp.

(i) Physical Methods: Hot water treatment of bulbs at 49°C for 60 min. eliminates root-knot nematode infection.

(ii) Cultural Methods: Soil application of neem cake at 1 t/ha is effective.

(iii) Host Resistance: Tuberose Cv. Shringar (Single type) was found resistant, while Suvasini (Double type) was tolerant to *M. incognita* (Nagesh *et al*., 1995).

(iv) Biological Methods: The experiment carried out for the biological control of root-knot nematode (*M. incognita*) in tuberose (*Polyanthes tuberosa*) showed that, *P. chlamydosporia* and *T. harzianum* treatments gave significantly higher control of nematodes than *P. lilacinus* (Shylaja, 2004).

(v) Integrated Methods: Integration of arbuscular mycorrhizal fungi (AMF) such as *Glomus mosseae* or *G. fasciculatum* with neem cake or/and aldicarb gave effective management of root-knot nematodes infecting tuberose (Khan and Parvatha Reddy, 1994). In tuberose, bulb treatment followed by spot soil drenching with *P. lilacinus* spores in 5% neem leaf extract significantly increased floral yield and reduced root infection by the nematode. The above treatment also reduced root galling and increased per cent egg parasitization by *P. lilacinus* (Nagesh *et al*., 1997).

Integration of *Paecilomyces lilacinus* with neem cake gave effective control of *M. incognita* on tuberose. Treatment of tuberose bulbs with neem cake extract mixed with *P. lilacinus* spores significantly reduced *M. incognita* infection and multiplication besides stimulating the plant growth. Soil application of *T. harzianum* (10^9 cfu/g) + *P. lilacinus* (10^6 cfu/g) gave effective control of root-knot nematodes. Soil application of neem cake enriched with *T. harzianum* at 1 t/ha is also effective.

Split application (at planting and 45 DAP) of neem cake at 1 kg/1.5 m^2 in combination with *P. lilacinus* (at 25 g/1.5 m^2 containing 18 x 10^8 spores/g) significantly reduced root galling, soil and root population of *M. incognita* and its multiplication rate. The above treatment also significantly increased number of spikes, spike length and weight, number of flowers/spike, parasitization of egg masses and root colonization by *P. lilacinus* (Nagesh *et al.*, 1998).

Bulb dip treatment with 5% neem leaf extract mixed with spores of *P. lilacinus* (2 x 10^4 spores/ml) for 30 min. combined with soil drench with 5% neem leaf extract (at 50 ml/plant) 30 DAP gave significant increase in plant growth, flower yield and parasitization of eggs and egg masses and least root galling and reproduction factor (Nagesh *et al.*, 1997).

Combined application of *P. chlamydosporia* + *T. harzianum* gave significantly higher control of *M. incognita* on tuberose (Shylaja *et al.*, 2004).

14.6.2. Bud and Leaf Nematode, *Aphelenchoides besseyi*

(i) Physical Methods: Cent percent mortality of *A. besseyi* was obtained by immersing tuberose bulbs in hot water bath maintained at 50°C for 30 minutes (Khan *et al.*, 2004).

(ii) Host Resistance: Tuberose cvs. Suhasini, Shringar, Prajwal and Hyderabad Double were tolerant to *A. besseyi* (Khan *et al.*, 2004).

(iii) Integrated Methods: Pre-soaking of tuberose bulbs in water overnight followed by hot water treatment at 50°C for 20 minutes + spraying of the crop twice with monocrotophos 36 EC at 0.15% (at sprouting and 30 days after first spray) in the first year and 3 sprays in second and third year crop at 15-20 days interval was found most effective in reducing % nematode infestation as well as disease indices (Khan *et al.*, 2004).

14.6.3. Root-knot nematode, *Meloidogyne incognita* and Wilt, *Fusarium oxysporum* f. sp. *dianthi* Disease Complex

(i) Integrated Methods: In course of the experiments carried out to evaluate combination of bio-agents for the control of wilt (*F.o.* f. sp. *dianthi*) and root-knot nematode (*M. incognita*) disease complex in tuberose, the best result was found in plants treated with *P. chlamydosporia* + *T. harzianum* (Shylaja, 2004).

14.7. CROSSANDRA

14.7.1. The Root-knot Nematode, *Meloidogyne incognita*

(i) Cultural Methods: Deep ploughing and fallowing in summer for about a month helps in reducing the nematode population. Planting crossandra after graminaceous crops should be avoided. Incorporation of FYM in soil and intercropping of crossandra with marigold or pangola grass would reduce the root-knot nematode infection.

Intercropping of crossandra with enemy plant like marigold significantly reduced the root-knot and lesion nematode population both in soil and roots (Khan and Parvatha Reddy, 1994). Application of farm yard manure has been found to reduce the lesion nematode population in crossandra.

(ii) Biological Methods: The parasitic fungus, *Paecilomyces marguandi* grown on paddy seeds at 2 g per kg soil gave effective control of root-knot nematode infecting crossandra (Khan and Parvatha Reddy, 1994).

(iii) Integrated Methods: Integrated management of root-knot nematodes infecting crossandra was achieved by rational combination of a biocontrol agent (*Trichoderma harzianum* or *Pochonia chlamydosporia*) with a nematicide (aldicarb) or oil cake (neem cake) (Khan and Parvatha Reddy, 1994). Integration of neem, karanj and castor cakes with AMF, *Glomus mosseae* significantly enhanced plant growth parameters and flower yield of crossandra, root colonization and sporulation of AMF. The above treatments also reduced root-knot nematode multiplication and root-galling (Nagesh *et al.*, 1998; Nagesh and Parvatha Reddy, 1996b). Combinations of *Verticillium lecanii* and *Paecilomyces lilacinus* (2 x 10^4 spores/ml each) with 5% neem leaf extract resulted in significantly higher plant growth parameters and crossandra flower yield (Nagesh and Parvatha Reddy, 1995). Root galling, nematode multiplication factor were least and the parasitization of eggs and egg masses was highest in *V. lecanii* + 5% neem leaf extract.

Incorporation *Verticillium lecanii* with neem cake facilitated the effective management of *M. incognita* on crossandra. Application of neem cake enriched with *Trichoderma harzianum* at 2 kg/m^2 (1 t/ha) in nursery beds gave effective control.

The root colonization of AM fungi increased significantly in the presence of neem cake which in turn improved their efficacy in reducing *M. incognita* population in crossandra roots.

Integration of a bioagent (*V. lecanii*), endomycorrhiza (*G. mosseae*) with botanicals improved the growth of crossandra and reduced the population of *M. incognita*. *G. mosseae* reduced the requirement of phosphatic fertilizer and favoured the antagonistic potential of *V. lecanii* against *M. incognita*.

Combined application of *P. lilacinus* and *Pasteuria penetrans* enhanced plant growth and flower yield of crossandra besides reducing root galling due to *M. incognita* (Nagesh *et al.*, 1995).

14.7.2. The Lesion Nematode, *Pratylenchus delattrei*

(i) Cultural Methods: Intercropping of crossandra with enemy plant like marigold significantly reduced the lesion nematode population both in soil and roots (Khan and Parvatha Reddy, 1994). Application of farm yard manure has been found to reduce the lesion nematode population in crossandra.

(ii) Biological Methods: Application of *T. viride* at 10 g/kg soil was found effective to maximize the flower yield and suppressing the lesion nematode population both in soil and roots (Jothi *et al.*, 2004).

14.7.3. The Needle Nematode, *Longidorus africanus*

(i) Cultural Methods: Pea, carrot, squash, spearmint and onion are reported to be poor hosts; while cauliflower, cabbage and radish as non-hosts. Either of these crops can be fitted as intercrops or in crop rotation scheme. Rotation of crossandra with spearmint which is a poor host of *L. africanus* would reduce the needle nematode population (Kolodge *et al.*, 1987).

Table 14.1. Biological control of plant parasitic nematodes in ornamental crops.

Crop	Nematode	Effective bioagent/s	Reference
Tuberose	*M. incognita*	*P. chlamydosporia, P. lilacinus, T. harzianum*	Shylaja, 2004
Crossandra	*M. incognita*	*P. lilacinus, V. lecanii*	Nagesh & Parvatha Reddy, 1995
		P. margaundi	Khan & Parvatha Reddy, 1994
		G. mosseae	Nagesh *et al.*, 1996b
Carnation	*M. incognita*	*P. lilacinus*	Shylaja, 2004

Chapter 15

MEDICINAL CROPS

15.1. BETEL VINE

15.1.1. The Root-knot Nematode, *Meloidogyne incognita*

(i) Physical Methods: White polyethylene mulching for 15 days on beds prepared for planting betel vine revealed high reduction in nematode populations due to increased soil temperature to 44.1°C (Sivakumar and Marimuthu, 1987).

(ii) Cultural Methods: Nematode-free planting material is a cardinal principle in the management of betel vine nematodes. Summer ploughing, summer ploughing and solarization or rice cropping have been beneficial. Application of FYM at 25 t/ha or neem cake at 1 t/ha before planting or crop rotation with rice helps to reduce the nematode population in soil. A pre-rice crop reduces populations of *M. incognita, Rotylenchulus reniformis* and *Helicotylenchus incisus* in the subsequent betel vine crop. Crop rotation with non-host crops like marigold, sesame and groundnut also suppress root-knot nematode populations. Support plants such as *Erythrina indica, Moringa oleifera* and *Bombax malbaricum* which are not susceptible to root-knot nematodes can be used in betel vine gardens.

In pots having 5 seedlings of *Tagetes errecta* around a betel vine plant, the population of *M. incognita* was reduced by 41.4% and root galling by 54% (Medhana *et al.*, 1985).

Soil application of neem, castor, karanj or groundnut cake all at 1 t/ha and poultry manure, saw dust or green leaves of *Calotropis* all at 2 t/ha to old betel vine gardens, suppressed the populations of root-knot and other nematodes and increased yields. Green leaf manuring with African marigold, neem or *Xanthium* have also been quite effective against root-knot, lesion and other nematodes.

(iii) Biological Methods: Three applications of *Paecilomyces lilacinus* grown on neem cake at 10 g/vine at 60 days interval gave a significant reduction in nematode infestation and increased vine growth and leaf yield under field conditions (Jonathan *et al.*, 2000). Nakat *et al.* (1995) reported that the application of *P. lilacinus* at 8 g/kg soil as the best treatment for the management of root-knot nematodes in betel vine.

(iv) Integrated Methods: *Trichoderma viride* mixed with mustard oil cake applied in three split doses reduced the nematode infestation and enhanced the leaf yield.

15.1.2. Root-knot Nematode, *Meloidogyne incognita* and Foot Rot, *Phytophthora capsici* Disease Complex

(i) Integrated Methods

- Summer ploughing and exposing the field to sunlight during May prior to sowing of *Sesbania* sp. (live standard) minimizes the initial load of inoculum of both the nematode and fungus.
- Selection of healthy seed vines from nematode-free and disease-free mother plants for planting.
- Dipping seed vines in 0.25% Bordeaux mixture solution for 5 min before planting.
- Application of FYM at 30 t/ha to promote multiplication of antagonistic microbes which in turn kills nematodes.
- Spot application of neem cake at 3 t/ha in 3 split doses – 1st dose at 45 DAP and remaining 2 splits at 45 days interval during North-East monsoon season (Oct.-Dec.).
- Rotation of betel vine crop with rice.

Combined application of *Pseudomonas* sp. (Pfbv 22) and *Bacillus* sp. (Bbv 57) gave significant reduction in nematode infestation (gall index, no. of egg laying females and soil population), wilt disease incidence and increase in leaf yield. The treatment also enhanced the biochemical markers responsible for induced systemic resistance such as peroxidase, polyphenol oxidase and phenylalanine ammonia lyase (Jonathan *et al.*, 2004).

15.2. ASHWAGANDHA, Withania somnifera

15.2.1. Root-knot Nematodes, *Meloidogyne* spp.

(i) Integrated Methods: Integration of neem cake with *T. harzianum*, vermicompost with *T. harzianum*, and cow urine with *T. harzianum* considerably reduced the root-knot nematode development and enhanced plant growth and yield. Maximum root-knot suppression was noticed in vermicompost with *T. harzianum* followed by mentha distillate with *Glomus aggregatum*. Highest increase in plant yield was recorded when the soil was amended with mentha/curry leaf distillates along with *T. harzianum*/ *Glomus aggregatum* (Pandey and Kalra, 2005b).

15.3. SARPAGANDHA

15.3.1. Root-knot Nematodes, *Meloidogyne* spp.

(i) Integrated Methods: Soil application of neem cake enriched with *T. harzianum* at 1 t/ha is recommended.

15.4. *COLEUS FORSKOHLII*

15.4.1. The Root-knot Nematode, *Meloidogyne incognita*

(i) Cultural Methods: Sree Bhadra, a high yielding variety of yam released by Central Tuber Crops Research Institute, Trivandrum, is identified as a resistant trap crop for the root-knot nematode. The nematodes were able to penetrate the root but giant cell formation was not induced and as a result the nematodes die inside the roots. Planting this variety in root-knot nematode infested field helped in clearing the field free of nematodes and giving good tuber yield. Subsequently growing of susceptible crop like coleus escaped nematode damage (Mohandas, 2001).

(ii) Biological Methods: Soil application of *Trichoderma harzianum* and *Pseudomonas fluorescens* each at 2.5 kg/ha recorded increased plant growth and reduced *M. incognita* population (Senthamarai *et al.*, 2006b).

(iii) Integrated Methods: Integration of soil solarization in the nursery for 15 days with 150 gauge LDPE film and application of *Paecilomyces lilacinus* + neem cake/ *P.lilacinus* + *Bacillus macerans* in the main field are the best treatments in increasing plant height (64.3 and 60.3 cm compared to 40.0 cm in control), number of leaves (593.3 and 583.3 compared to 310.0 in control), weight of tubers/plant (560.0 and 546.6 g compared to 350.0 g in control), yield (11.5 and 11.3 kg/plot compared to 6.9 kg/plot in control) and in reducing root galls (0.3 and 1.0 compared to 50.6 in control), nematode population in soil (25.0 and 30.0/100 ml soil compared to 196.6/100 ml soil in control) and roots (1.0 and 1.6/5 g roots compared to 79.0/5 g roots in control) (Nisha and Sheela, 2006).

15.5. KACHOLAM (Kaempferia galanga)

15.5.1. The Root-knot Nematode, *Meloidogyne incognita*

(i) Physical Methods: Hot water treatment of rhizomes at 55°C for 5 min was effective in reducing the nematode population (more than 74%) and giving 81% increase in yield (Nisha and Sheela, 2004).

(ii) Cultural Methods: Rhizome treatment with neem leaf extract (4%) + garlic extract (1%) was found effective.

(iii) Biological Methods: Rhizome treatment with *Pseudomonas fluorescens*, *Trichoderma viride* and AMF each at 3% w/w significantly reduced the nematode population (more than 74%) and improved plant growth and yield (Nisha and Sheela, 2004).

15.6. CHINESE POTATO (*Solenostemon rotundifolius*)

15.6.1. The Root-knot Nematode, *Meloidogyne incognita*

(i) Biological Methods: Dipping of plants in 3% solution of *Bacillus macerans* + soil drenching with 2% solution of *B. macerans* 7 days after planting recorded maximum reduction in nematode population and root-knot count (14 galls/g of roots) and gave maximum increase in yield (Sheela *et al.*, 2004).

15.7 HENBANE, *Hyoscyamus niger*

15.7.1. The Root-knot Nematode, *Meloidogyne incognita*

(i) Cultural Methods: It was observed that essential oils of palmarosa, citronella, basil and menthol mint were quite effective in reducing *M. incognita* population and improving plant growth. The oil of palmarosa was most effective (Pandey *et al.*, 2003).

(ii) Biological Methods: Negative impact of *Glomus aggregatum, G. fasciculatum* and *G. mosseae* on reproduction of *M. incognita* infecting *H. niger* has been observed (Pandey *et al.*, 1999). Significant reduction in *M. incognita* population was observed when henbane plants were inoculated with *Pseudomonas fluorescens* or *Glomus aggregatum* resulting in higher herbage yield and root colonization with the bioagents.

15.8. CHAMOMILE, *Matricaria recutita*

15.8.1. The Root-knot Nematode, *Meloidogyne incognita*

(i) Chemical Methods: Maximum reduction in root-knot severity and nematode population occurred with chemical activators such as 2-Chlorosalicylic acid, O-Actylsalicylic acid and 2-Chloronicotinic acid (Pandey and Kalra, 2005a).

(ii) Integrated Methods: Integration of *G. mosseae* and neem cake reduced the severity of root-knot disease and enhanced the growth, biomass and flower yield by 12 to 37% of chamomile.

15.9. BRAHMI, *Bacopa mannieri*

15.9.1. The Root-knot Nematode, *Meloidogyne incognita*

(i) Integrated methods: Neem compound, distillation waste of menthol mint, curry leaf, davana and vermicompost and bioinoculants such as *Glomus aggregatum* and *Trichoderma harzianum* alone or in combination were root-knot nematode suppressive and enhanced the growth and yield of brahmi.

15.10. *AMMI MAJUS*

15.10.1. The Root-knot Nematode, *Meloidogyne incognita*

(i) Cultural Methods: Treatment with *Costus speciosus* dried shoot powder at 10 g/kg soil showed the least nematode population (Rf=0.4) and root-knot index (RKI=0.6) followed by *Spilanthes acmella* (Rf=0.6, RKI=0.6), *C. speciosus* rhizome powder (Rf=0.5, RKI=1.0) and neem cake (Rf=0.5, RKI=1.0) (Pandey, 2002).

Table 15.1. Biological control of plant parasitic nematodes in medicinal crops.

Crop	Nematode	Effective bioagent/s	Reference
Betel vine	*M. incognita*	*P. lilacinus*	Jonathan *et al.*, 1995 Nakat *et al.*, 1995 Hazarika *et al.*, 1998
	M. incognita & *R. reniformis*	*P. lilacinus*	Jonathan *et al.*, 1995
	R. similis		Sosamma *et al.*, 1994
Henbane (*Hyoscyamus niger*)	*M. incognita*	*G. fasciculatum* *G. mosseae* *G. aggregatum*	Pandey *et al.*, 1999
Chamomile (*Matricaria recutita*)	*M. incognita*	*G. mosseae*	Pandey *et al.*, 2000

Chapter 16

AROMATIC CROPS

16.1. MINT

16.1.1. Root-knot Nematodes, *Meloidogyne incognita, M. hapla*

(i) Physical Methods: Hot water treatment of *Mentha arvensis* suckers at 48°C for 30 min gave complete eradication of *M. incognita* without any adverse effect to suckers and roots.

(ii) Cultural Methods: The nematodes which damage mints are generally endoparasitic (*Meloidogyne incognita* and *Pratylenchus thornei*) in nature and they can easily spread through suckers (propagating materials) from infested planting sites to uninfested planting sites. Therefore, it is essential and utmost important to use nematode-free planting material to check the spread of nematodes (Pandey *et al.*, 2003).

The severity of the nematode was reduced to a considerable extent in clayey soils. The high organic matter content in the soil suppressed the nematodes and enhanced the growth and oil yield. Soil application of neem/castor/mahua oil cakes at 1.0 g N/kg of soil was found effective against *M. incognita* and resulted in increase of oil yield and also plant growth parameters (Pandey *et al.*, 1992). Incorporation of neem cake 15 days before transplanting of *Mentha spicata* has been found the best approach for managing the population of *M. hapla* (Khan and Khanna, 1992).

Cultivation of some non-host crops in rotation with Japanese mint, water logging, rice cultivation during rainy season caused reduction in the population of root-knot nematodes. Crop rotation sequences of Menthol mint with maize, potato, mustard, pigeon pea, paddy and wheat were effective in reducing the root-knot nematode population and in increasing the yield (Table 16.1).

Table 16.1. Effective crop rotation sequences for the management of root-knot nematode in Menthol mint.

Crop rotation sequence	Root-knot index	Net benefit (in rupees)
Maize – potato – menthol mint	Mild	83,000
Paddy – potato – menthol mint	Mild	80,000
Maize – mustard – menthol mint	Mild	76,000
Pigeon pea – menthol mint	Mild	72,000
Paddy – menthol mint	Mild	75,000
Paddy – wheat – menthol mint	Mild	68,000

The application of hydro-distillation wastes of *Mentha arvensis, M. piperita, Cymbopogon martini* (palmarosa) and *C. winterianus* (citronella) in the soil (1 month before transplantation) was found effective in reducing the nematode population under field conditions.

Application of different oil seed-cakes improved the growth of Japanese mint under field conditions coupled with increased oil yield and reduced root-knot and lesion nematode population. However, best results were achieved when the soil was treated with neem cake. The use of vermicompost and different distillation wastes were found to enhance the growth and yield of different mint species and reduced nematode population significantly.

Maximum reduction in *M. incognita* population in Menthol mint was recorded in neem cake treated soil followed by mustard cake and *Trichoderma harzianum*. Higher herbage yield (21 t/ha) was observed in the soil treated with oil cakes in comparison to control (14.66 t/ha). The crop produced had significantly higher oil content (2.7%) and yield (151 kg/ha) in neem cake treatment followed by mustard cake (2.7% oil and 143.6 kg/ha yield), *T. harzianum* (2.5% oil and 116.4 kg/ha yield) and *Glomus aggregatum* (2.4% oil and 112 kg/ha yield). The untreated crop yielded 1.8% oil and 70.4 kg/ha yield (Pandey *et al*., 1998).

(iii) Host Resistance: The lowest root-knot (*M. incognita*) infection level was found on Kalka. Spear mint cvs. Neera and Arka were resistant to *M. incognita* (Pandey, 2003). Peppermint cv. Kukrail; Spearmint cvs. Arka, Neera and Neer-Kalka; and Bergamot mint cv. Kiran were found resistant to *M. incognita* (Pandey and Patra, 2001).

(iv) Biological Methods: Soil application of *Glomus fasciculatum, G. mosseae and G. aggregatum* enhanced herbage and oil yield and reduced the root-knot nematode population in the root rhizosphere.

(v) Integrated Methods: Application of carbofuran and neem cake in combination improved the growth of Japanese mint and oil yield and reduced root-knot nematode population very effectively.

The maximum reduction in *M. incognita* population was recorded in beds treated with mustard cake along with *T. harzianum* and produced significantly higher herbage and oil yields (Pandey, 2005a).

16.1.2. Lesion Nematodes, *Pratylenchus minyus, P. thornei, P. penetrans*

(i) Cultural Methods: Application of neem and mustard cakes at 1.5 t/ha reduced the lesion nematode population (*P. thornei*), improved the growth and oil yield of *M. piperita, M. citrata* and *M. spicata* (Shukla and Haseeb, 1996).

(ii) Host Resistance: Menthol mint cvs. Kosi and Kalka were resistant to *Pratylenchus thornei* and had least *M. incognita* infection. Bergamot mint cv. Kiran and Spear mint cv. Neer-Kalka were having least *P. thornei* infection.

16.2. BASIL, *Ocimum basilicum*

16.2.1. Root-knot Nematode, *Meloidogyne* spp.

(i) Cultural Methods: Neem cake was effective against *M. incognita* infecting basil (*Ocimum basilicum*) and increased plant growth and oil content. Application of mahua cake also increased the oil yield (Haseeb *et al.*, 1988).

16.3. JASMINE

16.3.1. The Root-knot Nematode, *Meloidogyne incognita*

(i) Integrated Methods: Cuttings or planting material should be raised in nematode-free or treated nursery beds. Application of neem cake enriched with *T. harzianum* at 1 t/ha is recommended. Application of phorate at 4 g a.i. per plant during May and September months and incorporation of 20 kg FYM increased flower yield of jasmine by 50% and reduced the root-knot nematode population by 70% (Sundarababu, 1992).

16.4. PATCHOULI

16.4.1. Root-knot Nematodes, *Meloidogyne* spp.

(i) Cultural Methods: Neem cake at 4 t/ha controlled the root-knot nematodes (Sarwar *et al.*, 1982). Planting of cultivated patchouli scion grafted on wild patchouli rootstock was effective for the management of root-knot nematodes (Krishna Prasad, 1978).

16.4.2. Root-knot Nematode and Root Rot Disease Complex

(i) Integrated Methods: Combined mortality due to root rot and root-knot nematode could be minimized by the application of *Trichoderma harzianum* + karanj cake at 5 t/ha.

Application of 150 kg N and 50 kg *Glomus fasciculatum*/ha enhanced fresh herbage (297.2 g/plant compared to 212.7 in control), dry herbage (92.0 g/plant compared to 64.0 in control) and oil yield (110.4 kg/ha compared to

28.0 kg/ha in control); increased mycorrhizal colonization (90.5% compared to 65.5% in control); reduced *M. incognita* population (145/200 ml soil compared to 395/200 ml soil in control) and recorded the highest cost benefit ratio (1:3.67) (Sumathi *et al.*, 2006).

16.5. DAVANA, *Artemisia pallens*

16.5.1. Root-knot Nematodes, *Meloidogyne* spp.

(i) Cultural Methods: Soil application of neem cake at 1 g N/kg soil significantly suppressed root-knot nematode population and increased oil yield (Haseeb and Butool, 1991). Better efficacy of neem cake over the other organic materials viz., mahua cake, castor cake and nematicides like aldicarb and carbofuran has been demonstrated on *M. incognita* infecting *A. pallens* (Pandey, 1994b). Increase in nematicidal efficacy of neem cake and FYM for 90 to 120 days in comparison to 60 days for fenamiphos has also been noticed (Anita and Vadivelu, 1997).

A considerable enhancement of herbage biomass, flower bud and oil yields were noticed when davana plants were treated with the distillation waste of palmarosa, citronella, curry leaf, menthol mint and vermicompost of pyrethrum, marigold, menthol mint and the bioagent, *T. harzianum*. Least nematode infections were recorded in distillation waste of curry leaf, palmarosa, lemon grass and vermicompost of pyrethrum, marigold and menthol mint (Pandey, 2005b).

Chapter 17

PLANTATION CROPS

17.1. COCONUT

17.1.1. The Burrowing Nematode, *Radopholus similis*

(i) Cultural Methods: Application of oil cakes, FYM and green foliage to the basins, growing of intercrops like cocoa that enriches the soil with sizeable quantities of shed foliage which help in the build up of beneficial organisms which may inhibit nematode multiplication. *Crotalaria juncea* may be grown as a green manure crop in the basins and interspaces and its incorporation helps in the inhibition of nematode multiplication. Avoiding banana as a shade crop in coconut nurseries and use of nematode-free planting material of coconut and other intercrops are other management approaches. Application of 50 kg cow dung/FYM, 2 kg neem/marotti cake and 25 kg green manure to the coconut basins in November is quite effective in bringing down the nematode population. Thirty per cent increase in yield has been recorded by application of *Hydnocarpus* sp. oil cake at 8 kg/palm/year in June-July and October-November to the burrowing nematode infested coconut palms (Koshy, 1986).

(ii) Biological Methods: Introduction of biocontrol agents like *Paecilomyces lilacinus* and AMF to nursery potting mixture helps in the reduction of nematode population and at the same time impart resistance/tolerance to nematode infestation (Koshy *et al.*, 1993). Among the AMF, *Acaulospora bireticulata* was found to be more effective in reducing the nematode population. Soil application of *Cylindrocarpon effusum* reduced the rate of multiplication of the nematode and damage to coconut seedlings (Sosamma and Koshy, 1978; Koshy and Sosamma, 1987).

(iii) Host Resistance: The dwarf cvs. Kenthali and Klappawangi and the hybrids such as Java Giant x Kulasekharam Dwarf Yellow, Java x Malayan Dwarf Yellow and San Ramon x Gangabondam recorded the least nematode multiplication and lesion indices (Sosamma *et al.*, 1980).

(iv) Integrated Methods: The following measures are suggested for an integrated management schedule for *R. similis* infestation on coconut palms:

- Application of cow dung, FYM, oil cakes and green manure to the basins. *Crotalaria juncea* may be cultivated in the basins and interspaces and used as a green manure.
- Avoiding banana as a shade crop in coconut nurseries.
- Use of nematode-free planting material of coconut and other intercrops.
- Use of less susceptible/tolerant cvs. such as Kenthali and Klappawangi or their hybrids in infested areas.

17.1.2. The Red Ring Nematode, *Rhadinaphelenchus cocophilus*

(i) Cultural Methods: 'Cut and burn' the infested trees is one of the oldest and most effective practices used to prevent further spread of the disease.

(ii) Chemical Methods: More recent methods involve the destruction of palms using arboricides, herbicides or phytotoxic nematicides applied to holes made in tree-trunks. Hoyle (1968) recommended the use of cacodylic acid ('Silvisar') to poison trees. The chemical prevented weevil breeding, killed the diseased trees within two weeks, but had no nematicidal effects. Victoria *et al.* (1970) used sodium and potassium arsenate to kill both diseased trees and nematodes.

Adult weevils are attracted to anaerobic fermentation products such as ethyl alcohol, n-butyl alcohol and to certain volatile esters extracted from diseased palm tissue (Griffith, 1987). This behaviour can be exploited to attract weevils to poisoned baits at 25/ha.

(iii) Biological Methods: Griffith (1977) reported the pathogenicity of a bacterium, *Micrococcus agilis* to the palm weevil and suggested its potential use as a biocontrol agent. The rhabditid nematodes of the genera *Steinernema* and *Heterorhabditis* can heavily parasitize the weevils by maintaining a high population of nematodes in the environment (Griffith and Koshy, 1990).

17.2. ARECANUT

17.2.1. The Burrowing Nematode, *Radopholus similis*

(i) Cultural Methods: Use of nematode-free planting materials of arecanut and other inter/mixed crops and avoiding *R. similis* susceptible inter/mixed crops like black pepper and banana in infested areas helps in reducing the nematode population.

(ii) Host Resistance: The arecanut hybrid VTL-11 x VTL-17 has been found to be highly resistant to *R. similis*. The cvs. Indonesia-6, Mahuva B and Andaman-5 are tolerant to *R. similis* and hence can be recommended for infested areas (Koshy *et al.*, 1979; Sundararaju and Koshy, 1982). The cvs.

Indonesia-6 and Singapore (VTL-17) are known to yield 50% more nuts over local South Kanara variety.

(iii) Integrated Methods

- Use of nematode-free planting material of arecanut and other inter/mixed crops.
- Avoiding *R. similis* susceptible inter/mixed crops like black pepper and banana in infested areas.

Integration of neem oil cake application at 1 kg/plant with phorate at 15 g a.i./plant gave effective control of *R. similis* in areca nut. In areca nut + banana + black pepper cropping system, integration of phorate at 15 g a.i./plant with neem oil cake at 500 g/plant was found most effective in reducing the nematode population (Sudha and Sundararaju, 1998).

17.3. COFFEE

17.3.1. The Lesion Nematode, *Pratylenchus coffeae*

(i) Cultural Methods: Nematode control in coffee is achieved through cultural practices like maintaining heavy shade, providing mulch to individual plants, application of FYM and giving frequent irrigation or providing nutrients through foliar spray. Soil application of coffee pulp at 18 kg/sq.m reduced the nematode population and increased the plant growth.

Lordello (1968) reported that replanting of old nematode-infested plantations preceded by fallow period of at least 2 years gave effective control of the lesion nematode.

(ii) Integrated Methods: It is advisable to change nursery site once every 7-8 years. Jungle soil and FYM should be dried and well sieved before use. The nursery site should also be dug and exposed to sun. The nematode-free plants should be used for planting. Affected plants should be pulled out and burnt. The planting site should be dug and the soil exposed to the sun for at least one summer season. Care should be taken to keep the site free from weeds. The site should be planted with either Robusta or Arabica –Robusta grafted plants. Grafting of Arabica scion on Robusta rootstock in 40 day-old 'topee stage' plants by using wedge-cleft method can be adopted (Kumar, 1988).

17.3.2. Root-knot Nematodes, *Meloidogyne exigua, M. coffeicola*

(i) Cultural Methods: At the crop innovation time, the infested coffee area may be turned to pasture. Most graminicolus plants are non-hosts to *M. exigua* and *M. coffeicola*. The decision of growing forage crops and turning back again the area to coffee will certainly reduce drastically the population of *Meloidogyne* species in 6-12 months.

Rotation of coffee with cotton, soybean and maize in *M. exigua* infested areas for one year significantly reduced the nematode population, after which the farmer can return to coffee cultivation. Coffee can also be rotated with groundnut, castor (both immune), *Styzolobium deeringiatum* and *Crotalaria spectablis* (both resistant).

ii) Host Resistance: *Coffea canephora* (line 2258 from Costa Rica) showed resistance to *M. exigua* and *M. incognita*. Planting of *Coffea arabica* var. Mundo Novo or Catuai Vermelho grafted on line 2258 in field showed good growth and production in Brazil.

17.4. TEA

17.4.1. Root-knot Nematodes, *Meloidogyne incognita* (in nurseries), *M. brevicauda* (in field)

Management in Nurseries

(i) Physical Methods: Soil solarization in nurseries is adequate to manage root-knot and other nematodes as the plants develop complete resistance by about 9-15 months (Gnanapragasam *et al.*, 1989).

(ii) Cultural Methods: Every effort should be made to raise the nurseries in nematode-free soil. Incorporation of organic matter into nematode infested soil is known to suppress the incidence of root-knot nematodes on tea.

The cultivation of marigold as a pre-plant crop considerably reduces the population of *Meloidogyne* spp. and *Pratylenchus loosi*. The reduction in the nematode population is significantly greater than when the land is left fallow. The interplanting of marigold in tea has resulted in a 7% increase in yield compared to 10% increase that followed soil fumigation with DBCP.

The rehabilitation crops such as Mana grass (*Cymbopogon confertiflorus*) or Guatemala grass (*Tripsacum laxum*) (which help to improve the soil) are resistant to *Meloidogyne* spp. The soil population of parasitic nematodes decline rapidly when these grasses are grown for one or more years before replanting tea.

(iii) Integrated Methods: Combination of neem cake at 5 g/seedling + carbofuran 3G at 1.5 g/seedling improved plant growth characters and reduced number of galls, egg masses and eggs/egg mass (Kalita and Bora, 2006).

Management Under Field Conditions

(i) Cultural Methods: The first step in the management of nematodes in replanted areas is by ensuring a very thorough uprooting of all old tea roots (up to pencil thickness). In Sri Lanka, such uprooted old tea lands are routinely reconditioned by planting deep-rooted grass species including Guatemala grass and Mana grass which are non-hosts to parasitic nematodes and bring down the residual nematode population (Kerr and Vythilingam, 1966).

Shade trees and green manure crops help to provide shade to the tea plants, aid in N fixation and provide regular supplies of mulch in the form of leaf litter and loppings which help in reducing the nematode population during their decomposition. The cultivation of shade trees and green manure crops susceptible to tea nematodes should be discouraged in tea plantations.

Plants such as marigolds, *Tagetes erecta* and *T. patula*, *Vetiveria zizanoides* and *Eragrostis curvula* have been found to help reduce nematode populations.

Large inputs of organic matter including cow dung and well decomposed plant residues have been reported to suppress nematode populations. In India, neem cake has been recommended to manage root-knot nematodes in the tea field (Muraleedharan, 1991). In Sri Lanka, castor oil cake has been proved to be the most effective one for suppressing nematodes, followed by mahua, neem and karanj cakes (Gnanapragasam, 1991).

17.4.2. The Lesion Nematode, *Pratylenchus coffeae*

(i) Cultural Methods: Mulching with Guatemala grass at 20 to 30 t/ha gave effective control of *P. coffeae* on tea.

Table 17.1. Biological control of plant parasitic nematodes in plantation crops.

Crop	Nematode	Effective bioagent/s	Reference
Coconut	*R. similis*	*G. mosseae, G. fasciculatum, G. macrocorpus, G. versiforme, Acaulospora bireticulata, Sclerocystis rubiformis, Scutellospora nigra*	Koshy *et al.*, 1999
Arecanut	*R. similis*	*P. lilacinus*	Sudha *et al.*, 2000

Chapter 18

SPICE CROPS

18.1. BLACK PEPPER

18.1.1. The Burrowing Nematode, *Radopholus similis*

(i) Cultural Methods: Use of healthy and nematode-free planting material can ensure better establishment, survival and higher yield. Soil solarization of nursery sites or potting mixture and random rotation of nursery sites are reported to be effective in black pepper. Mulching (with Guatemala grass) has an ameliorating effect on the symptoms of slow wilt/ pepper yellows (*R. similis*). Planting nematode-free rooted cuttings, uprooting of affected vines and replanting after a period of 9-12 months, use of non-living standards/supports, avoiding *R. similis* susceptible intercrops like banana, ginger, turmeric and exclusion of susceptible trees as standards for trailing the vines reduces the nematode population. Application of organic amendments like neem cake (200 g), green foliage (3-5 kg) or FYM (1 kg per vine) and earthing up in September-October are recommended. Growing marigold as a trap crop and uprooting the trap crop at flowering stage and burning reduce the incidence of nematodes. *R. similis* was not encountered when black pepper was intercropped with coffee (Venkitesan, 1976).

(ii) Biological Methods: *Pasteuria penetrans, Paecilomyces lilacinus* and *Pochonia chlamydosporia* are effective in suppressing the root-knot and burrowing nematodes. Five isolates of rhizobacteria (IISR-552, 528, 658, 853 and 859) were good growth promoters (26.7 to 55.6%) having dual nematicidal action (suppressing both *M. incognita* and *R. similis*).

(iii) Integrated Methods

- Planting of nematode-free rooted cuttings.
- Uprooting of affected vines and replanting after 9-12 months.
- Use of non-living standards for trailing pepper.
- Avoiding susceptible intercrops such as banana, ginger, turmeric etc.

18.1.2. The Root-knot Nematode, *Meloidogyne incognita*

(i) Cultural Methods: Live standards for black pepper such as *Artocarpus heterophyllus, A. hirsutus, Ailanthes malabarica, Mesopsis emini, Peltophorum pterocarpum, Swietenia macrophylla, Tamarindus indica, Garuga pinnata,* and *Macaranga peltata* are resistant to *M. incognita.*

Growing of non-host cover crops like Siratro (*Macroptilium atropurpureus*) in the interspaces and mulching with Guatemala grass (*Imperata cylindrica*) are recommended to reduce the root-knot nematode (*M. incognita*) population on black pepper in the Amazonian region (Ichinohe, 1980).

Application of neem cake reduces nematode populations followed by increase in yield.

(ii) Biological Methods: Application of *Paecilomyces lilacinus* and *Pochonia chlamydosporia* has potential to control nematodes which reduced the rate of infection to 15 and 12%, respectively. Effective strains of AMF (*Glomus* sp. and *Gigaspora* sp.) isolated from black pepper rhizosphere were very effective in suppression of root-knot nematodes.

Soil application of *Pochonia chlamydosporia* reduced nematode population and increased the yield (5.14 kg/vine) with cost: benefit ratio of 1:7.12.

Paecilomyces lilacinus, Pochonia chlamydosporia and *Bacillus* spp. gave good control of root-knot nematodes in black pepper.

Root-knot nematodes in pepper can be managed by the application of *Bacillus macerans* at 1.2 x 10^8 cells/vine just before the monsoon period.

AMF such as *Glomus fasciculatum, G. etunicatum and Acaulospora laevis* reduced root-knot nematodes and enhanced the growth of black pepper. Pre-inoculation of *Acaulospora laevis* in black pepper cv. Panniyur-1 reduced the root-knot index (0.75) followed by *Glomus mosseae/ Gigaspora margarita* (1.25) compared to control (3.5) (Anandraj *et al.*, 1991).

(iii) Host Resistance: The root-knot nematode infestation was less in cvs. Veliakaniakadan, Karimunda, Kalluvalli, Balankotta, Narayakodi and Padapan.

(iv) Integrated Methods: Black pepper vines combinedly inoculated with *P. lilacinus* and *P. penetrans* had put out 23-112 per cent increase in plant growth over control and were very effective in the management of root-knot nematodes (Sosamma and Koshy, 1995).

18.1.3. Nematodes, *Meloidogyne incognita/Radopholus similis* and Foot Rot, *Phytophthora capsicii* Disease Complex

(i) Integrated Methods: Integrated management of foot rot (*P. capsicii*) and nematodes (*M. incognita* and *R. similis*) on black pepper was achieved by (i) mixing AMF and *Trichoderma harzianum* in solarized nursery mixture to raise healthy and robust seedlings, (ii) application of *T.*

harzianum and FYM in planting pit, (iii) field application of neem cake at 1 kg/vine mixed with 50 g of *T. harzianum* during August.

18.2. CARDAMOM

18.2.1. Root-knot Nematodes, *Meloidogyne* spp.

(i) Physical Methods: Soil solarization for 40-45 days using 400 gauge transparent LDPE was proved ideal pre-sowing treatment for the management of nematodes and other soil-borne diseases in cardamom nurseries. Mean soil temperature during the solarization period increased by 8.7°C in solarized beds and the germination was enhanced by 25.5%, while weed growth was suppressed by 82%. *Pythium vexans* was totally eliminated, while populations of *Rhizoctonia solani, Phyllosticta elettaria* and nematodes were suppressed to varying levels. Solarization was also found to stimulate the growth of cardamom seedlings (Eapen, 1995).

(ii) Cultural Methods: Application of neem cake at 0.5 and 1.0 kg/plant twice a year reduces nematode populations followed by increase in yield (Ali, 1984).

(iii) Biological Methods: AMF such as *Glomus fasciculatum* and *Gigaspora margarita* were recorded to be effective in reducing the root-knot nematode population.

P. lilacinus suppressed *Meloidogyne* spp. population by 58.3-86.9 per cent in cardamom nursery. The number of quality seedlings for transplanting was greater in beds treated with this biocontrol agent (Eapen and Venugopal, 1995).

T. harzianum suppressed *Meloidogyne* spp. by 58.3-68.9 per cent in cardamom nursery and increased the number of quality seedlings for transplanting (Eapen and Venugopal, 1995).

(iv) Integrated Methods: Soil solarization of nursery beds and application of bioagents such as *P. lilacinus* or *Trichoderma* spp. improved growth of cardamom seedlings by suppressing root-knot nematode population (Eapen and Venugopal, 1995).

Integration of the following practices can help in the successful management of root-knot nematodes:

- Changing nursery sites frequently.
- Disinfecting nursery beds.
- Introduction of biocontrol agents at nursery level.
- Control of susceptible weeds.
- Exclusion of susceptible shade trees.
- Destruction of infested crop residues.
- Mulching with dead leaves.
- Application of neem cake.

18.2.2. Root-knot Nematode, *Meloidogyne* sp. and Rhizome Rot, *Rhizoctonia solani* Disease Complex

(i) Integrated Methods: *P. lilacinus* in combination with *Trichoderma* spp. suppressed *Meloidogyne* spp. and rhizome rot disease (*Rhizoctonia solani*) complex when incorporated in solarized cardamom nursery beds (Eapen and Venugopal, 1995).

18.3. GINGER AND TUEMERIC

18.3.1. Nematodes, *Meloidogyne incognita, Radopholus similis*

(i) Physical Methods: Use of nematode-free rhizomes for fresh planting, washing seed material to free of soil and drying in shade before planting helps to reduce the inoculum. Hot water treatment of rhizomes at 50-55°C for 10 min or 45°C for 30 min eliminates nematode infection. Soil solarization of beds for 40 days during summer reduces nematode population in soil.

(ii) Cultural Methods: Washing turmeric seed material free of soil and shade drying before planting helps in reducing the nematode population.

Crop rotation with cereals and millets should be practiced, preferably with rice at least once in 3 years. For ginger cultivation in Fiji, a ginger-taro-fallow rotation has been recommended. Intercropping of ginger with maize, capsicum and mulching with FYM are effective in reducing the nematode population. Deep summer ploughing with furrow turner ploughs should be practiced. Application of well decomposed FYM/compost at 25-30 t/ha or neem cake at 2 t/ha and mulching with green leaves at 10-12 t/ha at planting and repeating the mulching during the growth period helps in reducing nematode multiplication (Kaur, 1987). Using ginger and turmeric as intercrops in *R. similis* infested coconut and areca nut based farming systems should be avoided.

(iii) Biological Methods: Complete suppression of nematodes could be achieved by the application of *Fusarium oxysporum* (isolate IISR-11) at 50 g/bed of 3 sq. m. at the time of planting. *Paecilomyces lilacinus* and *Aspergillus nidulans* suppressed nematode population.

Turmeric rhizome treatment with *P. fluorescens* at 10 g/kg increased plant growth characters (58 and 31% increase in pseudo stem height and number of tillers over control) and decreased nematode population (40%) and gall indices (25%) (Seenivasan *et al.*, 2001).

(iv) Host Resistance: Turmeric cvs. Kodur, Chayapasupu, Duggirala, Guntur-1, Guntur-9, Rajampet, Sugandham and Appalapadu are resistant to *M. incognita* (Gunasekharan *et al.*, 1987).

(v) Integrated Methods: Soil application of *T. harzianum* + neem cake at 1 t/ha is recommended.

Summer ploughing at an interval of 15 days and use of hot water treated ginger rhizomes as seed proved promising in reducing the nematode population and in increasing the yield up to 65%.

18.4. CORIANDER

18.4.1. The Root-knot Nematode, *Meloidogyne incognita*

(i) Biological Methods: *P. lilacinus* was effective in increasing the plant growth parameters and in reducing root galling and egg mass production due to *M. incognita* in coriander (Sobita Devi and Pandey, 2001).

Table 18.1. Biological control of plant parasitic nematodes in spice crops.

Crop	Nematode	Effective bioagent/s	Reference
Black pepper	*M. incognita*	*P. lilacinus* *P. fluorescens*	Sosamma & Koshy, 1995
		P. penetrans	Eapen *et al.*, 1997
		G. mosseae	Sosamma & Koshy, 1995 Sivaprasad & Sheela, 1999
		G. mosseae, G. fasciculatum, Acaulospora laevis, Gigaspora margarita	Anandraj *et al.*, 1991
Cardamom	*Meloidogyne* spp.	*P. lilacinus, Trichoderma* spp.	Eapen & Venugopal, 1995
Turmeric	*M. incognita*	*P. fluorescens*	Seenivasan *et al.*, 2001
Coriander	*M. incognita*	*P. lilacinus*	Midha *et al.*, 2001

Chapter 19

TUBER CROPS

19.1. YAM

19.1.1. Root-Knot Nematodes, *Meloidogyne* spp.

(i) Physical Methods: Hot water treatment of corms at 51°C for 30 min. eliminates root-knot nematode infection.

(ii) Cultural Methods: The root-knot nematode can be controlled by growing sweet potato cv. Shree Bhadra as a trap crop.

(iii) Host Resistance: Yam cvs. Sree Latha and Sree Kirthi were found resistant to root-knot nematodes.

19.2. SWEET POTATO

19.2.1. Root-Knot Nematodes, *Meloidogyne* spp.

(i) Physical Methods: Hot water treatment at 45°C for 30 hr, 46.7°C for 65 min or 50°C for 3-5 min eliminates root-knot nematode infection without seriously impairing the viability of roots. Hot air treatment of roots, 5-8 cm in diameter, at 50°C for 4-8 hrs eliminates most root-knot nematodes without significantly affecting their keeping qualities.

(ii) Cultural Methods: The clean cuttings of seed roots from the apex are less prone to root-knot damage when planted out than cuttings from the basal end.

Groundnut or maize in rotation with sweet potato decreases numbers of *M. incognita* to low levels.

(iii) Host Resistance: Sweet potato cvs. Heartgold, Nemagold, Redmar, Jersey types and Kyushu No. 52 were reported to be resistant to root-knot nematodes (Giamalava *et al.*, 1960). The cv. Jasper is moderately resistant, while cv. Painter shows some tolerance.

(iv) Integrated Methods: Control of root-knot damage to sweet potatoes involves integration of at least 3 methods:

- The selection of seed roots that are either free of nematodes or freed of nematodes by hot water treatment.
- Planting of the seed roots into beds of sand or coarse textured soil that is either nematode-free or, if infested, is pre-plant fumigated.
- Transplanting of the clean 'slips' into nematode-free soil or pre-plant fumigated soil.

19.2.2. The Reniform Nematode, *Rotylenchulus reniformis*

(i) Physical Methods: Hot water treatment of sweet potatoes at 50°C for 3 to 5 minutes eliminated the reniform nematode (Martin, 1970).

(ii) Host Resistance: The sweet potato variety Goldrush is less susceptible to *R. reniformis* than other varieties, but is highly susceptible to root-knot nematodes.

19.3. CASSAVA

19.3.1. Root-Knot Nematodes, *Meloidogyne* spp.

(i) Host Resistance: Cassava cvs. Sree Sahya and Narayaniyakappa were found resistant to root-knot nematodes.

19.3.2. The Lesion Nematode, *Pratylenchus brachyurus*

(i) Host Resistance: A number of cassava varieties (Atitogen, Ba Pou II, Agba Tiega, Agba Boquia, Sodjievi) show some resistance to the lesion nematode.

Chapter 20

FUTURE OUTLOOK

In looking to the future, what are some of the things we need to do from a realistic and practical stand point of view?

We should continue research on nematode management to include efforts to obtain (a) nematicides which are more effective, can be used at lower dosage, and cause less pollution; (b) resistant cultivars for major crops and at least for the most important nematode species; and knowledge necessary to utilize the benefits and capabilities offered by integrated nematode management systems.

A second area of concern is to advance biological control to its fullest potential, utilizing biological control agents plus organic amendments, sewage, sludge, etc. This would include cultural methods for biocontrol parasites, understand their nature, make them practical, and genetically modify the biocontrol agents to improve their effectiveness.

The present research thrust on biological control of nematode pests need to be intensified. Many fungal and bacterial bioagents have been found effective under *in vitro* conditions. Efforts be made to test them under field conditions. Further, our efforts should be directed towards developing mass production technologies for effective bioagents and easy field application techniques.

In recent years, it has been observed that the combination of fungal bioagents perform better in reducing nematode population and increasing the plant health. Thus the combination of highly toxic fungus, *Aspergillus niger* (killed most of the infective second stage juveniles) and an egg parasite, *Cladosporium oxysporum* (invaded and killed the eggs in egg sac) both at half the doses reduced significantly more *M. incognita* population and exhibited better plant growth than when either of the fungal bioagents in brinjal (Goswami and Singh, 2001).

A third area would be, through surveys and identifications, determine more accurately the numbers and kinds of nematodes which actually exist and pinpoint their distribution. Success of crop rotations and resistant cultivars depend upon knowing what nematodes occur in which areas and fields.

A fourth area would be to develop a better understanding of the biological aspects of all our more important plant nematodes. This would include the molecular genetics of closely related species and races; host-parasite relationships and histochemical details which may provide clues to the nature of plant resistance to nematodes; ecological factors, such as temperature, moisture and soil types which may influence nematode development and survival which in turn affect control measures; and make full use of current biotechnologies as appropriate in this and other nematology research.

With regard to viruses transmitted by soil-borne nematodes, it is advisable to explore the possibility of finding dorylaimid nematodes from the rhizosphere and their role in transmitting soil-borne viruses mainly in crops like fruits, vegetables and ornamentals.

Our future research programmes need to be more based towards development of sustainable and non-chemical methods for the management of nematode pests.

There is a need for development of cropping systems based nematode management technologies.

Another promising area of research in nematode management will be to identify and develop new generation chemicals which may not kill the nematodes but either change their behaviour or delay the development and force them into the inactive stages or alert the plant to defend the nematode attack.

Phytoalexins effectively induce plant resistance mechanisms to nematodes, particularly to sedentary ones. However, the mechanism by which phytoalexins are accumulated in plants is not known. Further research in accurately elucidating the effect of potential phytoalexins on the nematodes is urgently needed.

Although valuable improvement has been obtained in development of nematode resistant horticultural crop cultivars, there is a need for breeding varieties with combined resistance to different nematode species and to other organisms such as soil-borne fungi and bacteria which cause wilt and root rot.

The development of transgenic plants with durable nematode resistance keeping pace with the evolution of resistance breaking populations should be considered.

Once the INM technologies are developed, it is necessary that they should be put to practice under field conditions. The implementation of these programmes will require scientists/trained technicians and efficient extension media. In this direction, national institutes and SAU's can play an important role by holding training courses and carrying out problem oriented researches. Since integrated approach may involve more than one discipline, it is therefore, essential that professional scientists must collaborate in its planning and execution.

We have made much progress in nematology; but we still have a long way to go in the battle against these destructive and insidious plant nematodes. Through our continued dedication and strong research efforts, greater knowledge and better understanding of nematodes will be achieved. This will result in further reductions in losses caused by nematodes and a significant increase of food and fiber in our countries.

Chapter 21

SOURCES OF CRITICAL INPUTS FOR INTEGRATED NEMATODE MANAGEMENT

21.1. ARBUSCULAR MYCORRHIZAL FUNGI

Table 21.1. Trade names and source of availability of Arbuscular Mycorrhizal Fungi

Arbuscular Mycorrhizal Fungi	Trade Name	State	Source of Availability
Glomus mosseae	Symbion VAM	Karnataka	Agri Technol Infmn Centre, IIHR, B'lore
		Maharashtra	Bharatiya Agro Industries Foundation, Pune
		Tamil Nadu	T Stanes & Co, Coimbatore
G. fasciculatum	Symbion VAM	Karnataka	Agri Technol Infmn Centre, IIHR, B'lore
		Maharashtra	Bharatiya Agro Industries Foundation, Pune
		Tamil Nadu	T Stanes & Co, Coimbatore
G. aggregatum		Maharashtra	Bharatiya Agro Industries Foundation, Pune
Gigaspora margarita		Maharashtra	Bharatiya Agro Industries Foundation, Pune
Arbuscular mycorrhizal fungi	Multiplex Trishul	Karnataka	Karnataka Agro Chemicals, B'lore

21.2. ANTAGONISTIC FUNGI

Table 21.2. Trade names and sources of availability of antagonistic fungi.

Antagonistic Fungi / Trade Name	State	Source of Availability
Arthrobotrys cladodes var. *macroides*	Karnataka	Project Directorate of Biol Control, Bangalore
A. oligospora	Karnataka	Project Directorate of Biol Control, Bangalore
Aspergillus niger (Kalisena SD, Kalisena SL, Sanjeevni)	Gujarat	Cadila Pharmaceuticals Ltd, Ahmedabad
	New Delhi	Plant Path Divn, Indian Agri Res Inst, New Delhi
Dactylella brochophaga	Karnataka	Project Directorate of Biol Control, Bangalore
Gliocladium catenulatum	Karnataka	Project Directorate of Biol Control, Bangalore
G. deliquescens	Karnataka	Project Directorate of Biol Control, Bangalore
G. roseum	Karnataka	Project Directorate of Biol Control, Bangalore
Pochonia chlamydosporia (Bionema, Biovert, Multiplex Varsha)	Karnataka	Project Directorate of Biol Control, Bangalore Agri Technol Infmn Centre, IIHR, Bangalore Central Sericultural Res & Trng Inst, Mysore Karnataka Agro Chemicals, B'lore
Paecilomyces lilacinus (Bio-Nemator, Biocon, Multiplex Niyantran, Abtec Paecilomyces)	Karnataka	Project Directorate of Biol Control, Bangalore Agri Technol Infmn Centre, IIHR, Bangalore
Trichoderma aureoviride	Karnataka	Project Directorate of Biol Control, Bangalore
T. hamatum	Karnataka	Project Directorate of Biol Control, Bangalore
T. harzianum (Binab-T, Ecoderma, F-Stop, Supravit, Tricodex, Trichoderma)	Andhra Pradesh	ADA (BCL), P M Palem ADA (BCL), Kakinada ADA (BCL), Nidadavole ADA (BCL), Ibrahimpatnam ADA (BCL), Ongole ADA (BCL), Nellore

	DDA (FTC), Nandyal
	ADA (BCL), Anantapur
	DDA (FTC), Rajendranagar
	ADA (Oilseeds) (BCL), Mahabubanagar
	ADA (BCL), Nalagonda
	ADA (BCL), Mulugu Road
	ADA (BCL), Karimnagar
	ADA (BCL), Adilabad
	Pest Control (India) Pvt Ltd, Hyderabad
Assam	State Biocontrol Laboratory, Guwahati
	Pest Control (India) Pvt Ltd, Guwahati
Bihar & Jarkhand	Pest Control (India) Pvt Ltd, Jamshedpur
Gujarat	Pest Control (India) Pvt Ltd, Vadodara
Haryana	State Biocontrol Laboratory, Sirsa
	Pest Control (India) Pvt Ltd, Chandigarh
Karnataka	Project Directorate of Biol Control, Bangalore
	Agri Technol Infmn Centre, IIHR, Bangalore
	Central IPM Centre, Bangalore
	Central Sericultural Res & Trng Inst, Mysore
	Coffee Research Sub Station, Chethalli
	Central Coffee Res Inst, Entomol Divn, Chikmagalur
	Univ of Agri Sci, Entomol Dept, Dharwad
	Inst of Pulses & Oilseeds Res, Gulbarga
	Regional Res Stn, UAS, Raichur
	Agri Res Stn, UAS, Bijapur
	Kadur Agro Pvt Ltd, Bangalore
	Margo Pvt Ltd, Bangalore
	Pest Control (India) Pvt Ltd, Bangalore
Kerala	Travancore Organic Fertilizers, Kottayam
	VRM Biotech Products, Changanacherry
Madhya Pradesh	Pest Control (India) Pvt Ltd, Bhopal
Maharashtra	Bio Era Technologies, Nagpur
	Hoechst Schering Agr Envt, Mumbai
	KNS Biotech, Nanded
	Nathkrupa Biocontrol Lab, Nagpur
	Om Agro Organics, Yavatmal
	Pest Control (India) Pvt Ltd, Mumbai-63
	Pest Control (India) Pvt Ltd, Mumbai-59

	Pest Control (India) Pvt Ltd, Thane
	Pest Control (India) Pvt Ltd, New Mumbai
	Pest Control (India) Pvt Ltd, Nashik
	Pest Control (India) Pvt Ltd, Pune-411 033
	Pest Control (India) Pvt Ltd, Pune-411 030
	Sanvardhini Agro Pvt Ltd, Satara
	Soman Biofertilizers, Pune
	Vidyas Biotech Lab, Nagpur
	West Coast Herbochem P Ltd, Mumbai
New Delhi	Biotech Intern Ltd, New Delhi
	Pest Control (India) Pvt Ltd, New Delhi
Orissa	Pest Control (India) Pvt Ltd, Bhubaneswar
Punjab	Pest Control (India) Pvt Ltd, Chandigarh
Rajasthan	Biocontrol Lab, ARS, RAU, Sri Ganganagar
	JDA (Pl Path), Biocontrol Lab, Durgapura
	DDA (Agro), IPM Lab, Ajmeer
	DDA (Agro), IPM Lab, Malikpur
	DDA (Agro), IPM Lab, Rampura
	DDA (Agro), IPM Lab, Cjjatrapura
	DDA (Agro), IPM Lab, Chittorgarh
	DDA (Agro), IPM Lab, Hanumangarh
	DDA (Pl Path), IPM Lab, Banaswara
	Pest Control (India) Pvt Ltd, Jaipur
Tamil Nadu	Centre for Pl Prot Studies, TNAU, Coimbatore
	Bio-control Lab, Papparapatty
	Bio-control Lab, Coimbatore
	Bio-control Lab, Salem
	Bio-control Lab, Villupuram
	Bio-control Lab, Trichy
	Bio-control Lab, Panjupettai
	Bio-control Lab, Kuruppanaichenpalyam
	Bio-control Lab, Vinayagapuram
	Bio-control Lab, Tirunelveli
	Bio-control Lab, Tanjavur
	Bio-control Lab, Namakkal
	Basarass Biocontrol Res, Vallavar
	Greentech Agro Prod P Ltd, Coimbatore
	Jeypee Bio Techs, Virudhunagar
	Pest Control (India) Pvt Ltd, Chennai
	Rajendra Foundn for Agri Res, Chittar
	Shri Durga Agro Services, Coimbatore
	Shrishti Bioprod P Ltd, Coimbatore

		Sun Agro Ind India, Chennai
		TN Co-op Sugar Fedn Ltd, Chengalpattu
		TAP Industries, Chennai
	Uttarakhand	GB Pant Uni Agri & Technol, Pl Path Dept, Pantnagar
		Pest Control (India) Pvt Ltd, Dehradun
	Uttar Pradesh	Crop Health Bioprod Res Centre, Ghaziabad
		Pest Control (India) Pvt Ltd, Bareilly
		Pest Control (India) Pvt Ltd, Lucknow
		Pest Control (India) Pvt Ltd, Kanpur
	West Bengal	Pest Control (India) Pvt Ltd, Kolkata
T. koningii	Karnataka	Project Directorate of Biol Control, Bangalore
T. longibactriatum	Karnataka	Project Directorate of Biol Control, Bangalore
T. polysporum	Karnataka	Project Directorate of Biol Control, Bangalore
T. pseudokoningii (Nursery Guard)	Karnataka	Project Directorate of Biol Control, Bangalore
		Central Sericultural Res &Trng Inst, Mysore
T. virens (Gligard, WRC-GV)	Karnataka	Project Directorate of Biol Control, Bangalore
T. viride (Bio-Cure, Bio-Cure F, Agroderma, Antogon TV, Bioderma, Dermapak, Ecofit, Monitor, Multiplex Nisarga, Rakshak, Trichosan, Trichoft, Trichoderm, Trichonik, Tricho-X, Trichodermin)	Andhra Pradesh	ADA (BCL), P M Palem
		ADA (BCL), Kakinada
		ADA (BCL), Nidadavole
		ADA (BCL), Ibrahimpatnam
		ADA (BCL), Ongole
		ADA (BCL), Nellore
		DDA (FTC), Nandyal
		ADA (BCL), Anantapur
		DDA (FTC), Rajendranagar
		ADA (Oilseeds) (BCL), Mahabubanagar
		ADA (BCL), Nalagonda
		ADA (BCL), Mulugu Road
		ADA (BCL), Karimnagar
		ADA (BCL), Adilabad
		Pest Control (India) Pvt Ltd, Hyderabad
	Assam	State Biocontrol Laboratory, Guwahati
		Pest Control (India) Pvt Ltd, Guwahati
	Bihar & Jarkhand	Pest Control (India) Pvt Ltd, Jamshedpur

Gujarat	Pest Control (India) Pvt Ltd, Vadodara
Haryana	State Biocontrol Laboratory, Sirsa
	Pest Control (India) Pvt Ltd, Chandigarh
Karnataka	Project Directorate of Biol Control, Bangalore
	Coffee Research Sub Station, Chethalli
	Central Coffee Res Inst, Entomol Divn, Chikmagalur
	Kadur Agro Pvt Ltd, Bangalore
	Karnataka Agro Chemicals, B'lore
	Margo Pvt. Ltd, Bangalore
	Pest Control (India) Pvt Ltd, Bangalore
	Travancore Org Fert Co P Ltd, Bangalore
Kerala	Travancore Organic Fertilizers, Kottayam
	VRM Biotech Products, Changanacherry
Madhya Pradesh	Pest Control (India) Pvt Ltd, Bhopal
Maharashtra	Bio Era Technologies, Nagpur
	Hoechst Schering Agr Envt, Mumbai
	KNS Biotech, Nanded
	Nathkrupa Biocontrol Lab, Nagpur
	Om Agro Organics, Yavatmal
	Pest Control (India) Pvt Ltd, Mumbai-63
	Pest Control (India) Pvt Ltd, Mumbai-59
	Pest Control (India) Pvt Ltd, Thane
	Pest Control (India) Pvt Ltd, New Mumbai
	Pest Control (India) Pvt Ltd, Nashik
	Pest Control (India) Pvt Ltd, Pune-411 033
	Pest Control (India) Pvt Ltd, Pune-411 030
	Sanvardhini Agro Pvt. Ltd., Satara
	Soman Biofertilizers, Pune
	Vidyas Biotech Lab, Nagpur
	West Coast Herbochem P Ltd, Mumbai
New Delhi	Biotech Intern Ltd, New Delhi
	Pest Control (India) Pvt Ltd, New Delhi
Orissa	Pest Control (India) Pvt Ltd, Bhubaneswar
Punjab	Pest Control (India) Pvt Ltd, Chandigarh
Rajasthan	Biocontrol Lab, ARS, RAU, Sri Ganganagar
	JDA (Pl Path), Biocontrol Lab, Durgapura
	DDA (Agro), IPM Lab, Ajmeer
	DDA (Agro), IPM Lab, Malikpur
	DDA (Agro), IPM Lab, Rampura
	DDA (Agro), IPM Lab, Cjjatrapura

		DDA (Agro), IPM Lab, Chittorgarh
		DDA (Agro), IPM Lab, Hanumangarh
		DDA (Pl Path), IPM Lab, Banaswara
		Pest Control (India) Pvt Ltd, Jaipur
	Tamil Nadu	Centre for Pl Prot Studies, TNAU, Coimbatore
		Bio-control Lab, Papparapatty
		Bio-control Lab, Coimbatore
		Bio-control Lab, Salem
		Bio-control Lab, Villupuram
		Bio-control Lab, Trichy
		Bio-control Lab, Panjupettai
		Bio-control Lab, Kuruppanaichenpalyam
		Bio-control Lab, Vinayagapuram
		Bio-control Lab, Tirunelveli
		Bio-control Lab, Tanjavur
		Bio-control Lab, Namakkal
		Basarass Biocontrol Res, Vallavar
		Greentech Agro Prod P Ltd, Coimbatore
		Jeypee Bio Techs, Virudhunagar
		Pest Control (India) Pvt Ltd, Chennai
		Rajendra Foundn for Agri Res, Chittar
		Shri Durga Agro Services, Coimbatore
		Shrishti Bioprod P Ltd, Coimbatore
		Sun Agro Ind India, Chennai
		TN Co-op Sugar Fedn Ltd, Chengalpattu
		TAP Industries, Chennai
		T Stanes & Co, Coimbatore
	Uttarakhand	GBPant Univ Agri & Technol, Pl Path Dept, Pantnagar
		Pest Control (India) Pvt Ltd, Dehradun
	Uttar Pradesh	Crop Health Bioprod Res Centre, Ghaziabad
		Pest Control (India) Pvt Ltd, Bareilly
		Pest Control (India) Pvt Ltd, Lucknow
		Pest Control (India) Pvt Ltd, Kanpur
	West Bengal	Pest Control (India) Pvt Ltd, Kolkata
Verticillium lecanii	Karnataka	Project Directorate of Biol Control, Bangalore
		Pest Control (India) Pvt Ltd, Bangalore
	Maharashtra	Sio Agro Research Lab, Mumbai
	Tamil Nadu	Sakthi Biocontrol & Biofert Centre, Kallipatti Gobi

V. suchlosporium	Karnataka	Project Directorate of Biol Control, Bangalore
Arthrobotrys sp. + *P lilacinus* + *Verticillium* sp. (Nemastin)	Maharashtra	Kumar Krishimitra Bioproducts, Pune
Aspergillus niger + *T. viride* + *P. lilacinus* (Multiplex Shock)	Karnataka	Karnataka Agro Chemicals, B'lore
P. chlamydosporia + *B. pumilis* (Bacillus Plus-P)	Karnataka	Agri Technol Infmn Centre, IIHR, Bangalore
P. chlamydosporia + *B. subtilis* (Bacillus Plus-S)	Karnataka	Agri Technol Infmn Centre, IIHR, Bangalore
P. chlamydosporia + *P. lilacinus* (Biovert Plus)	Karnataka	Agri Technol Infmn Centre, IIHR, Bangalore
P. chlamydosporia + *P. fluorescens* (Pseudomonas Plus)	Karnataka	Agri Technol Infmn Centre, IIHR, Bangalore
P. chlamydosporia + *T. harzianum* (Trichorich)	Karnataka	Agri Technol Infmn Centre, IIHR, Bangalore
P. chlamydosporia + *T. viride* (Trichovert)	Karnataka	Agri Technol Infmn Centre, IIHR, Bangalore
T. harzianum + *T. viride* (Ecoderma, NIPROT)	Karnataka	Bio Control Res Lab, PCI Pvt Ltd, Bangalore Margo Pvt Ltd, Bangalore
T. harzianum + *T. viride* + *T. virens* (Combat)	Maharashtra	Kumar Krishimitra Bioproducts, Pune

21.3. BACTERIAL BIOPESTICIDES

Table 21.3. Trade names and sources of availability of bacterial biopesticides.

Bacterial Biopesticides/ Trade Name's	State	Source of Availability
Bacillus sphaericus	New Delhi	Biotech Intern Ltd, New Delhi
	Tamil Nadu	Sinagro Industries India, Chennai
B. subtilis (Kill Dew DP Epic, Kodiac)	Maharashtra	Kumar Krishimitra Bioproducts, Pune
B. thuringiensis var. *kurstaki* (Biolep, Biosap, Delfin, Dipel, Bibot, Spicturin, Halt, Multiplex Bt. K)	Andhra Pradesh	Indian Inst of Chem Technol, Hyderabad
	Karnataka	Karnataka Agro Chemicals, B'lore

	Madhya Pradesh	Indore Biotech Inputs & Res (Pvt) Ltd, Indore
	Maharashtra	Lupin Laboratories (I) P Ltd, Mumbai Sandoz (I) Ltd, Mumbai Wockhardt Ltd, Mumbai
	New Delhi	Biotech Intern Ltd, New Delhi Hindustan Insecticides, New Delhi
	Tamil Nadu	Tuticorn Alkali Chem & Fert Ltd, Chennai
Pseudomonas fluorescens (Bio-Cure B, Bioshield, Multiplex Sparsha, Tracomonas, Sudocel, Dagger-G)	Assam	State Biocontrol Laboratory, Guwahati
	Haryana	State Biocontrol Laboratory, Sirsa
	Karnataka	Agri Technol Infmn Centre, IIHR, Bangalore Karnataka Agro Chemicals, B'lore Travancore Org Fert Co P Ltd, Bangalore
	Kerala	Plantrich Chem & Fert Ltd, Kottayam
	New Delhi	Biotech Intern Ltd, New Delhi
	Tamil Nadu	Bio-control Lab, Papparapatty Bio-control Lab, Coimbatore Bio-control Lab, Salem Bio-control Lab, Villupuram Bio-control Lab, Trichy Bio-control Lab, Panjupettai Bio-control Lab, Kuruppanaichenpalyam Bio-control Lab, Vinayagapuram Bio-control Lab, Tirunelveli Bio-control Lab, Tanjavur Bio-control Lab, Namakkal Basarass Biocontrol Res, Vallavar Greentech Agro Prod P Ltd, Coimbatore Jeypee Bio Techs, Virudhunagar Rajendra Foundn for Agri Res, Chittar Sun Agro Ind India, Chennai T Stanes & Co, Coimbatore
B. pumilis + P. chlamydosporia (Bacillus Plus-P)	Karnataka	Agri Technol Infmn Centre, IIHR, Bangalore

B. subtilis + *P. chlamydosporia* (Bacillus Plus-S)	Karnataka	Agri Technol Infmn Centre, IIHR, Bangalore
P. fluorescens + *P. chlamydosporia* (Pseudomonas Plus)	Karnataka	Agri Technol Infmn Centre, IIHR, Bangalore

21.4. NEEM-BASED AND OTHER BOTANICALS

Table 21.4. Source of availability of neem–based & other botanicals

Neem-Based & other Botanicals	State	Source of Availability
Achook	Maharashtra	Bahar Agrochemicals, Ratnagiri Mahagrapes, Pune
Aphidin	Maharashtra	Eco-Max Agrosystems Ltd, Mumbai
Avana	Tamil Nadu	EID Parry (I) Ltd, Bioprod Divn, Chennai
Aza	Andhra Pradesh	SOM-IPM Systems (I) Ltd, Hyderabad
Azadit	Rajasthan	Pesticides India Ltd, Udaipur
Biogold	Tamil Nadu	SPIC, Bio-Tech Divn, Chennai
Bioneem	Maharashtra	Ajay Biotech Lab (P) Ltd, Pune
Bionema	Uttar Pradesh	Bioved Res Soc, Allahabad
Biopest	Uttar Pradesh	Bioved Res Soc, Allahabad
Biosol	Tamil Nadu	AV Thomas Pvt Ltd, Chennai
Eco-Neem	Karnataka	Murkumbi Bio Agro P Ltd, Belgaum Pest Control (India) Pvt Ltd, Bangalore
Eco-Neem Plus	Karnataka	Murkumbi Bio Agro P Ltd, Belgaum Pest Control (India) Pvt Ltd, Bangalore
Ecotin	Karnataka	Monarch Biofert & Res Centre, Bangalore
Field Marshal	Gujarat	Ganesh Biocontrol System, Gondal
Fortuna Aza	Andhra Pradesh	Fortune Bio-tech Limited, Secunderabad
Jai Neem	Haryana	Jai Chemicals, Faridabad
Jawan	Maharashtra	Mc DA Agro Pvt Ltd, Mumbai
Jeevan Crop Protector	Maharashtra	Mc DA Agro Pvt Ltd, Mumbai
Juerken	Tamil Nadu	Madurai Chem & Agro Ind Ltd, Madurai
Limonool	Karnataka	Bio-Multi Tech Pvt Ltd, Bangalore
Margocide-CK	Karnataka	Monofix Agro Products Ltd, Hubli
Margosan-O	Karnataka	Monofix Agro Products Ltd, Hubli
Multineem	Karnataka	Karnataka Agro Chemicals, B'lore
Multiplex Mahaan	Karnataka	Karnataka Agro Chemicals, B'lore
Multiplex Urea Coat	Karnataka	Karnataka Agro Chemicals, B'lore
Neemactin	Maharashtra	Wockhardt Ltd, Mumbai

Neemak	Karnataka	West Coast Herbochem P Ltd, Bangalore
	Maharashtra	West Coast Herbochem P Ltd, Mumbai
Neemarin	New Delhi	Biotech Intern Ltd, New Delhi
Neemark	Karnataka	West Coast Herbochem P Ltd, Bangalore
Neemasol	Tamil Nadu	EID Parry (I) Ltd, Pesticides Divn, Chennai
Neemax	Maharashtra	Eco-Max Agrosystems Ltd, Mumbai
Neemazal-F	Tamil Nadu	EID Parry (I) Ltd, Bioprod Divn, Chennai
Neemazal-T/S	Tamil Nadu	EID Parry (I) Ltd, Bioprod Divn, Chennai
Neemocide	West Bengal	Vinayak Fats & Proteins P Ltd, Kolkata
Neemol	Andhra Pradesh	Ramson Laboratories, Vijayawada
	Tamil Nadu	Agro Links, Trichy
Neemolin	Maharashtra	Khatau Agro Tech, Mumbai
Neemosan	Tamil Nadu	Agronule Industries, Trichy
Neemox	Maharashtra	Prakash Farm Chemicals, Mumbai
Neempourn	Tamil Nadu	Prabhakar Oil Mills, Trichy
Neemta	Gujarat	AJ Chemicals, Ahmedabad
Neethrin	Uttar Pradesh	Amitul Agro Chem Pvt Ltd, Gorakhpur
Neem Base	New Delhi	Indian Agri Res Inst, Agrochem Divn, New Delhi
Neem Based pesticides	Maharashtra	Bahar Agrochemicals, Ratnagiri
		KNS Biotech, Nanded
		Om Agro Organics, Yavatmal
	Tamil Nadu	Ecobiocides & Botanicals Pvt Ltd, Theni
		Jeypee Bio Techs, Virudhunagar
Neem Cake	Uttarakhand	Sri Ram Solvent Extraction Plant, Jaspur
Neem Gold	Tamil Nadu	EID Parry (I) Ltd, Bioprod Divn, Chennai
		SPIC, Bio-Tech Divn, Chennai
Neem Guard	Maharashtra	Gharda Chemicals Pvt Ltd, Mumbai
Neem Hit	Maharashtra	Skylark Agro Chemicals, Thane
Neem Oil	Karnataka	Agro Extracts Ltd, Bangalore
	Maharashtra	Mc DA Agro Pvt Ltd, Mumbai
		National Organic Chem Ind Ltd, Mumbai
		Sio Agro Research Lab, Mumbai
	Uttarakhand	Sri Ram Solvent Extraction Plant, Jaspur
Neem Plus	Maharashtra	BD Keathen & Co, Mumbai
Neem Rich	Maharashtra	National Chem Lab, Pune
Neem Seed	Maharashtra	Cheminova Agro-Tech, Nandubar
Neem Soap	Karnataka	Agri Technol Infmn Centre, IIHR, Bangalore

Neem Top	Tamil Nadu	Sri Krishna Co, Coimbatore
Nim-76	New Delhi	Defence Inst of Physiol, New Delhi
Nimba	New Delhi	Indian Agri Res Inst, Entomol Divn, New Delhi
Nimbasal	New Delhi	Nimba Foods & Agrochem, New Delhi Pesto-Chem India Ltd, New Delhi
Nimbicidine	Tamil Nadu	T Stanes & Co, Coimbatore
Nimbitor	Maharashtra	Zandu Pharmaceutical Works, Mumbai
Nimbosol	Tamil Nadu	Victoria Laboratories, Salem
Nimitox	Karnataka	Rallis India Ltd, Bangalore
Nimlin	Maharashtra	Sunline Agro Chem P Ltd, Dhulia
Nimbo Bas	Tamil Nadu	Basarass Biocontrol Res, Vallavar
Ozoneem	Haryana	Ozone Biotech, Faridabad
Phytowin	Tamil Nadu	Phyto Products Ltd, Thiruthuripoondi
Pongamia oil-based biopesticide	Tamil Nadu	Jeypee Bio Techs, Virudhunagar
Pongamia Soap	Karnataka	Agri Technol Infmn Centre, IIHR, Bangalore
Rakshak	Karnataka	Murkumbi Bio Agro P Ltd, Belgaum Margo Pvt Ltd, Bangalore
RD-9 Repellin	Andhra Pradesh	ITC Ltd, Hyderabad
Replin 555	Tamil Nadu	Micro Chemicals Ltd, Madurai
Shaktiman	Gujarat	Khetiwadi Corner, Vadodara Krishna Biotech Pvt Ltd, Veravel Oceon Agro India Ltd, Baroda
Sukrina	Tamil Nadu	Canster Chemicals Ltd, Chennai
Suneem	Maharashtra	Sunida Exports Ltd, Mumbai
Swaticure	Maharashtra	Swathi Industries Ltd, Mumbai
Tric	Uttar Pradesh	Amitul Agro Chem Pvt Ltd, Gorakhpur
Tricure	Karnataka	Murkumbi Bio Agro P Ltd, Belgaum Margo Pvt Ltd, Bangalore
Vapacide	Andhra Pradesh	Indian Inst of Chem Technol, Hyderabad

21.5. FULL ADDRESSES OF SOURCES OF CRITICAL INPUTS

21.5.1. ANDHRA PRADESH

Central Government Organizations

Indian Institute of Chemical Technology (IICT), Uppal Road, Hyderabad-500 007.

State Government Organizations

ADA (BCL), P.M. Palem, Visakhapatnam District, Phone: 0891-2781283, Fax: 0891-2504139.

ADA (BCL), Kakinada, East Godhavari District, Phone: 0884-378794, Fax: 0884-2384238.

ADA (BCL), Nidadavole, West Godhavari District, Phone: 08812-222794, Fax: 08812-244014.

ADA (BCL), Ibrahimpatnam, Krishna District, Phone: 08672-2881968, Fax: 08672-252483.

ADA (BCL), Ongole, Prakasham District, Phone: 08592-233126, Fax: 08592-280046.

ADA (BCL), Nellore, Nellore District, Phone: 0861-2326655, Fax: 0862307132.

DDA (FTC), Nandyal, Kurnool District, Phone: 08518-248090, Fax: 08518-277754.

ADA (BCL), Anantapur, Anantapur District, Phone: 08554-231743, Fax: 08554-275984.

DDA (FTC), Rajendranagar, Rangareddy District, Phone: 040-24015188, Fax: 044-24547507.

ADA (Oilseeds) (BCL), Mahabubanagar, Mahabubanagar District, Phone: 08542-242624, Fax: 08542-242398/242203.

ADA (BCL), Nalagonda, Nalagonda District, Phone: 08682-244560, Fax: 08682-232981.

ADA (BCL), ARS Campus, Mulugu Road, Warangal District, Phone: 0870-2543433, Fax: 0870-2511100.

ADA (BCL), Karimnagar, Karimnagar District, Phone: 0878-262161, Fax: 0878-2240343.

ADA (BCL), Adilabad, Adilabad District, Phone: 08732-226454, Fax: 08732-227311.

Private Organizations

Fortune Bio-tech Limited, 14, Ishaq Colony, 108, Bazar Road, Secunderabad-500 051, Phone: 040-819000, 841519, Fax: 040-843945.

ITC Ltd., ILTD Division, IBD 6-3-1110, Amrutha Hall, Begumpet, Hyderabad-500 016.

Pest Control (India) Pvt. Ltd., Flat No. 213, Bhanu Enclave, Sanjeeva Reddy Nagar, Hyderabad-500 038, Phone: 040-2381 3851, 2370 1411, Fax: 040-2370 4305, E-mail: hyderabad@pcil.co.in (Contact person: Mr. S. Srinivasa Reddy, Mobile: 31017824).

Ramson Laboratories, 29-14-53, Prakasam Road, Vijayawada-520 002.

SOM-IPM Systems (India) Ltd., 7-1-80/202, Kemson Apartments, Ameerpet, Hyderabad-500 016, E-mail: somphyto/Hyderabad@dartmaildartnet.com.

21.5.2. ASSAM

State Government Organizations

State Biocontrol Laboratory, C/O Director of Agriculture, Ulubari, Guwahati 781007, Phone: 0361-2332215, Fax: 0361-2332796, E-mail: assamagri@sify.com.

Private Organizations

Pest Control (India) Pvt. Ltd., Latasil Lamb Road, Behind Nice Tele Services, Guwahati-781 001, Phone: 0361-263 1230, E-mail: bedanta.baruah@pcil.co.in (Contact person: Mr. Bedanta Baruah, Mobile: 98640 54187).

21.5.3. BIHAR & JHARKHAND

Private Organizations

Pest Control (India) Pvt. Ltd., "Madhukunj", 3rd Floor, Q Road, Bistupur, Jamshedpur-831 001, Phone: 0657-242 3158, 228 8580, Fax: 0657-242 3158, E-mail: kailash.sharma@pcil.co.in (Contact person: Mr. Kailash Sharma, Mobile: 3105493).

21.5.4. GUJARAT

Private Organizations

A.J. Chemicals, Kisan Brothers Pvt. Ltd., New Cloth Market, Ahmedabad-380 002.

Cadila Pharmaceuticals Ltd., Ahmedabad.

Ganesh Biocontrol System, Ganesh Chamber, Gondala Road, Gondal 360311, Phone: 02825-221610, 228912, E-mail: ganeshbiocontrol@yahoo.com.

Khetiwadi Corner, Shiyabave, Vadodara-390 001.

Krishna Biotech Pvt. Ltd., 8-B, National Highway, Veravel-360 002, Rajkot District.

Oceon Agro India Ltd., Baroda.

Pest Control (India) Pvt. Ltd., 118-119, VIP View, VIP Road, Karelibaug, Vadodara-390 018, Phone: 0265-236 3546, 236 1027, Fax: 0265-236 1027, E-mail: vadodara_branch@pcil.co.in (Contact person: Mr.Y. Ramprasad, Mobile: 3132536).

21.5.5. HARYANA

State Government Organizations

State Biocontrol Laboratory, Sirsa, Phone: 01662-226572, Fax: 01662-225713.

Private Organizations

Jai Chemicals, 14/1, Mathura Road, Faridabad-121 003.

Ozone Biotech, Plot No. 6, Site No. 2, Rajdhani Land and Finance, 14/3 Mathura Road, Faridabad-121 001, Phone: 0129-5047602, Fax: 0129-5047604, E-mail: ozonebiotech@rediffmail.com.

Pest Control (India) Pvt. Ltd., SCO No. 9, Cabin No. 5 (Basement), Sector-26, Madhya Marg, Chandigarh-160 019, Phone: 0172-2790230, E-mail: psd_chandigarh@pcil.co.in (Contact person: Mr. Sunil Awasti, Mobile: 0141 312 2470).

21.5.6. KARNATAKA

ICAR Institutes

Agricultural Technology Information Centre (ATIC), Indian Institute of Horticultural Research (IIHR), Hessaraghatta Lake, Bangalore-560 089.

Project Directorate of Biological Control (PDBC), P.B.No. 2491, H.A. Farm Post, Bellary Road, Bangalore-560 024.

Central Government Organizations

Central Integrated Pest Management Centre, Whitefield, Bangalore.

Central Sericultural Research Training Institute (CSR&TI), Mysore 570 008.

Coffee Research Sub Station, Coffee Board, Chethalli-571 248, Kodagu District, Phone: 08276-266726, 266292, E-mail: ddcrss@sancharnet.in.

Central Coffee Research Institute (CCRI), Divn. of Plant Pathology, Coffee Research Station-577 117, Chikmagalur District, Phone: 08265-543008, Fax: 08265-543143, E-mail: crsento@yahoo.com.

State Agricultural Universities

Department of Agricultural Entomology, University of Agricultural Sciences (UAS), Krishinagar, Dharwad-580 005.

Institute of Pulses & Oilseeds Research, Aland Road, University of Agricultural Sciences, Gulbarga-581 101.

Agricultural Research Station (ARS), UAS, Bijapur.

Regional Research Station (RRS), UAS, Raichur.

Private Organizations

Agro Extracts Ltd., Plot No. 16, Phase II, Peenya Industrial Area, Bangalore-560 058.

Bio-Control Research Laboratories, Pest Control (India) Pvt. Ltd., No. 36/2, Sriramanahalli, Rajanakunte P.O., Post Box No. 6426, Yelahanka, Bangalore-560 064, Phone: 080-2246 8839/40/41/42, Fax: 080-2846 8838, E-mail: bcrl@pcil.co.in (Contact person: Mr. N. Venkatesh, Mobile: 3188 2116).

Bio-Multi Tech Pvt. Ltd., Ganganagar, Bangalore-560 030.

Kadur Agro & Video Vision Production Pvt. Ltd., R.V. Vidyaniketan, Mylasandra P.O., Bangalore-560 078.

Karnataka Agro Chemicals, 180, Ist Main Road, Mahalakshmi Layout, Bangalore-560 086, Phone: 080-2349 7464, 2349 4406, 2349 7360, Fax: 080-2349 0647, E-mail: multiplex@vsnl.com.

Margo Pvt. Ltd., 344/8, IVth Main, Sadashivnagar, Bangalore-560 080, Phone: 080-23613051/2/3, Fax: 080-23613055, E-mail: margo@vsnl.com, info@margobiocon.com, msm.christopher@margobiocon.com.

Monarch Bio-Fertilizers & Research Centre, 7/32, 47th Cross, Ist Main, 8th Block, Jayanagar, Bangalore-560 082.

Monofix Agro Products Ltd., VG Limbikai Building, Gokul Road, Hubli-580 030.

Murkumbi Bio Agro Pvt. Ltd., BC 105, Havelock Road, Camp, Belgaum-590 001.

Pest Control (India) Pvt. Ltd., B-385, Industrial Shed, KISSDC Ist Stage, Bangalore-560 058, Phone: 080-2372 4707, Fax: 080-2372 4708, E-mail: psd_bangalore@pcil.co.in (Contact Person: Mr. Hemant Kumar, Mobile: 080 3186 7949).

Rallis India Ltd., Agrochemical Station, Plot No. 21 & 22, Phase II, Peenya Industrial Area, P.B. No. 5813, Bangalore-560 058, E-mail: rallis.research@gems.vsnl.net.in.

Travancore Organic Fertilizers Co. Pvt. Ltd., Biotech Division, 1609, MVR Layout, Kalyan Nagar, Bangalore-560 043.

West Coast Herbochem Pvt. Ltd., 105/B, Industrial Suburb, IInd Stage, Rajajinagar, IIIrd Cross Road, Bangalore-560 022.

21.5.7. KERALA

Private Organizations

Plantrich Chemical & Fertilizers Ltd., Industrial Estate, Manarcad P.O., Kottayam-686 019, Phone: 91481-2371477/877, E-mail: plantrich@bioplantrich.com.

Travancore Organic Fertilizers, Kottayam-686 001.

VRM Biotech Products, Southern Fertilizers & Chemicals, Salim Complex, Changanacherry-686 101, Kottayam.

21.5.8. MADHYA PRADESH

Private Organizations

Indore Biotech Inputs & Research (Pvt.) Ltd., 6, Sikh Mohalla Main Road, Opposite Kothari Market, Indore-452 007.

Pest Control (India) Pvt. Ltd., 109, Malaviya Nagar, Bhopal-462 003, Phone: 0755-255 0090, 255 4767, Fax: 0755-255 4767, E-mail: salim.naushad@pcil.co.in (Contact person: Mr. S.M. Naushad, Mobile: 3139019).

21.5.9. MAHARASHTRA

Private Organizations

Ajay Biotech India Ltd., 100/3, Kalpana Apartments, Dr. Katkar Marg, Erandawane, Pune-411 040.

Ajay Biotech Laboratories (Pvt.) Ltd., A-12/5, Kubera Park, Kanchawa, Pune-411 040.

Bahar Agrochemicals, E/24, MIDC, Lote Parashuram Road, Ratnagiri-22.

BD Keathen & Co., Bandra (East), Mumbai-440 029.

Bharatiya Agro Industries Foundation, Pune.

Bio Era Technologies, 12, Sri Ram Apartments, Dindayal Nagar, Ring Road, Nagpur-412 022.

Cheminova Agro-Tech, Plot No. 2, Gat N 352, Shahada Dondaicha Road, Mohide Shivar, Nandubar -425 409, Shahada District.

Eco-Max Agrosystems Ltd., Good Value Marketing Company, Mumbai-20.

Gharda Chemicals Pvt. Ltd., 48, Hill Road, Mumbai-440 050.

Hoechst Schering Agr. Environment, Hoechst Centre, 54/A, Andheri-Kurla Road, Andheri West, Mumbai-400 093.

Khatau Agro Tech., Khatau House, Moghul Lane, Mahim, Mumbai-440 016.

KNS Biotech, Shivaneri Nagar, Rampur Road, Degloor, Nanded-431 717.

Kumar Krishimitra Bioproducts (I) Pvt. Ltd., 12, Ganeshwadi, Prin K.R. Kanitkar Path, Off F.C. Road, Pune-411 004.

Lupin Laboratories (I) Pvt. Ltd., 259, CST Road, Kalima, Santa Cruz (East), Mumbai-440 098.

Mc DA Agro Pvt. Ltd., Mumbai-440 001.

Nathkrupa Biocontrol Laboratories, 55, Sriram Nagar, Near Wireless Station, Pawan Bhumi Marg, Wardha Road, Nagpur-440 002.

National Organic Chemical Industries Ltd., 951/A, Appasaheb Marathi Marg, Prabhadevi, Mumbai-440 029.

Om Agro Organics, Samarthwadi, Yavatmal-445 002.

Pest Control (India) Pvt. Ltd., 2nd & 5th Floor, Jagdamba House, Next to Annapurna Theatre, P.B.No. 9060, Goregaon (East), Mumbai-400 063, Phone: 022-2686 5550, 5699 0252, Fax: 022-2686 5555, 5699 0256, E-mail: nikhil.chatterjee@pcil.co.in (Contact person: Mr. Nikhil Chatterjee, Mobile: 3898 1560).

Pest Control (India) Pvt. Ltd., 127, Keytuo Industrial Estate, 220, Kondivita Road (off Andheri-Kurla Road), Andheri (East), Mumbai-400 059, Phone: 022-2821 3546/7, 2821 7385, 2832 8481, Fax: 022-2822 1647, E-mail: nanda.khamkar@pcil.co.in (Contact person: Mr. N.B. Khamkar, Mobile: 022 3210 2063).

Pest Control (India) Pvt. Ltd., Krishna Niwas, Ist Floor, Kaduva Lane, Near Collectorate, Thane-400 601, Phone: 022-2534 1131, 2536 1506, Fax: 022-2538 5142, E-mail: siddharth.sarangpani@pcil.co.in (Contact person: Mr. S. Sarangpani, Mobile: 022-3210 1586).

Pest Control (India) Pvt. Ltd., Shop Nos. 11 & 12, Vindhya Commercial Complex, Sector 11, CBD Belapur, New Mumbai-400 614, Phone: 022-2757 2225, 2757 1840, 2757 1837, Fax: 022-2757 1837, E-mail: priyendra.khona@pcil.co.in (Contact person: Mr. Priyendra Khona, Mobile: 022 3210 1587).

Pest Control (India) Pvt. Ltd., 11 & 12, Wastu Park 'A', Ashwin Sector, New CIDCO, Nashik-422 009, Phone: 0253-2393887, 2390807, Fax: 0253-2390807, E-mail: suresh.bramhankar@pcil.co.in (Contact Person: Mr. S.A. Bramhankar, Mobile: 0253 3106357).

Pest Control (India) Pvt. Ltd., Flat No. 5, Sandesh Building, Opp. Gokhale Park, Chinchwad, Pune-411 033, Phone: 020-2745 2250, Fax: 020-2745 2250, E-mail: suneel.joshi@pcil.co.in (Contact Person: Mr.S.P. Joshi, Mobile: 020 3101 0545).

Pest Control (India) Pvt. Ltd., Manashanti Apartments, 504-B, Shaniwar Peth, Pune-411 030, Phone: 020-2448 2756, 2448 2914, Fax: 020-2745 2250, E-mail: pcilpune@vsnl.net (Contact Person: Mr.B.B. Haldar, Mobile: 0202449 4607).

Prakash Farm Chemicals, Gulmohar Cross, Road No. 9, Ville Parle West, Mumbai-440 049.

Sandoz (I) Ltd., Agro Division, Sandoz House, Dr. Annie Besant Road, Worli, Mumbai-400 018.

Sanvardhini Agro Pvt. Ltd., M-14, Addl. MIDC, Satara-415 004.

Sio Agro Research Laboratories, Dhiraj Apartments, Mulund (West), Mumbai-440 080.

Skylark Agro Chemicals, W-92, MIDC, Phase-II, Dombivali (East), Thane-421 204.

Soman Biofertilizers, 594, Sadashiv Peth, 101, Express Towers, Laxmi Road, Pune-411 030.

Sunida Exports Ltd., Bandra (East), Mumbai-440 029.

Sunline Agro Chemicals Pvt. Ltd., P.B. No. 73, Sakri Road, Dhulia.

Swathi Industries Ltd., Morray House, Bandra, Mumbai-440 050.

Vidyas Biotech Laboratories, 33, Savita Vihar, Somalwada, Wardha Road, Nagpur-440 015.

West Coast Herbochem Pvt. Ltd., 23A/101, Nalanda, C.S.T. Road, Kurla West, Mumbai-400 070.

Wockhardt Ltd., Ready Money Terrace, 2nd Floor, 167, Dr. Annie Besant Road, Worli, Mumbai-400 018.

Zandu Pharmaceutical Works Ltd., 70, Gokhale Road (South), Mumbai 25.

21.5.10. NEW DELHI

ICAR Institutes

Division of Agrochemicals, Indian Agricultural Research Institute (IARI), New Delhi-110 012.

Division of Entomology, Indian Agricultural Research Institute (IARI), New Delhi-110 012, E-mail: iari@dBt.ernet.in

Division of Plant Pathology, Indian Agricultural Research Institute (IARI), New Delhi-110 012.

Central Government Organizations

Defence Institute of Physiology & Applied Sciences, Delhi Cantonment, New Delhi-110 022.

Private Organizations

Biotech International Ltd., VIPPS Centre, 2 Local Shopping Centre, Block-EFGII, Masjid Moth, Greater Kailash-II, New Delhi-110048, Phone: 011-29220546/47/7359, Fax: 011-29229166/3083, e-mail: info@biotech-int.com

Hindustan Insecticides Ltd., Scope Complex, Core-6, II Floor, Lodi Road, New Delhi-110 003.

Nimba Foods & Agrochemicals Division, 14-A/10, Western Extension, Pusa Road, New Delhi-110 055.

Pest Control (India) Pvt. Ltd., 7, Jantar Mantar Road, Post Box No. 581, New Delhi-110 001, Phone: 011-2336 8772, Fax: 011-2336 8771, E-mail: pcilmkt@bolnet.in / psd_delhi@pcil.co.in (Contact person: Mr. Naresh Patti, Mobile: 011 3257 9897).

Pesto-Chem India Ltd., CM-3, Ansal Dilkush, Industrial Complex, GT Karnal Road, Azadpur, New Delhi-110 033.

21.5.11. ORISSA

Private Organizations

Pest Control (India) Pvt. Ltd., 440, Saheed Nagar, Bhubaneswar-751 007, Phone: 0674-, 2540235, 2544274, Fax: 0674-2544274, E-mail: anjan_samantray@yahoo.com (Contact Person: Mr. Anjan Samantray, Mobile: 98611 49280).

21.5.12. PUNJAB

Private Organizations

Pest Control (India) Pvt. Ltd., SCO No. 9, Cabin No. 5 (Basement), Sector-26, Madhya Marg, Chandigarh-160 019, Phone: 0172-2790230, Mobile: 0172 3105264, E-mail: psd_chandigarh@pcil.co.in.

21.5.13. RAJASTHAN

State Agricultural Universities

Biocontrol Laboratory, Agricultural Research Station (ARS), Rajasthan Agricultural University (RAU), Sri Ganganagar-335 001.

State Government Organizations

Deputy Director of Agriculture (Agronomy), ATC, Regional IPM Lab, Tabiji Farm, Ajmeer, Phone: 0145-2440652.

Deputy Director of Agriculture (Agronomy), ATC, Regional IPM Lab, Malikpur, Bharatpur, Phone: 05644-223337.

Deputy Director of Agriculture (Agronomy), ATC, Regional IPM Lab, Rampura, Jodhpur, Phone: 02962-222006.

Deputy Director of Agriculture (Agronomy), ATC, Regional IPM Lab, Cjjatrapura, Bundi, Phone: 0747-2442368.

Deputy Director of Agriculture (Agronomy), ATC, Regional IPM Lab, Farm Chittorgarh, Phone: 01472-241319.

Deputy Director of Agriculture (Agronomy), ATC, Regional IPM Lab, Krishi Farm, Hanumangarh Town, Phone: 01552-231291.

Deputy Director of Agriculture (Plant Pathology), IPM Laboratory, Dahod Road, Banaswara, Phone: 02962-243054.

Joint Director of Agriculture (Plant Pathology), State Biocontrol Lab, Durgapura, Jaipur, Phone: 0141-2550830.

Private Organizations

Pest Control (India) Pvt. Ltd., Shankar Sadan, Ist Floor, Adinath Marg, Opp. SMS Hospital, C-Scheme, Jaipur-302 004, Phone: 0141-236 8004, 237 8346, Fax: 0141-237 8346, E-mail: pcijpr@datainfosys.net (Contact person: Mr. Rajan Prakash, Mobile: 0141 312 2470).

Pesticides India Ltd., PO Box No. 20, Udaisagar Road, Udaipur-313 001.

21.5.14. TAMIL NADU

State Agricultural Universities

Centre for Plant Protection Studies (CPPS), Tamil Nadu Agricultural University (TNAU), Coimbatore- 3, Phone: 0422-431222, Fax: 0422-431672.

State Government Organizations

Bio-control Lab., State Seed Farm, Papparapatty Lab., Pennagaram Taluk, Dharmapuri District.

Bio-control Lab., JDA Office Complex, Tadagam Road, Coimbatore-641 013, Phone: 0422-2430766.

Bio-control Lab., ADA Office Complex, Seelanai-Chanpatty, Salem, Phone: 0427-2280214.

Bio-control Lab., Collectorate Office Complex, Villupuram-605 602, Phone: 04146-222291.

Bio-control Lab., ADA Office Complex, Puttur, Trichy-620 017, Phone: 0431-2793241.

Bio-control Lab., JDA Office Complex, Kancheepuram, At Panjupettai-651 501, Kancheepuram District, Phone: 04112-222977.

Bio-control Lab., State Seed Farm Complex, Kuruppanaichenpalyam Post, Bhavani, Erode.

Bio-control Lab., ADA Office Complex, Vinayagapuram, Therkkutheru Post, Melur Taluk, Madurai District.

Bio-control Lab., JDA Office, Tirunelveli Complex, Nirubar Colony, Tirunelveli-626 002.

Bio-control Lab., JDA Office, Thaniavur Complex, Kattuthottam Marianaman Koil Post, Tanjavur.

Bio-control Lab., Panchayat Union Complex, Namakkal-637 001.

Private Organizations

Agro Links, Trichy.

Agronule Industries, Trichy-632 217.

Agro Research Laboratories, 10/224, Avinashi Road, Coimbatore-641 018.

A.V. Thomas Pvt. Ltd., Anna Salai, Chennai-600 032.

Bacto Agro Culture Pvt. Ltd., Bharatha Nagar, Kalapally-643 253, The Nilgiris.

Basarass Biocontrol Research, 3/204, Main Road, Eraiyur Pennadanam RS & PO, SA, Vallavar-606 111, Cuddalore Dist., Phone: 4143-222403/037, Fax: 44-24867867, E-mail: basaras@vsnl.net.

Canster Chemicals Ltd., Adayar, Chennai-600 020.

Ecobiocides and Botanicals Pvt. Ltd., 2 East Market Street, Periyakulam Road, Theni-625 531, Phone: 4546-252020, Fax: 4546-251010, Mobile: 094433 42020, E-mail: ecobiocides@hotmail.com, Web site: www.azgro.com.

E.I.D. Parry (India) Ltd., Bio-Products Division, 234, NSC Bose Road, Chennai-600 001.

E.I.D. Parry (India) Ltd., Pesticides Division, "Dare House", P.B.No. 12, Chennai-600 001.

Green Tech Agro Products Pvt. Ltd., 47-D, Parsen Residency, Rajaji Road, Ramnagar, Coimbatore-641 009, Phone: 422-230107, Fax: 422-235237.

Jeypee Bio Techs, 25, Chinnaiah School Street, Virudhunagar-626 001.

Madurai Chemicals & Agro Industries Ltd., Rajaji Road, Madurai-625 104.

Micro Chemicals Ltd., Rajaji Road, Madurai-625 104.

Pest Control (India) Pvt. Ltd., 28, Errabalu Street, Chennai-600 001. Phone: 044-2522 6745/4266, Fax: 044-2522 2815, E-mail: psd_chennai@pcil.co.in (Contact person: Mr. V.M. Sagar, Mobile: 044 3106 1225).

Phyto Products Ltd., Thiruthuripoondi, S.A. District.

Prabhakar Oil Mills, Trichy-632 217.

Rajendra Foundation for Agricultural Research & Rural Development, Chittar-638 311,

Kesarimangalam Post, Bhavani Taluk, Erode Dist., Phone: 04256-39258, 39858.

Sakthi Biocontrol & Bio-fertilizer Centre, 9/11 Sathy Main Road, Kallipatti Gobi-638 505, Erode Dist., Phone: 04285-263464, 04554-231832.

Shri Durga Agro Services, 10/1, Kongu Nagar, Kalveerapalyam, Bharathiyar University Post, Coimbatore-641 046.

Shrishti Bio Products Pvt. Ltd., 331, Chinnasamy Naidu Road, New Siddapudur, Coimbatore-641 044.

Sinagro Industries India, 178, 9th Street, Mangalanagar, Porur, Chennai-600 116.

Southern Petrochemicals Industries Corporation Ltd. (SPIC), Bio-Tech Division, Agro Industrial Complex, 97, Mount Road, Chennai-600 032.

Sri Krishna Company, Race Course Road, Coimbatore-641 018.

Sun Agro Industries India, 178, 9th Cross Street, Mangala Nagar, Porur, Chennai-600 016.

Tamil Nadu Co-operative Sugar Federation Ltd., Main Biocontrol Research Laboratories, 2E/1, Rajeshwari Vedachalam Street, Chengalpattu-603 001.

TAP Industries, 7, Guna Complex, Tirumogeer Road, Y. Othathadai, Chennai-600 107.

T. Stanes & Co., 8/23-24, Race Course Road, Post Box No. 3709, Coimbatore-641 018.

Tuticorn Alkali Chemicals & Fertilizers Ltd., 553, Anna Salai, Teynampet, Chennai-600 018.

21.5.15. UTTARAKHAND

State Agricultural Universities

Department of Plant Pathology, GB Pant University of Agriculture & Technology (GBPUA&T), Pantnagar-263 145.

Private Organizations

Pest Control (India) Pvt. Ltd., 37/1, Ballapur Road, Near Hotel Surabhi, Dehradun-248 001, Phone: 0135-275 8568 (Contact person: Mr. Randeep Singh).

Sri Ram Solvent Extraction Plant (Neem), Jaspur.

21.5.16. UTTAR PRADESH

Private Organizations

Amitul Agro Chem Pvt. Ltd., New Snehapuri Colony P.O., Gorakhpur-273 001.

Bioved Research Society, 103/42, MLN Road, Allahabad-211 002.

Crop Health Bio-Product Research Centre, R-12/50, Rajanagar, Ghaziabad-201 003.

Pest Control (India) Pvt. Ltd., Ratanadeep Complex, Ist Floor, Chauki Chauraha, Post Box 89, Bareilly-243 001, Phone: 0581-247 5032 / 245 3059, E-mail: sandeepkumar@pcil.co.in (Contact person: Mr. Sandeep Kumar, Mobile: 0581 310 3763).

Pest Control (India) Pvt. Ltd., 306, Lekhraj Khazana, Faizabad Road, Indira Nagar, Lucknow-226 016, Phone: 0522-238 7360 / 234 2606, E-mail: girish.lohumi@pcil.co.in (Contact person: Mr. G.C. Lahumi, Mobile: 0522 324 4071).

Pest Control (India) Pvt. Ltd., Ramchandani Palace, IInd Floor, 183, The Mall, Post Box No. 271, Kanpur-208 001, E-mail: sanjay.mishra@pcil.co.in (Contact person: Mr. S.K. Mishra, Mobile: 0512 311 2751).

21.5.17. WEST BENGAL

Private Organizations

Pest Control (India) Pvt. Ltd., 23, Mirza Ghalib Street, Post Box No. 9001, Kolkata-700 016, Phone: 033-2252 1184/2824, Fax: 033-2252 3652, E-mail: kolkata_branch@pcil.co.in (Contact person: Mr. Sanjay Choubey, Mobile: 033 3107 5887).

Vinayak Fats & Proteins Pvt. Ltd., Agri. In-put Division, 206, Bentinck Chambers, 37A, Bentinck Street, Kolkata-700 069.

REFERENCES

Aalten, P.M., Vitour, D., Blanvillain, D., Gowen, S.R. and Sutra, L. 1998. Effect of rhizosphere fluorescent *Pseudomonas* strains on plant-parasitic nematodes *Radopholus similis* and *Meloidogyne* spp. *Letters in Applied Microbiology* **27**: 357-361.

Abid, M. and Maqbool, M.A. 1991. Effect of bare-root dip treatment in oil-cakes and neem leaf extract on the root-knot development and growth of tomato. *Pakistan Journal of Nematology* **9**: 13-16.

Abuzar, S. and Haseeb, A. 2006. Efficacy of carbofuran, bavistin, neem, *Trichoderma harzianum* and *Aspergillus nizer* against *Meloidogyne incognita* and *Fusarium oxysporum* f. sp. *vasinfectum* disease complex on okra. *Indian Journal of Nematology* **36**: 299-300.

Acharya, A., Dash, H.C. and Padhi, N.N. 1987. Pathogenic association of *Meloidogyne incognita* with *Sclerotium rolfsii* and *Xanthomonas bataticola* on betel vine. *Indian Journal of Nematology* **17**: 105-106.

Acharya, A. and Padhi, N.N. 1988. Effect of neem oilcake and saw dust against root-knot nematode, *Meloidogyne incognita* on betel vine (*Piper betle* L.). *Indian Journal of Nematology* **18**: 137-139.

Adams, S. 1997. Seeing red colored mulch starves nematodes. *Agriculture Research*, October, p. 18.

Ahmed, J.A. and Choudhury, B.N. 2004. Management of root-knot nematode, *Meloidogyne incognita* using organic amendments in French bean. *Indian Journal of Nematology* **24**: 137-139.

Ahmed, M.A. 1970. *Studies on the Root-Lesion Nematode*, Pratylenchus delattrei *Luc, 1958*. Ph.D. thesis, University of Madras, Madras.

Ahuja, S. 1983a. Studies on the variability in infection of root-knot nematode *Meloidogyne incognita* on potato. *Third Nematology Symposium,* Himachal Pradesh Agricultural University, Solan, p. 4.

Ahuja, S. 1983b. Comparative efficacy of some nematicides against root-knot nematode, *Meloidogyne incognita* in tomato, okra, brinjal and cauliflower. *Third Nematology Symposium,* Himachal Pradesh Agricultural University, Solan, p. 43.

Ahuja, S. and Mukhopadhyaya, M.C. 1984. Efficacy of some agricultural residues as compared to nematicides/insecticides in the control of root-knot nematode

(*Meloidogyne incognita*) infestation in okra (*Abelmoschus esculentus*). *Proceedings of the Symposium on Utilization of Agricultural Wastes*, pp. 131-135.

Ahuja, S. and Mukhopadhyaya, M.C. 1987. Split application of nematicides to control root-knot nematode (*Meloidogyne incognita*) and shoot and fruit borer (*Leucinodes orbonalis* Guen.) in eggplant (*Solanum melongena*) seed crop. *Capsicum Newsletter* No.6: 81-82.

Akhtar, M. 1993. Evaluation of nematicidal potential in some plants against plant parasitic nematodes on tomato. *Journal of Bangladesh Academy of Science* **17**: 125-127.

Akhtar, M. 1997. Biological control of plant parasitic nematodes by neem products in agricultural soils. *Applied Soil Ecology* **7**: 219-223.

Akhtar, M. and Alam, M.M. 1990. Evaluation of nematicidal potential in some plants against root-knot nematode on tomato and chilli. *International Nematology Network Newsletter* **7**(3): 10-12.

Akhtar, M. and Alam, M.M. 1991. Integrated control of plant parasitic nematodes on potato with organic amendments, nematicide and mixed cropping with mustard. *Nematologia Mediterranea* **19**: 169-171.

Akhtar, M. and Alam, M.M. 1993. Control of plant parasitic nematodes by 'Nimin'- an urea-coating agent and some plant oils. *Journal of Plant Disease and Protection* **100**: 337-342.

Akhtar, M. and Mahmood, M.M. 1993b. Control of plant parasitic nematodes with 'Nimin' and some plant oils by bare root-dip treatment. *Nematologia Mediterranea* **21**: 89-92.

Akhtar, M. and Mahmood, M.M. 1993a. Effect of *Mononchus aquaticus* and organic amendments on *Meloidogyne incognita* development on chilli. *Nematologia Mediterranea* **21**: 251-252.

Akhtar, M. and Mahmood, M.M. 1994a. Potentiality of phytochemicals in nematode control - A review. *Bioresource Technology* **48**: 189-201.

Akhtar, M. and Mahmood, M.M. 1994b. Control of root-knot nematode by bare root-dip in undecomposed and decomposed extracts of neem cake and leaf. *Nematologia Mediterranea* **22**: 55-57.

Akhtar, M. and Mahmood, M.M. 1994c. Prophylactic and therapeutic use of oilcakes and leaves of neem and castor extracts for the control of root-knot nematode on chilli. *Nematologia Mediterranea* **22**: 107-109.

Akhtar, M. and Mahmood, M.M. 1994d.. Nematode populations and short-term tomato growth in response to neem-based products and other soil amendments. *Nematropica* **22**: 169-173.

Akhtar, M. and Mahmood, M.M. 1995a. Control of root-knot nematode *Meloidogyne incognita* in tomato plants by seed coating with Achook and neem oil. *International Pest Control* **37**: 86-87.

Akhtar, M. and Mahmood, M.M. 1995b. Evaluation of neem-based product against root-knot nematode *Meloidogyne incognita*. *Annals of Applied Biology* (Suppl. 126) **16**: 6-7.

Akhtar, M. and Mahmood, M.M. 1996a. Integrated nematode control in potato, *Solanum tuberosum*. *International Pest Control* **38**: 62-64.

Akhtar, M. and Mahmood, M.M. 1996b. Effect of plant based product – Nimin and some plant oils on root-knot and ectoparasitic nematodes. *Nematologia Mediterranea* **24**: 3-5.

Akhtar, M. and Mahmood, M.M. 1996c. Evaluation of nematicidal properties of a biopesticide against root-knot nematode (*Meloidogyne incognita*). *Annals of Applied Biology* (Suppl. 127) **17**: 104-105.

Akhtar, M. and Mahmood, M.M. 1997a. Control of root-knot nematode *Meloidogyne incognita* in tomato plants by seed coating with Suneem and neem oil. *Journal of Pesticide Science* **22**(1): 37-38.

Akhtar, M. and Mahmood, M.M. 1997b. Impact of organic inorganic management and plant based products on plant parasitic and microbivorous nematode communities. *Nematologia Mediterranea* **25**: 21-23.

Akhtar, M., Wahi, A.H. and Alam, M.M. 1992. Control of root-knot nematode with bare-root dip in leaf extracts of Persian lilac and calotropis. *Current Nematology* **3**: 41-42.

Alam, M.M. 1987. Pollution free control of plant parasitic nematodes by soil amendment with plant wastes. *Biological Wastes* **22**: 75-79.

Alam, M.M. 1989. Control of root-knot and stunt nematodes with horn meal, bone meal and oilseed cakes. *Indian Journal of Nematology* **19**: 166-170.

Alam, M.M. 1990. Control of plant parasitic nematodes with organic amendments and nematicides in nurseries of annual plants. *Journal of Bangladesh Academy of Science* **14**: 107-113.

Alam, M.M. 1993. Bioactivity against nematodes. In *Neem Research and Development* (N.S. Randhawa and B.S. Parmar, eds.), pp.123-143. Society of Pesticide Science, India.

Alam, M.M., Ahmad, M. and Khan, A.M. 1980. Effect of organic amendments on the growth and chemical composition of tomato, eggplant and chilli and their susceptibility to attack by *Meloidogyne incognita*. *Plant and Soil* **57**: 231-236.

Alam, M.M., Khan, A.M. and Saxena, S.K. 1973a. Efficacy of 'Vydate' oxamyl for the control of root-knot nematode, *Meloidogyne incognita* on eggplant and okra. *Indian Journal of Nematology* **3**: 148-152.

Alam, M.M., Khan, A.M. and Saxena, S.K. 1973b. Some new host records of root-knot nematode, *Meloidogyne incognita* (Kofoid & White, 1919) Chitwood, 1949. *Indian Journal of Nematology* **3**: 159-160.

Alam, M.M., Khan, A.M. and Saxena, S.K. 1974. Reaction of some cultivated varieties of egg plant, pepper and okra to the root-knot nematode, *Meloidogyne incognita*. *Indian Journal of Nematology* **4**: 64-68.

Alam, M.M., Khan, A.M. and Saxena, S.K. 1977. Influence of different cropping sequences on soil population of plant parasitic nematodes. *Nematologia Mediterranea* **5**: 71-78.

Alam, M.M., Khan, A.M. and Saxena, S.K. 1982. Relative toxicity of decomposed and undecomposed oilcakes to plant-parasitic nematodes. *Acta Botonica Indica* **10**: 124-127.

Alam, M.M., Saxena, S.K. and Khan, A.M. 1973. Control of root-knot nematode, *Meloidogyne incognita* (Kofoid & White, 1919) Chitwood, 1949 on tomato and egg plant with V. C. -13 and Basamid liquid as bare-root dip. *Indian Journal of Nematology* **3**: 154-156.

Alam, M.M., Saxena, S.K. and Khan, A.M. 1977. Influence of interculture of marigold and margosa with some vegetable crops on plant growth and nematode population. *Acta Botanica Indica* **5**: 33-39.

Alam, M.M., Saxena, S.K. and Khan, A.M. 1979. Suitability of crops to the reniform, stunt and root-knot nematodes. In *Recent Researches in Plant Science* (S.S. Bir, ed.), pp. 630-634. Kalyani Publishers, New Delhi.

Alam, M.M., Siddiqi, Z.A., Khan, A.M. and Saxena, S.K. 1980. Effect of different cropping sequences on the population of plant parasitic nematodes. *Indian Journal of Nematology* **10**: 35-39.

Al-Hazmi, A.S. 1985. Interaction of *Meloidogyne incognita* and *Macrophomina phaseolina* in root disease of French bean. *Phytopathology Zeist* **113**: 311-316.

Ali, S.S. 1984. *Proc. of Sixth Symp. on Plantation Crops*, Kottayam, Kerala, India, pp. 215-223.

Ali, S.S. 1986. Proc. of Second Group Discussion on the Nematological Problems of Plantation Crops, Central Coffee Res. Stn., Balehonnur, pp. 10-12.

Ali, S.S. and Venugopal, M.N. 1992. *Nematologia Mediterranea* **20**: 65-66.

Ali, S.S. and Venugopal, M.N. 1993. *Current Nematology* **4**: 19-24.

Anandraj, M., Ramana, K.V. and Sarma, Y.R. 1991. In *Mycorrhizal Symbiosis and Plant Growth* (D.J. Bagyaraj and A. Manjunath, eds.), pp. 110-112. Univ. of Agri. Sci., Bangalore.

Anandraj, M., Venugopal, M.N., Veena, S.S., Kumar, A. and Sarma, Y.R. 2001. Ecofriendly management of diseases of spices. *Indian Spices* **38** (3): 28-31.

Anita, B. and Subramanian, S. 1998. Management of the reniform nematode *Rotylenchulus reniformis* in tomato. In *Nematology - Challenges & Opportunities in 21st Century* (Usha K. Mehta, ed.), pp. 249-250. Sugarcane Breeding Inst., Coimbatore.

Anita, B. and Vadivelu, S. 1997. Management of root-knot nematode, *Meloidogyne hapla* on scented geranium, *Pelorgonium graveolans*. *Indian Journal of Nematology* **27**: 123-125.

Anon. 1966. Annual Report of the Research Department for 1964-65, Indian Coffee Board, Bangalore, pp. 61-72.

Anon, 1985. Kufri Swarna, a High Yielding Late Blight and Cyst Nematode Resistant Potato Cultivar Suitable for Cultivation in Tamil Nadu Hills. No. 21(E), Central Potato Research Institute, Shimla, 4 pp.

Anon, 1989a. Biennial Report (1987-89) of AICRP on Plant Parasitic Nematodes with Integrated Approach for Their Control, Indian Agri. Res. Inst., New Delhi, 80pp.

Anon, 1989b. Consolidated Biennial Report (1987-89) of AICRP on Plant Parasitic Nematodes with Integrated Approach for Their Control, Dept. of Nematology, Haryana Agri. Univ., Hisar, 35 pp.

Anon, 1989c. QRT Report of All India Co-ordinated Research Project on Plant Parasitic Nematodes with Integrated Approach for their Control (1984-88). Indian Council of Agricultural Research, New Delhi, 103 pp.

Anon. 1990a. Final Report of Research Scheme on Nematode Problems in Some Selected Temperate Fruit Crops and Their Management. Y. S. Parmar University of Horticulture & Forestry, Solan, 54 pp.

Anon. 1990b. Annual Report of the Central Tuber Crops Research Institute, Trivandrum.

Anon. 1993a. *QRT Report of AICRP on Nematodes (1989-93)*, Dept. of Nematol., Haryana Agri. Univ., Hisar, India, 149 pp.

Anon. 1993b. Biennial Report (1991-93) of AICRP on Plant Parasitic Nematodes with Integrated Approach for Their Control, Dept. of Nematology, Haryana Agri. Univ., Hisar, 80 pp.

Anusuya, K. and Vadivelu, S. 2002. Mutualistic symbiosis of VAM fungi on plant growth promoting rhizobacteria (PGPR) on micropropagated carnation. *J. Microbiology World* 4: 105-107.

Atilano, R.A., Menge, J.A. and van Gundy, S.D. 1982. Interaction between *Meloidogyne arenaria* and *Glomus fasciculatum* in grapes. *Journal of Nematology* **13**: 52-57.

Atwal, A.S. and Mangar, A. 1969. Repellent action of root exudates of *Sesamum orientale* against the root-knot nematode, *Meloidogyne incognita* (Heteroderidae : Nematoda). *Indian Journal of Entomology* **31**: 286.

Ayala, A., Roman, J. and Gonzalez, J. 1967. Pangola grass as a rotation crop for pineapple nematode control. *Journal of Agriculture Science, University of Puerto Rico* **51**: 94-96.

Ayyar, P.N.K. 1926. A preliminary note on the root nematode, *Heterodera radicicola* Mitter and its economic importance in South India. *Science and Culture* **22**: 391-393.

Ayyar, P.N.K. 1933. Some experiments on the control of root-gall nematode (*Heterodera radicicola* Greeff) in South India. *Madras Agricultural Journal* **21**: 97-107.

Ayyar, P.N.K. 1934. Further experiments on the root gall nematode, *Heterodera marioni*. *Indian Journal of Agricultural Science* **3**: 1064-1071.

Baghel, P.P.S. 1995. nematode problems in citrus. In *Nematode Pest Management-An Appraisal of Eco-friendly Approach* (G. Swarup, D.R. Dasgupta and J.S. Gill, eds.), pp. 203-210.

Baghel, P.P.S. and Bhatti, D.S. 1983a. Evaluation of pesticides for the control of phytonematodes on citrus. *Third Nematology Symposium*, Himachal Pradesh Agricultural University, Solan, pp. 38-39.

Baghel, P.P.S. and Bhatti, D.S. 1983b. Relative efficacy of nematicides for control of phytonematodes on grapevine varieties. *Third Nematology Symposium*, Himachal Pradesh Agricultural University, Solan, p. 39.

Baghel, P.P.S., Bhatti, D.S. and Jalali, B.L. 1990. *Proceedings of the National Conference on Mycorrhiza*, Haryana Agri. Univ., Hisar, p. 118.

Baghel, P.P.S. and Gupta, D.C. 1986. Effect of intercropping on root-knot nematode (*Meloidogyne javanica*) infecting grapevine (var. Parlette). *Indian Journal of Nematology* **16**: 283-285.

Bagyaraj, D.J., Manjunath, A. and Reddy, D.D.R. 1979. Interaction of vesicular arbuscular mycorrhiza with root-knot nematodes in tomato. *Plant and Soil* **51**: 397-403.

Bajaj, K.L. and Mahajan, R. 1977. Phenolic constituents of tomato resistant and susceptible to *Meloidogyne incognita* (Kofoid & White) Chitwood. *Nematologia Mediterranea* **5**: 329-333.

Bala, S.K. and Sukul, N.C. 1987. Systemic nematicidal effect of eugenol. *Nematropica* **17**: 219-222.

Balasubramanian, M. and Rangaswami, G. 1964. Studies on host range and histopathology of root-knot infections caused by *Meloidogyne javanica*. *Indian Phytopathology* **17**: 126-132.

Balasubramanian, P. and Ramakrishnan, C. 1983. *Nematologia Mediterranea* **11**: 203-204.

Bansal, R.K. and Verma, V.K. 2002. Antagonistic efficacy of *Azatobacter chroococcum* against *Meloidogyne javanica* infecting brinjal. *Indian Journal of Nematology* **32:** 132-134.

Barber, C.A. 1901. *Department Land Records and Agriculture*. Madras Agri. Branch 2, Bull. No. 45, pp.227-234.

Barbercheck, M.E. and Broembsen, S.L. von. 1986. *Plant Disease* **70**: 945-950.

Barone, A. Ritter, E., Schanchtschabel, U., Debner, T., Salamini, I. and Gebhardt, C. 1990. Localization by restriction fragment length polymorphism mapping of a major dominant gene conferring resistance to the potato cyst nematode, *Globodera rostochiensis*. *Molecular Genetics* **224**: 177-182.

Bhatnagar, A. and Goswami, B.K. 1987. Comparative efficacy of neem and groundnut oilcakes with aldicarb against *Meloidogyne incognita* in tomato. *Review de Nematologia* **10**: 467-470.

Bhattacharya, D. and Goswami, B.K. 1988. Effect of oil-cake used alone and in combination with aldicarb on *Meloidogyne incognita* infecting tomato. *Nematologia Mediterranea* **16**: 139-142.

Bergeson, G.B. 1963. Influence of *Pratylenchus penetrans* alone and in combination with *Verticillium albo-utrum* on growth of peppermint. *Phytopathology* **53**: 1164-1166.

Bergeson, G.B. 1975. The effect of *Meloidogyne incognita* on the resistance of four musk melon varieties to *Fusarium* wilt. *Plant Disease Reporter* **59**: 410-413.

Berkeley, M.J. 1855. Vibrio forming excrescences on the roots of cucumber plants. *Garden Chronicle* **14**: 220.

Bhagawati, B and Goswami, B. K. 2000. Interaction of *M. incognita* and *Fusarium oxysporum* f. sp. *lycopersici* on tomato. *Indian Journal of Nematology* **30**: 93-94.

Bhagawati, B., Goswami, B. K. and Singh, C.S. 2000. Management of disease complex of tomato caused by *M. incognita* and *Fusarium oxysporum* f. sp. *lycopersici* through bioagents. *Indian Journal of Nematology* **30**: 16-22.

Bhattacharya, D. and Goswami, B.K. 1987. A study on the comparative efficacy of neem and groundnut oilcakes against root-knot nematode, *Meloidogyne incognita* as influenced by microorganisms on sterilized and unsterilized soil. *Indian Journal of Nematology* **17**: 81-83.

Bhattacharya, R.K. and Rao, V.N.M. 1984. Effect of soil covers and soil moisture regimes on nematode population in soil and in root of banana. *Journal of Research Assam Agricultural University* **5**:206-209.

Bhatti, D.S. and Jain, R.K. 1977. Estimation of loss in okra, tomato and brinjal yield due to *Meloidogyne incognita. Indian Journal of Nematology* **7**: 37-41.

Bhosle, B.B., Sehgal, M., Puri, S.N., Sardana, H.R. and Singh, D.K. 2006. Efficacy of organic amendment in management of root-knot nematode, *Meloidogyne incognita* on okra. *Indian Journal of Nematology* **36**: 37-40.

Bindra, O.S., Chhabra, H.K., Chadha, K.L. and Mehrotra, N.K. 1967. A study on the correlation of citrus nematode population with decline of citrus. *Journal of Research, Punjab Agricultural University* **4**: 543-546.

Bird, G.W. 1981. Integrated nematode management for plant protection. In *Plant Parasitic Nematodes, Vol. 3* (B.M. Zuckerman & R.A. Rohde, eds.), pp. 355-375. Academic Press, New York.

Borah, A. and Phukan, P.N. 2000. Effect of VAM fungus, Glomus fasciculatum and root-knot nematode, Meloidogyne incognita on brinjal. Journal of the Agricultural Sciences Society of North-Eastern India **13**: 212-214.

Borah, A. and Phukan, P.N. 2004. Comparative efficacy of *Glomus fasciculatum* with neem cake and carbofuran for the management of *Meloidogyne incognita* on brinjal. *Indian Journal of Nematology* **24**: 129-132.

Borkakaty, D. 1993. Management of Root-knot Nematode, Meloidogyne incognita (Kofoid and White, 1919) Chitwood, 1949 on Brinjal by Integrated Approach. M. Sc. (Agri.) thesis, Assam Agri. Univ., Jorhat.

Boruah, H.P.D. and Kumar, B.S.D. 2002. Biological activity of secondary metabolites produced by a strain of *Pseudomonas fluorescens. Folia Microbiologica* **47**: 359-363.

Bourne, J. M., Kerry, B.R. and de Leij, F.A.A.M. 1994. Methods for the study of *Verticillium chlamydosporium* in the rhizosphere. Workshop:Organisms and Methods for Biological control of Nematodes. *Journal of Nematology* **26** (Suppl.): 587-591.

Bowman, P. and Bloom, J.R. 1966. Breaking the resistance of tomato varieties to *Fusarium* wilt by *Meloidogyne incognita. Phytopathology* **56**: 871.

Butler, E.J. 1906. Report of root-knot nematodes on black pepper in Kerala. *Agricultural Journal of India* **1**: 25-36.

Cannayane, I. and Sivakumar, C.V. 2001. Bare root dip treatment with *Paecilomyces lilacinus* against *Meloidogyne incognita* in tomato. *Annals of Plant Protection Sciences* **9**: 351-354.

Caperton, C.M., Martyn, R.D. and Starr, J.L. 1986. Effects of *Fusarium* inoculum density and root-knot nematodes on wilt resistance on summer squash. *Plant Disease Reporter* **70**: 207-209.

Castillo, M.B. 1985. Some studies on the use OF organic soil amendments for nematode control. *Philippine Agriculturist* **68**: 76-93.

Caubel, G. and Samson, R. 1984. Effect of the stem nematode, *Ditylenchus dipsaci* on the development of 'Café au lait' bacteriosis in garlic (*Allium sativum*) caused by *Pseudomonas fluorescens. Agronomie* **4**: 311-313.

Cayrol, J.C. 1974. *Proc. Symp. XII Intnl. De Nematologia*, Society of European Nematologists, Grenada, Spain, pp 21-22.

Chahal, P.P.K. and Chahal, V.P.S. 1988. Biological control of root-knot nematode with *Azatobacter chroococcum* on brinjal. In *Advances in Plant Nematology*, (M.A. Maqbool, A. Morgan Golden, A. Gaffar and L.R. Krusberg, eds.), pp. 257-263. Natl. Nematol. Centre, Univ. of Karachi, Karachi, Pakistan.

Chahal, P.P.K. and Chahal, V.P.S. 2003. *Azotobacter* for the control of plant parasitic nematodes and for other benefits. In *Nematode Management in Plants* (P.C. Trivedi, ed.), pp. 241-250. Scientific Publishers (India), Jodhpur.

Chahal, P.P.K. and Chhabra, H.K. 1984. Interaction of *Meloidogyne incognita* with *Rhizoctonia solani* on tomato. *Indian Journal of Nematology* **14**: 56-57.

Chahal, V.P.S. and Chahal, P.P.K. 1991. Control of *Meloidogyne incognita* with *Bacillus thuringiensis*. In *Plant Soil Interaction* (R.J. Wright *et al.*, eds.), pp. 677-680.

Chahal, V.P.S. and Chahal, P.P.K. 2003. *Bacillus thuringiensis* for the control of *Meloidogyne incognita*. In *Nematode Management in Plants* (P.C. Trivedi, ed.), pp. 251-257. Scientific Publishers (India), Jodhpur.

Chandawani, G.H. and Reddy, T.S.N. 1967. The host range of root-knot nematode, *Meloidogyne javanica* in tobacco nurseries at Rajahmundry, Andhra Pradesh. *Indian Phytopathology* **20**: 383-384.

Channabasappa, B.S., Krishnappa, K. and Reddy, B.M.R. 1995. Utilization of ecofriendly biological agents and biocomponents in the integrated management of *Radopholus similis* on banana. *Natl. Symp. on Nematode problems of India – An Appraisal of the Nematode Management with Ecofriendly Approaches and Biocomponents,* Indian Agri. Res. Inst., New Delhi.

Charles, J.S.K. and Venkitesan, T.S. 1984. New hosts of *Heterodera oryzicola* Rao & Jayaprakash, 1978 in Kerala, India. *Indian Journal of Nematology* **14**: 181-182.

Charles, J.S.K. and Venkitesan, T.S. 1993. *Status Report on the Nematological Investigations on Banana in Kerala Agicultural University*. Kerala Agicultural University, Vellanikkara, 43 pp.

Charles, J.S.K., Venkitesan, T.S. and Thomas, Y. 1983. Field screening of banana against the burrowing nematode, *Radopholus similis* (Cobb) Thorne. *Third Nematology Symposium*, Himachal Pradesh Agricultural University, Solan, p. 28.

Charles, J.S.K., Venkitesan, T.S., Thomas, Y. and Varkey, P.A. 1985. Correlation of plant growth components to bunch weight in banana infested with burrowing nematode, *Radopholus similis* (Cobb). *Indian Journal of Nematology* **15** : 186-190.

Chhabra, H.K. and Bindra, O.S. 1974. Screening of citrus rootstocks against the citrus nematode, *Tylenchulus semipenetrans* Cobb, 1913 in the Punjab. *Indian Journal of Horticulture* **31**: 194-195.

Chhabra, H.K. and Sharma, J.K. 1981. Combined effect of *Meloidogyne incognita* and *Rhizoctonia bataticola* on pre-emergence damping-off of okra and brinjal. *Science and Culture* **47**: 256-257.

Chhabra, H.K., Sidhu, A.S. and Singh, I. 1977. *Meloidogyne incognita* and *Rhizoctonia solani* interaction on okra. *Indian Journal of Nematology* **7**: 54-57.

Choudhury, B.C. 1981. Root-knot nematode problem in various crop plants in Bangladesh. In *Proc. of the IIIrd Intn.* Meloidogyne *Project Res. and Planning Conf. on Root-knot Nematodes,* Region VI, Jakarta, Indonesia, pp. 142-145.

Cobb, N.A. 1914. *Nematodes and Their Relationships.* USDA Year Book, pp. 457-490.

Connick, W.J. Jr. 1988. Formulation of living biological control agents with alginate. *ACS Symposium Series* No. **371**: 241-250.

Conroy, J.J. and Green, R.J., Jr. 1974. Interactions of the root-knot nematode *Meloidogyne incognita* and the stubby root nematode *Trichodorus christiei* with *Verticillium albo-utrum* on tomato at controlled inoculum densities. *Phytopathology* **64**: 1118-1121.

Conroy, J.J., Green, R.J., Jr. and Ferris, J.M. 1972. Interaction of *Verticillium albo-utrum* and the root lesion nematode, *Pratylenchus penetrans*, in tomato roots at controlled inoculum densities. *Phytopathology* **62**: 362-366.

Cook, R.J. and Baker, K.F. 1983. *The Nature and Practice of Biological Control of Plant Pathogens.* American Phytopathological Society, St. Paul, Minnesota.

Corbett, D.C.M. and Hide, G.A. 1971. Interactions between *Heterodera rostochiensis* Woll. and *Verticillium dahliae* Kleb. on potatoes and the effect of CCC on both. *Annals of Applied Biology* **68**: 71-80.

Cronin, D., Moenne-Loccoz, Y., Fenton, A., Dunne, C., Dowling, D.N. and O'-Gara, F. 1997. Role of 2,4-diacetylphloroglucinol in the interactions of the bio-control pseudomonad strain F113 with the potato cyst nematode *Globodera rostochiensis. Applied and Environmental Microbiology* **63**: 1357-1361.

Dahiya, R.S., Bansal, R.K. and Paruthi, I.J. 1998. Efficacy of mustard cake and poultry manure for the management of *Meloidogyne javanica* in bottle gourd and okra. *Proc. of Natl. Symp. on Rational Approaches in Nematode Management for Sustainable Agriculture,* Nematol. Soc. of India, New Delhi, pp. 100-102.

Dalal, M.R. and Vats, R. 1998. Estimation of loss in pea (*Pisum sativum* L.) due to *Rotylenchulus reniformis.* In *Nematology – Challenges and Opportunities in 21st Century,* (Usha K. Mehta, ed.), pp. 8-9. Sugarcane Breeding Institute, Coimbatore.

Darekar, K.S. and Mahse, N.L. 1988. Assessment of yield losses due to root-knot nematode, *Meloidogyne incognita* Race 3 in tomato, brinjal and bitter gourd. *Internatioanl Nematology Network Newsletter* **5**(4): 7-9.

Darekar, K.S. and Patil, B.D. 1981. Reaction of some cowpea varieties to root-knot nematode. *Indian Journal of Nematology* **11**: 83-84.

Darker, K.S. and Patil, B.D. 1982. Reaction of some grapevine varieties to root-knot and citrus nematode. *Indian Journal of Nematology* **12**: 390-392.

Das, A.K. 1994. *Assessment of Yield Loss Due to* Meloidogyne incognita *on French Bean and its Management.* M. Sc. (Agri.) thesis, Assam Agri. Univ., Jorhat.

Das, B.K. and Padhi, N.N. 1990. Response of root-knot nematode infecting tomato to seed treatment by some neem products. *Orissa Journal of Agricultural Research* **2**: 148-150.

Das, N. and Sinha, A.K. 2005. Integrated management of root-knot nematode (*Meloidogyne incognita*) on okra (*Abelmoschus esculentus* (L.) Moench). *Indian Journal of Nematology* **35**: 175-182.

Dasgupta, D.R. and Ganguly, A.K. 1986. Plant resistance to nematodes. In *Plant Parasitic Nematodes of India, Problems and Progress* (G. Swarup & D.R. Dasgupta, eds.), pp. 458-475. Indian Agri. Res. Inst., New Delhi, India.

Dasgupta, D.R. and Seshadri, A.R. 1971. Races of the reniform nematode, *Rotylenchulus reniformis* Linford and Oliverra, 1940. *Indian Journal of Nematology* **1**: 21-24.

Davide, R.G. and Marasigan, L.Q. 1985. Yield loss assessment and evaluation of resistance of banana cultivars to the nematodes *Radopholus similis* Thorne and *Meloidogyne incognita* Chitwood. *Philippine Agriculturist* **68**: 333-349.

Davide, R.G. and Zorilla, R.A. 1986. Evaluation of a fungus, *Paecilomyces lilacinus* for the biological control of root-knot nematodes *Meloidogyne incognita* on okra. *International Nematology Network News Letter* **3**(3): 32-33.

Davide, R.G. and Zorilla, R.A. 1995. *Biocontrol* **1**: 63-75.

Davis, R.A. and Jenkins, W.R. 1963. Effects of *Meloidogyne* spp. and *Tylenchorhynchus claytoni* on pea wilt incited by *Fusarium oxysporum* f. sp. *pisi* race 1. *Phytopathology* **53**: 745.

De Leij, F.A.A.M., Kerry B.R. and Dennehy, J.A. 1993. *Verticillium chlamydosporium* as a biological control agent for *Meloidogyne incognita* and *M. hapla* in pot and micro-plot test. *Nematologica* **39**:115-126.

De Moura, R.M., Echandi, E. and Powell, N.T. 1975. Interaction of *Corynebacterium michiganense* and *Meloidogyne incognita* on tomato. *Phytopathology* **65**: 1332-1335.

Den-Belder, E., Fitters, P.F.L. and Waalnijk, C. 1994. PN : EP 623672; 09.11.94, Res. Inst. Plant Prot.

D'Errico, F.P. and Maio, S.K. 1980. Effect of some materials on tomatoes in field preliminary experiment. *Nematologia Mediterranea* **8**: 107-110.

Deshmukh, M.R., Karmakar, S.P. and Patil, S.G. 2004. Screening of grape varieties for resistance to root-knot nematode (*Meloidogyne incognita*). *Indian Journal of Nematology* **24**: 224-226.

Devarajan, K. and Rajendran, G. 2001. Effect of fungus *Paecilomyces lilacinus* (Thom.) Samson on the burrowing nematode *Radopholus similis* (Cobb) Thorne in banana. *Pest Management in Horticultural Ecosystems* **7**: 171-173.

Devarajan, K. and Rajendran, G. 2002. Effect of fungal egg parasite, *Paecilomyces lilacinus* (Thom.) Samson on *Meloidogyne incognita* in banana. *Indian Journal of Nematology* **32**: 98-101.

Devi, Geetanjali. 1993. Pathogenicity, Crop Loss Assessment and Management of Meloidogyne incognita (Kofoid and White, 1919) Chitwood, 1949 on Carrot (Dacus carota L.). M. Sc. (Agri.) thesis, Assam Agri. Univ., Jorhat.

Dhawan, S.C. and Sethi, C.L. 1977. Inter-relationship between root-knot nematode, *Meloidogyne incognita* and little leaf of brinjal. *Indian Phytopathology* **30**: 55-63.

Doraiswamy, K., Sundaram and Manian, K. 1979. Root-knot nematode control in scented geranium (*Pelargonium graveolens* L.). *Madras Agricultural Journal* **66**: 205-206.

DuCharme, E.P. and Suit, R.F. 1956. Immunity of the lychee from the burrowing nematode. *Proceedings of the Florida State Horticultural Society* **68**: 270-272.

Dwivedi, K., Upadhyay, K.D. and Ahmed, F. 2001. Integrated management of root-knot nematode infecting tomato (*Lycopersicon esculentum* Mill.). *Natl. Cong. on Centenary of Nematol. in India – Appraisal & Future Plans*, Indian Agri. Res. Inst., New Delhi, p.166.

Eapen, S.J. 1995. *Investigations on Plant Parasitic Nematodes Associated with Cardamom*. Final Report of the Project Nema I (813), Indian Inst. of Spices Research, Calicut, 39 pp.

Eapen, S.J., Ramana, K.V. and Sarma, Y.R. 1997. Evaluation of *Pseudomonas fluorescens* isolates for control of *Meloidogyne incognita* in black pepper (*Piper nigrum* L.). In *Biotechnology of Spices, Medicinal & Aromatic Plants* (S. Edison, K.V. Ramana, B. Sasikumar, K.N. Babu and S.J. Eapen, eds.), pp. 129-133. Indian Inst. of Spices Res., Calicut

Eapen, S.J. and Venugopal, M.N. 1995. Field evaluation of *Paecilomyces lilacinus* and *Trichoderma* spp. in cardamom nurseries for the control of root-knot nematodes and rhizome rot disease. *National Symp. on Nematode Problems of India – An Appraisal of the Nematode Mangmt. with Eco-friendly Approaches and Biocomponents*. Indian Agri. Res. Inst., New Delhi.

Ekanayake, H.M.R.K. and Jayasundara, N.J. 1994. Effect of *Paecilomyces lilacinus* and *Beauvaria bassiana* in controlling *Meloidogyne incognita* on tomato in Sri Lanka. *Nematologia Mediterranea* **22**: 87-88.

El-Goorni, M.A., Abo-Eldahar, M.E. and MeNair, F.F. 1974. Interaction between root-knot and *Pseudomonas marginata* on gladiolus corms. *Phytopathology* **64**: 271-272.

Ellenby, C. 1945. Control of potato root eelworm, *Heterodera rostochiensis* by allyl isothiocynate, the mustard oil of *Brassica nigra*. *Annals of Applied Biology* **32**: 237-239.

Ellenby, C. 1951. Mustard oils and control of the potato root eelworm, *Heterodera rostochiensis* Wollenweber: Further field and laboratory experiments. *Annals of Applied Biology* **38**: 859-875.

Fassuliotis, G. and Bhatt, D.P. 1982. Potential of tissue culture for breeding root-knot nematode resistance into vegetables. *Journal of Nematology* **14**: 10-14.

Fassuliotis, G. and Rau, G.J. 1969. The relationship of *Meloidogyne incognita acrita* to the incidence of cabbage yellows. *Journal of Nematology* **1**: 219-222.

Fassuliotis, G. and Skucus, G.P. 1969. The effect of pyrralizidine alkaloid ester and plants containing pyrrolizidine on *Meloidogyne incognita acrita*. *Journal of Nematology* **1**: 287-288.

Fassuliotis, G. and Sparrow, A.H. 1955. Preliminary report of x-ray studies on the golden nematode. *Plant Disease Reporter* **39**: 572.

Faulkner, L.R. and Skotland, C.B. 1965. Interaction of *Verticillium dahliae* and *Pratylenchus miniyus* in *Verticillium* wilt of peppermint. *Phytopathology* **55**: 583-586.

Ferraz, S. and Lear, B. 1976. Interaction of four plant parasitic nematodes and *Fusarium oxysporum* f. sp. *dianthi* on carnation. *Experientiae* **22**: 272-277.

Ford, H.W. 1967. Burrowing nematode resistant rootstocks as biological barriers in citrus groves. *Proceedings of the Florida State Horticultural Society* **80**: 56-59.

Ford, H.W. and Feder, W.A. 1969. *Proceedings of the First International Citrus Symposium*, Riverside, California, USA **2**: 941-948.

Fritzsche, R., Pelcz, J., Ottel, G. and Thiele, S. 1983. Interaction between Meloidogyne incognita and Fusarium oxysporum f. sp. cucumerinum in greenhouse cucumbers. Tagungsbericht Akademie der Landwirschaftwissenschaften der Deutschen Demokratischen Republik **261**: 685-690.

Ganguly, S. and Dasgupta, D.R. 1981. Protein pattern in resistant and susceptible tomato varieties inoculated with the root-knot nematode, *Meloidogyne incognita. Indian Journal of Nematology* **11**: 180-188.

Garabedian, S. and Van Gundy, S.D. 1983. Use of avermectins for the control of *Meloidogyne incognita* on tomatoes. *Journal of Nematology* **15**: 503-510.

Gautam, A., Siddiqui, Z.A. and Mahmood, I. 1995. Integrated management of *Meloidogyne incognita* on tomato. *Nematologia Mediterranea* **23**: 245-247.

Geoghegan, I., Robertson, W., Birch, N. and Gatehouse, A.M.R. 1995. PN : WO 9526634 : 12.10.95, Axis- Genet, Cambridge, UK.

George, J., Saji, K., Sasikumar, K.V. and Krishnamoorthy, B. 2001. Spice varieties of IISR. *Indian Spices* **38** (3): 13-16.

Giamalava, M.J., Martin, W.J. and Hernandez, T.P. 1960. *Phytopathology* **50**: 575.

Giannakou, I.O., Pembroke, B., Gowen, S.R. and Doloumpala, S. 1999. Effects of fluctuating temperatures and different host plants on development of *Pasteuria penetrans* in *Meloidogyne javanica. Journal of Nematology* **31**: 313-318.

Gilbert, J.C. and Mc Guire, D.C. 1956. Inheritance of resistance to severe root-knot from *Meloidogyne incognita* in commercial type tomatoes. *Proceedings of Annals of Horticultural Science* **68**: 437-442.

Gill, J.S. 1981. Fighting the foliar nematode of chrysanthemum. *Indian Horticulture* **25**: 21-22.

Gill, J.S. and Sharma, N.K. 1976. Additional hosts of the foliar nematode, *Aphelenchoides ritzemabosi* from India. *Indian Journal of Nematology* **6**: 169-171.

Gill, J.S. and Walia, R.K. 1980. Control of chrysanthemum foliar nematode, *Aphelenchoides ritzemabosi* on *Zinnia elegans. Indian Journal of Nematology* **10**: 86-88.

Gintis, B.O., Morgan-Jones, G. and Rodriguez-Kabana, R. 1983. Fungi associated with several developmental stages of *Heterodera glycines* from an Albama soybean field soil. *Nematologica* **13**: 181-200.

Gnanapragasam, N.C. 1991. Influence of soil amendments in reducing pathogenicity to tea by the root lesion nematode, *Pratylenchus loosi. Proc. Intn. Symp. Tea Sci.*, Shizuoka, Japan, pp. 684-688.

Gnanapragasam, N.C., Dharmasena, W.A.M. and Liyanage, D.D. 1989. Development of age resistance in young tea to different species of *Meloidogyne*. *International Nematology Network Newsletter* **6**: 4.

Goeldi, E.A. 1887. Relatorio sobre a molestia do cafeeiro na provincial do Rio de Janeiro. *Apparently an advanced separate of Arch. Mus. Nac. Rio de Janeiro* **8**: 7-121, 1892.

Golden, J.K. and Van Gundy, S.D. 1975. A disease complex of okra and tomato involving the nematode, *Meloidogyne incognita* and the soil inhabiting fungus, *Rhizoctonia solani*. *Phytopathology* **65**: 265-273.

Good, J.M. 1964. Effect of soil application and sealing methods on the efficacy of row application of several soil nematicides for controlling root-knot nematodes, weeds and *Fusarium* wilt. *Plant Disease Reporter* **48**: 199-203.

Gopinath, K.V., Nanje Gowda, D.N. and Nagesh, M. 2002. Management of root-knot nematode *Meloidogyne incognita* on tomato using bioagent *Verticillium chlamydosp-orium*, neem cake, marigold and carbofuran. *Indian Journal of Nematology* **32**: 179-181.

Goswami, B.K. and Chenulu, V.V. 1974. Interaction of root-knot nematode, *Meloidogyne incognita* and tobacco mosaic virus in tomato. *Indian Journal of Nematology* **4**: 69-80.

Goswami, B.K., Haque, M.M., Mishra, S.D. and Mojumder, V. 1988. Comparative efficacy of Hind-O-Meal, Karanj, castor oilseed cakes and carbofuran against root-knot nematode in cowpea. *Annals of Agricultural Research* **10**: 316-318.

Goswami, B.K. and Singh, H. 2001. Biomanagement of disease complex caused by *Meloidogyne incognita* and *Fusarium oxysporum* on tomato. *Natl. Cong. on Centenary of Nematol. in India – Appraisal & Future Plans*, Indian Agri. Res. Inst., New Delhi, p.169.

Goswami, B.K. and Vijayalakshmi, K. 1987. Effects of period of decomposition of oilseed cakes in soil on *Meloidogyne incognita* juveniles. *Indian Journal of Nematology* **17**: 84-86.

Goswami, B.K. and Swarup, G. 1971. Effect of oilcake amended soil on the growth of tomato and root-knot nematode population. *Indian Phytopathology* **24**: 291-294.

Govindaiah, Dandin, S.B., Philip, T. and Datia, R.K. 1990. Effect of marigold (*Tagetes patula*) intercropping against *Meloidogyne incognita* infecting mulberry. *Indian Journal of Nematology* **20**: 96-99.

Gowda, D.N. 1972. Studies on comparative efficacy of oilcakes on control of root-knot nematode, *Meloidogyne incognita* Chitwood on tomato. *Mysore Journal of Agricultural Sciences* **6**: 524-525.

Gowda, D.A., Swamy, B.C.N. and Setty, K.G.H. 1982. Nematophagous fungi from cultivated soils in Karnataka. *Journal of Soil Biology and Ecology* **2**: 95-96.

Gowdar, S.B. and Kulkarni, S. 1999. Compatibility effect of antagonists and seed dressers against *Fusarium udum* – the causal agent of pigeon pea wilt. *Karnataka Journal of Agriculture Science* **12**: 197-199.

Grewal, P.S. 1989. Effect of leaf-matter incorporation on *Aphelenchoides composticola* (Nematode), mycofloral composition, mushroom compost quality and yield of *Agaricus bisporus*. *Annals of Applied Biology* **115**: 299-312.

Grewal, P.S. and Grewal, S.K. 1988. Solar heat pasteurization of casing soil and chicken manure for mushroom cultivation. *Mushroom Journal of Tropics* **8**: 35-45.

Grewal, P.S. and Sohi, H.S. 1988. A new and cheaper technique for the rapid multiplication of *Arthrobotrys conoides* and its potential as a bio-nematicide in mushroom culture. *Current Science* **57**: 44-46.

Griffith, R. 1977. A species of bacterium pathogenic to the palm weevil, *Rhynchophorus palmarum*. 1976 Ann. Rept., Red Ring Res. Divn., Minst. of Agri., Land and Fisheries, Trinidad and Tobago, pp. 31-35.

Griffith, R. 1987. Red ring disease of coconut palm. *Plant Disease* **71**: 193-196.

Griffith, R. and Koshy, P.K. 1990. Nematode parasites of coconut and other palms. In *Plant Parasitic Nematodes in Subtropical and Tropical Agriculture* (M. Luc, R.A. Sikora & J. Bridge, eds.), pp. 363-386. CAB International, Wallingford, Oxon, U.K.

Gunasekaran, C.R. 1979. Control of nematodes associated with patchouli. *Indian Journal of Nematology* **9**: 87.

Gunasekaran, C.R., Vadivelu, S. and Jayaraj, S. 1987. Experiments on nematodes of turmeric – A review. In *Proc. of the Third Group Discussions on the Nematological Problems of Plantation Crops*. Sugarcane Breeding Institute, Coimbatore, pp. 45-46.

Gupta, D.C. and Mukhopadhyaya, M.C. 1971. Effect of N, P and K on the root-knot nematode, *Meloidogyne javanica* (Treub) Chitwood. *Science and Culture* **37**: 246-247.

Gupta, J.C. and Atwal, A.S. 1972. Biology and ecology of *Hoplolaimus indicus* (Hoplolaiminae : Nematoda). 3. Influence of population density on tomato and citrus plants. *Proceedings of XIth International Symposium on Nematology*, Reading, U. K., p.26.

Hackney, R.W. and Dickerson, O.J. 1975. Marigold, castor bean and chrysanthemum as controls of *Meloidogyne incognita* and *Pratylenchus alleni*. *Journal of Nematology* **7**: 84-90.

Handa, D.K. and Mathur, B.N. 1981. Control of root-knot of chillies with nematicides. *Second Nematology Symposium*, Tamil Nadu Agricultural University, Coimbatore, p. 48.

Haq, S., Singh, R.P. and Saxena, S.K. 1985. Control of the root-knot nematode *Meloidogyne incognita* on tomato with soft coal fly ash. *International Nematology Network Newsletter* **2**(3): 4-6.

Haque, M.M. and Gaur, H.S. 1985. Effect of multiple cropping sequences on the dynamics of nematode population and crop performance. *Indian Journal of Nematology* **15**: 262-263.

Hare, W.W. 1951. *Phytopathology* **41**: 16.

Harinath Naidu, P., Haritha, V. and John Sudheer, M. 2006. Screening of vegetable crop varieties for resistance to root-knot nematode (*Meloidogyne incognita*). *Indian Journal of Nematology* **36**: 293-294.

Harrison, A.L. and Young, P.A. 1941. Effect of root-knot nematode on tomato wilt. *Phytopathology* **31**: 749-752.

Hartman, R.W. 1976. Breeding for nematode resistance in vegetables. *Sabrao Journal* **8**: 1-10.

Hasan, A. 1985. Synergistic interaction between *Pythium aphanidermatum* and *Rhizoctonia solani* with *Meloidogyne incognita* on chilli. *Nematologica* **31**: 210-217.

Hasan, A. and Khan, M.N. 1985. The effect of Rhizoctonia solani, Sclerotium rolfsii and Verticillium dahliae on the resistance of Meloidogyne incognita. Nematologia Mediterranea **13**: 133-136.

Hasan, N. 1988. Plant parasitic nematodes associated with herbaceous forages and their management. Ann. Rept., Indian Grassland and Fodder Res. Inst., Jhansi, India, pp.88-90.

Hasan, N. and Jain, R.K. 1992. Field application of *Paecilomyces lilacinus* in combination with certain organic matters for controlling *Meloidogyne incognita* infecting cowpea followed by Lucerne. *First Afro-Asian Nematol. Symp.*, Aligarh Muslim Univ., Aligarh, pp. 10-11.

Hasan, N. and Jain, R.K. 1998. Nematological research in Uttar Pradesh: An overview. In *Phytonematology in India* (P.C. Trivedi, ed.), pp. 150-169.

Haseeb, A. 1992. Nematode pests of medicinal and aromatic plants. In *"Nematode Pests of Crops"* (D. S. Bhatti and R. K. Walia, eds.), pp. 239-249. CBS Publishers & Distributors, Delhi, 381 pp.

Haseeb, A. and Butool, F. 1991. Evaluation of different pesticides and oilcake on the control of root-knot nematode (*Meloidogyne incognita*) infesting davana (*Artemisia pallens*). *Current Nematology* **2**: 201-204.

Haseeb, A., Butool, F. and Pandey, R. 1988. Influence of different intial population densities of *Meloidogyne incognita* on the growth and oil yield of *Ocimum basilicum* cv. French. *Indian Journal of Mycology and Plant Pathology* **6**: 176-179.

Haseeb, A., Kumar, V., Shukla, P.K. and Ahmad, A. 2006. Effect of different bioinoculants, organic amendments and pesticides on the management of *Meloidogyne incognita* – *Fusarium solani* disease complex on tomato cv. K-25. *Indian Journal of Nematology* **36**: 65-69.

Haseeb, A. and Pandey, R. 1987. Incidence of root-knot nematodes in medicinal and aromatic plants - new host records. *Nematropica* **17**: 209-212.

Haseeb, A. and Pandey, R. 1989a. Root-knot disease of Henbane, *Hyscyamus niger* - a new disease record. *Tropical Pest Management* **35**: 212-213.

Haseeb, A. and Pandey, R. 1989b. Observations on *Meloidogyne* spp. affecting Japanese mint - new disease records. *Nematropica* **19**: 93-97.

Haseeb, A. and Pandey, R. 1990. Root-knot nematodes - a constraint to cultivation of Davana, *Artemisia pallens*. *Tropical Pest Management* **36**: 317-319.

Haseeb, A., Pandey, R. and Husain, A. 1988. A combination of nematicides and oilseed cakes for control of *Meloidogyne incognita* on *Ocimum basilicum*. *Nematropica* **18**: 65-69.

Haseeb, A., Shukla, P.K., Kumar, V. and Khan, R.U. 2004. Comparative efficacy of bio-control agents, neem seed powder and pesticides against *Meloidogyne incognita* and *Fusarium oxysporum* on brinjal. *National Symposium on*

Paradigms in Nematological Research for Biodynamic Farming, Univ. of Agri. Sci., Bangalore, p. 80.

Hashmi, S. and Hashmi, G. 1990. Crop rotation and *Meloidogyne javanica* in vegetable production. *International Nematology Network Newsletter* **7** (1): 19-21.

Hassan, A. 1988. Interaction between *Pratylenchus coffeae* and *Pythium aphanidermatum* and/or *Rhizoctonia solani* on chrysanthemum. *Zeist Phytopathology* **123** : 227-232.

Hassan, M.G. and Sobita Devi. 2004. Efficacy of bioagents against root-knot nematode, (*Meloidogyne incognita*) of tomato. *National Symposium on Paradigms in Nematological Research for Biodynamic Farming*, Univ. of Agri. Sci., Bangalore, p. 88.

Hawn, E.J. 1963. Transmission of bacterial wilt of alfalfa by *Ditylenchus dipsaci* (Kuhn). *Nematologica* **8**: 65-68.

Hazarika, B.P. and Roy, A.K. 1974. Effect of *Rhizoctonia solani* on the reproduction of *Meloidogyne incognita* on eggplant. *Indian Journal of Nematology* **4**: 246-248.

Hazarika, K., Dutta, P.K. and Saikia, L. 1998. Field efficacy of *Paecilomyces lilacinus* against root-knot nematode in betel vine. *Journal of Phytological Research* **11**: 171-173.

Hemavathi, M.R. 2002. *Evaluation of Bioagents,* Pasteuria penetrans *and* Glomus fasciculatum *in the Management of Root-Knot Nematode,* Meloidogyne incognita *Infesting Tomato.* M.Sc. (Agri.) thesis, Univ. of Agri. Sci., Bangalore, 102 pp.

Hewitt, W.B., Raski, D.J. and Goheen, A.C. 1958. Nematode vector of soil-borne fanleaf virus of grapevines. *Phytopathology* **48**: 586-595.

Hide, G.A. and Corbett, D.C.M. 1973. Controlling early death of potatoes caused by *Heterodera rostochiensis* and *Verticillium dahliae*. *Annals of Applied Biology* **5**: 461-462.

Hide, G.A. and Corbett, D.C.M. 1974. Field experiments in the control of *Verticillium dahliae* and *Heterodera rostochiensis* on potatoes. *Annals of Applied Biology* **78**: 295-307.

Hirano, K. and Kawamura, T. 1972. *Incidence of complex disease caused by* Pratylenchus penetrans *or* P. coffeae *and* Fusarium *wilt fungus in tomato seedlings.* Tech. Bull., Faculty of Hort., Chiba Univ. **20**: 37-43.

Hoyle, J.C. 1968. A simple method for poisoning coconut trees. *Journal of Agricultural Science Trinidad* **68**: 459-465.

Ibrahim, I.K.A. and Rezk, M.A. 1988. The root-knot nematodes – a major problem in crop production in Egypt. In *Advances in Plant Nematology* (M.A. Maqbool, A.M. Golden, A. Gaffar and L.R. Krusberg, eds.), pp. 81-98. Natl. Nematol. Res. Centre, Univ. of Karachi, Karachi, Pakistan.

Ichniohe, M. 1980. *Ann. Rept.* Soc. of Plant Prot., North Japan **31**: 1-8.

Inomoto, M.M. and Monteiro, A.R. 1989. Thermal treatment of suckers of 'Giant Cavendish' banana for the eradication of plant-parasitic nematodes. *Nematologia Brasileira* **13**: 139-150.

Jacob, A.J. and Illaque, M.M. 1998. Effect of neem products on the suppression of root-knot nematode *Meloidogyne incognita* in tomato. *Third International Symposium of Afro-Asian Society of Nematologists*, Coimbatore, India, P. 85.

Jacobsen, B.J., MacDonald, D.H. and Bissonnette, H.L. 1979. Interaction between *Meloidogyne hapla* and *Verticillium albo-utrum* in soils. *Phytopathology* **69**: 978-981.

Jaffee, B.A., Gaspard, J.T., Ferris, H. and Muldoon, A. 1988. *Soil Biology and Biochemistry* **20**: 631-636.

Jagdale, G.B., Pawar, A.B. and Darekar, K.S. 1985. Studies on control of root-knot nematode on betel vine. *International Nematology Network Newsletter* **2**: 10-13.

Jain, R.K. and Bhatti, D.S. 1978. Bare root treatment with systemics for controlling root-knot nematode in tomato transplants. *Indian Journal of Nematology* **8**: 19-24.

Jain, R.K. and Bhatti, D.S. 1981. Systemics as seed treatment for the control of *Meloidogyne javanica* in okra. *Second Nematology Symposium,* Tamil Nadu Agricultural University, Coimbatore, p. 44.

Jain, R.K. and Bhatti, D.S. 1983. Chemical control of *Meloidogyne javanica* in tomato through nursery bed treatment. *Third Nematology Symposium*, Himachal Pradesh Agricultural University, Solan, pp. 37-38.

Jain, R.K. and Bhatti, D.S. 1985. Effect of summer ploughing alone and in combination with some other effective practices on the incidence of root-knot nematode (*Meloidogyne javanica*) on tomato (Cultivar HS 101). *Indian Journal of Nematology* **15**: 262.

Jain, R.K. and Bhatti, D.S. 1987. Population development of root-knot nematode (*Meloidogyne javanica*) and tomato yield as influenced by summer ploughing. *Tropical Pest Management* **33**: 122-125.

Jain, R.K. and Bhatti, D.S. 1989. An integrated approach for the management of root-knot nematode (*Meloidogyne javanica*) in tomato. *Indian Journal of Nematology* **19**: 159-161.

Jain, R.K., Bhatti, D.S., Bhutani, R.D. and Kalloo, G. 1983. Screening of germplasm of some vegetable crops for resistance to root-knot nematode *Meloidogyne incognita. Indian Journal of Nematology* **13**: 212-215.

Jain, R.K. and Gupta, D.C. 1990. Integrated management of root-knot nematode (*Meloidogyne javanica*) infecting tomato and okra. *Indian Journal of Nematology* **20**: 14-18.

Jain, R.K. and Gupta, D.C. 1991. Integrated management of root-knot nematode, *Meloidogyne javanica* infecting *Solanum melongena. Afro-Asian Journal of Nematology* **1**: 156-160.

Jain, R.K., Paruthi, I.J. and Gupta, D.C. 1988. Control of root-knot nematode (*Meloidogyne javanica*) in tomato through nursery bed treatment alone and in combination with application of carbofuran at transplanting. *Indian Journal of Nematology* **18**: 340-341.

Jain, R.K., Paruthi, I.J. and Gupta, D.C. 1990. Influence of intercropping on the incidence of root-knot nematode (*Meloidogyne javanica*) in brinjal. *Indian Journal of Nematology* **20**: 236-238.

Jain, R.K., Paruthi, I.J., Gupta, D.C. and Dhankar, B.S. 1986. Appraisal of losses due to root-knot nematode, *Meloidogyne javanica* on okra under field conditions. *Tropical Pest Management* **32**: 341-342.

Jain, R.K. and Sethi, C.L. 1988. Interaction between vesicular arbuscular mycorrhiza and *Meloidogyne incognita, Heterodera cajani* on cowpea as influenced by time of inoculation. *Indian Journal of Nematology* **18**: 263-268.

Jakobsen, I. 1999. Effects of *Pseudomonas fluorescens* DF 57 on growth and P uptake of two arbuscular mycorrhizal fungi in symbiosis with cucumber. *Mycorrhiza* **8**: 319-334.

Jatala, P. 1986. Biological control of nematodes. *Annual Review of Phytopathology* **24**: 453-489.

Jatala, P., Kaltenbach, R., Devaux, A.J. and Campos, R. 1980. Field application of *Paecilomyces lilacinus* for controlling *Meloidogyne incognita* on potatoes. *Journal of Nematology* **12**: 226-227.

Jatala, P., Martin, C., Mendoza, H.A., Hidalgo, O.A. and Rincon, R.H. (eds.). 1990. Role of nematodes in the expression of *Pseudomonas solanacearum* and strategies for screening and breeding for combined resistance. *Advances en el Mejoramiento Genetico de la Papa en los Paises del cono sur.*, pp.187-189.

Jayakumar, J., Rajendran, G. and Ramakrishnan, S. 2004. Management of reniform nematode, Rotylenchulus reniformis in okra through Streptomyces avermitilis. National Symposium on Paradigms in Nematological Research for Biodynamic Farming, Univ. of Agri. Sci., Bangalore, p. 98.

Jayakumar, J., Rajendran, G. and Ramakrishnan, S. 2005. bio-efficacy of avermectin against root-knot nematode, *Meloidogyne incognita* in tomato. *Indian Journal of Nematology* **35**: 151-156.

Jayaraman, V., Rajendran, G. and Muthukrishnan, T.S. 1975. Occurrence of root-knot nematodes in *Polianthes tuberosa* L. in Tamil Nadu. *Indian Journal of Nematology* **5**: 101-102.

Jenkins, W.R. and Coursen, B.W. 1957. The effect of root-knot nematodes, *Meloidogyne incognita acrita* and *M. hapla*, on *Fusarium* wilt of tomato. *Plant Disease Reporter* **41**: 182-186.

Johnson, A.W. and Littrell, R.H. 1969. Effect of *Meloidogyne incognita, M. hapla* and *M. javanica* on the severity of *Fusarium* wilt of chrysanthemum. *Journal of Nematology* **1**: 122-125.

Johnson, L.F. 1963. Temperature as a factor in the control of tomato root-knot with oat straw. *Phytopathology* **53**: 879.

Johnson, N. 2000. Studies on Nematode Pests of Cut Flowers with Special Reference to Carnation, Cystera, Gladiolus and Asiatic Lily. Ph. D. thesis, Tamil Nadu Agri. Univ., Coimbatore, 160 pp.

Jonathan, E.I. 1994. *Studies on the Root-knot Nematodes (*Meloidogyne incognita*) (Kofoid and White, 1919) Chitwood, 1949, in Banana (*Musa *sp.) cv. Poovan (AAB)*. Ph. D. Thesis, Tamil Nadu Agri. Univ., Coimbatore, Tamil Nadu, India.

Jonathan, E.I., Arulmozhiyan, R., Muthusamy, S. and Manuel, W.W. 2000. Field application of *Paecilomyces lilacinus* for the control of *Meloidogyne incognita* on betel vine, *Piper betle*. *Nematologia Mediterranea* **28**: 131-133.

Jonathan, E.I. and Cannayane, I. 2002. Field application of biocontrol agents for management of spiral nematode, *Helicotylenchus multicinctus* in banana (Abstr.). *Global Conference on Banana and Plantain,* Bangalore, p. 193.

Jonathan, E.I., Cannayane, I., Bommiraju, P., Amutha, G. and Samiyappan, R. 2004. Biological control of root-knot nematode, *Meloidogyne incognita* and *Phytophthora* wilt complex in betel vine with plant growth promoting rhizobacteria. *National Symposium on Paradigms in Nematological Research for Biodynamic Farming*, Univ. of Agri. Sci., Bangalore, p. 73-74.

Jonathan, E.I., Nagalakshmi, S. and Padmanabhan, D. 1990. Estimation of yield losses in betel vine due to *Meloidogyne incognita. International Nematology Network Newsletter* **7** (4): 26.

Jonathan, E.I., Padmanabhan, D. and Ayyamperumal, A. 1995. Biological control of root-knot nematode on betel vine, *Piper betel* by *Paecilomyces lilacinus. Nematologia Mediterranea* **23**: 191-193.

Jonathan, E.I. and Rajendran, G. 2000. Assessment of avoidable yield loss in banana due to root-knot nematode, *Meloidogyne incognita. Indian Journal of Nematology* **30**: 162-164.

Jones, F.C.E. 1961. The potato root eelworm *Heterodera rostochiensis* Woll. in India. *Current Science* **30**: 187.

Jones, F.C.E. 1969. Integrated control of the potato cyst nematode. *Proceedings of Fifth British Insecticide and Fungicide Conference* **3**: 646-656.

Jones, J.P. and Overman, A.J. 1976. Tomato wilts, nematodes and yields as affected by soil reaction and persistent contact nematicide. *Plant Disease Reporter* **60**: 913-917.

Jothi, G., Ramakrishnan, S. and Sundara Babu, R. 2004. Integrated management of *Pratylenchus delattrei* in crossandra. *National Symposium on Paradigms in Nematological Research for Biodynamic Farming,* Univ. of Agri. Sci., Bangalore, p. 97.

Jothi, G. and Sundarababu, R. 2000. Interaction effect of VAM and root-knot nematode on the growth and nutrient content of brinjal. *International Journal of Tropical Plant Diseases* **18**: 91-100.

Jothi, G. and Sundarababu, R. 2001. Management of root-knot nematode in brinjal by using VAM and crop rotation with green gram and pearl millet. *Journal of Biological Control* **15**: 77-80.

Jung, C., Clauussen, U., Harsthemke, B., Her, F. and Herman, R.G. 1992. A DNA library from an individual *Beta patellaris* chromosome conferring nematode resistance obtained by micro dissection of meotic metaphase chromosome. *Plant Molecular Biology* **20**: 503-511.

Kalaiarasan, P., Sivakumar, M. and Subramanian, S. 2006. Reaction of pulse genotypes for resistance to pigeonpea cyst nematode, *Heterodera cajani* Koshy, 1967. *Indian Journal of Nematology* **36**: 291-292.

Kalita, S. and Bora, B.C. 2006. Effect of nematicides and organic amendments in the management of *Meloidogyne incognita* in nursery tea. *Indian Journal of Nematology* **36**: 148-149.

Kamra, A. and Gaur, H.S. 1998. Control of nematodes, fungi and weeds in nursery beds by soil solarization. *Indian Journal of Nematology* **28**: 46-52.

Kandasamy, K. and Sivakumar, C.V. 1981. Effect of *Meloidogyne arenaria* on the germination of chilli (*Capsicum annum* Linn.) seeds. *Second Nematology Symposium*, Tamil Nadu Agricultural University, Coimbatore, p.16.

Kantharaju, V., Prakash Babu, N., Krishnappa, K. and Reddy, B.M.R. 2002. Biodiversity of indigenous isolates of VAM, *Glomus fasciculatum* against *Meloidogyne incognita* race-1 on tomato (Abstr.). *National Symposium on Bio-diversity and Management of Nematodes in Cropping Systems for Sustainable Agriculture,* Jaipur, p.3.

Kanwar, R.S. 1990. Studies on Cropping Systems for Management of Root-knot Nematode, Meloidogyne javanica (Treub, 1985) Chitwood, 1949. Ph. D. thesis, Haryana Agri. Univ., Hisar, 118 pp.

Kanwar, R.S. and Bhatti, D.S. 1990. Yield performance of resistant and susceptible tomato cultivars in monoculture and related reproduction of *Meloidogyne javanica. International Nematology Network Newsletter* **7**(3): 16-18.

Kanwar, R.S. and Bhatti, D.S. 1992. Controlling root-knot nematode, *Meloidogyne javanica* through crop rotation. *Current Nematology* **3**: 65-70.

Karuna, K., Krishnappa, K. and Reddy, B.M.R. 2001. Utilization of *Pasteuria penetrans* in the management of root-knot nematode on brinjal. *Natl.Conf. on Centenary of Nematology in India- Appraisal and Future Plans,* Indian Agri. Res. Inst., New Delhi, p. 142.

Karuna, K., Ravichandra, N.G., Krishnappa, K. and Shreenivasa, K.R. 2004. Management of root-knot nematode, *Meloidogyne incognita* on banana by biocontrol agents. *National Symposium on Paradigms in Nematological Research for Biodynamic Farming,* Univ. of Agri. Sci., Bangalore, p. 70.

Kaur, D.J. 1987. Ph. D. thesis, Himachal Pradesh Univ., Shimla, India, 198 pp.

Kerr, A. and Vythilingam, M.K. 1966. Replanting eelworm infested areas. *Tea Quarterly* **37**: 67-72.

Kerry, B.R., Kirkwood, I.A., De Leij, F.A.A.M., Barba, J., Leidjens, M.B. and Brookes, P.C. 1993. Growth and survival of *Verticillium chlamydosporium* Goddard, a parasite of nematodes in soil. *Biocontrol Science and Technology* **3**: 355-365.

Khan, A.M., Adhami, A., Siddiqi, Z.A. and Saxena, S.K. 1966. Effect of different oilcakes on hatching of larvae and on the development of root-knot caused by *Meloidogyne incognita* (Kofoid and White) Chitwood. In *Plant Disease Problems* (S.P. Raychaudhuri *et al*., eds.), pp. 582-588. Indian Phytopathological Society, Indian Agricultural Research Institute, New Delhi, India.

Khan, A.M., Alam, M.M., Siddiqi, Z.A. and Saxena, S.K. 1976. Efficacy of nematicides and oilcakes for the control of phytoparasitic nematodes in nurseries of perennial plants. *Geobios* **3**: 78-81.

Khan, A.M., Saxena, S.K. and Khan, M.W. 1971. Interaction of *Rhizoctonia solani* Kuhn and *Tylenchorhynchus brassicae* Siddiqi, 1961 in pre-emergence damping-off of cauliflower seedlings. *Indian Journal of Nematology* **1**: 85-86.

Khan, A.M., Saxena, S.K. and Siddiqi, Z.A. 1971a. Efficacy of *Tagetes erecta* in reducing root infesting nematodes of tomato and okra. *Indian Phytopathology* **24**: 166-169.

Khan, A.M., Saxena, S.K. and Siddiqi, Z.A. 1971b. Reaction of some cultivated and wild species of *Lycopersicon* and cucurbits to *Meloidogyne incognita* (Kofoid and White) Chitwood. *Indian Phytopathology* **24**: 197-199.

Khan, A.M., Saxena, S.K., Siddiqi, Z.A. and Upadhyay, R.S. 1975. Control of nematodes by crop rotation. *Indian Journal of Nematology* **5**: 214-221.

Khan, F.A. 1990. Nematicidal potentials of some naturally-growing medicinal plants against *Pratylenchus zeae*. *Revue de Nematologie* **13**: 463-465.

Khan, F.A. and Khan, A. M. 1973. Studies on the reniform nematode, *Rotylenchulus reniformis* I. Host range and population changes. *Indian Journal of Nematology* **3**: 24-30.

Khan, M.R. and Ghadipur, M.H. 2004. Management of root-knot disease of some solanaceous vegetables by soil application of fly ash. *National Symposium on Paradigms in Nematological Research for Biodynamic Farming,* Univ. of Agri. Sci., Bangalore, p. 92.

Khan, M.R. and Tarannum, Z. 1999. Effects of field application of various micro-organisms on *Meloidogyne incognita* on tomato. *Nematologia Mediterranea* **27**: 233-238.

Khan, M.R., Shit, S., Pal, A.K. and Biswas, A.B. 2004. Studies on tuberose foliar nematode, *Aphelenchoides besseyi* in West Bengal. *National Symposium on Paradigms in Nematological Research for Biodynamic Farming*, Univ. of Agri. Sci., Bangalore, p. 135-136.

Khan, M.W. and Esfahani, M.N. 1990. Efficacy of *Paecilomyces lilacinus* for controlling *Meloidogyne javanica* on tomato in greenhouse in India. *Pakistan Journal of Nematology* **8**: 95-100.

Khan, M.W., Khan, A.M., Haider, S.R. and Ashraf, S.S. 1988. Identity of races of *Meloidogyne incognita* in Western region of Uttar Pradesh, India. *Nematologica* **34**: 115-116.

Khan, R.M. and Parvatha Reddy, P. 1992. Nematode pests of ornamental crops. In *Nematode Pests of Crops* (D. S. Bhatti and R. K. Walia, eds.), pp. 250-257. CBS Publishers & Distributors, Delhi, 381 pp.

Khan, R.M. and Parvatha Reddy, P. 1994. Nematode problems of ornamental crops and their management. In *Floriculture-Technology, Trades and Trends* (J. Prakash and K.R. Bhandary, eds.), pp.468-472. Oxford & IBH Publishing Co. Pvt. Ltd., New Delhi, 667 pp.

Khan, T., Khan, S.T., Fazal, M. and Siddiqi, Z.A. 1997. Biological control of *Meloidogyne incognita* and *Fusarium solani* disease complex in papaya using *Paecilomyces lilacinus* and *Trichoderma harzianum*. *International Journal of Nematology* **7**: 127-132.

Khan, T.A. 1991. *Studies on a Disease Complex of Papaya caused by* Meloidogyne incognita *and* Fusarium solani. Final Tech. Rept., DST Project, Aligarh Muslim Univ., Aligarh, 47 pp.

Khan, T.A. and Husain, S.I. 1988. Studies on the efficacy of seed treatment with pesticides, oilcakes, neem leaf and culture filtrate of *Paecilomyces lilacinus* for the control of diseases caused by the presence of *Rotylenchulus reniformis*,

Meloidogyne incognita and *Rhizoctonia solani* either individually or concomitantly on cowpea. *Indian Journal of Nematology* **18**: 192-198.

Khan, T.A. and Khan, S.T. 1998. Nematode diseases of papaya. In *Nematode diseases in Plants* (P.C. Trivedi, ed.), pp. 344-360. CBS Publishers and Distributors, New Delhi.

Khan, T.A., Khan, S.T. and Saxena S.K. 1992. Effect of interaction of *Meloidogyne incognita* and *Fusarium solani* on the relative response of papaya cvs. *First Afro-Asian Nematology Symposium*, Aligarh Muslim Univ., Aligarh, p. 14.

Khan, T.A., Nisar, S. and Husain, S.I. 1991. Effect of water soluble extracts of certain oilcakes on the development of root-knot disease of okra (*Abelmoschus esculentus* var. Sevendhari). *Current Nematology* **2**: 167-170.

Khanna, A.S. and Sharma, N.K. 1990. Efficacy of nematophagous fungi against myceliophagus nematode *in vitro*. *Proc. National Seminar on Bioagents in Nematode Management*, Indian Agri. Res. Inst., New Delhi, pp. 83-84.

Khanna, A.S. and Sharma, N.K. 1994. Management of *Aphelenchoides composticola* infesting white button mushroom by neem cake. *Indian Journal of Hill Farming* **7**: 114-115.

Kiran Kumar, K.C., Krishnappa, K., Karuna, K., Ravichandra, N.G. and Shreenivas, K.R. 2004. Evaluation of proposed management package of growth, development and yield of tomato infested by root-knot nematode. *National Symposium on Paradigms in Nematological Research for Biodynamic Farming,* Univ. of Agri. Sci., Bangalore, p. 94.

Kirkpatrik, J.D., Mai, W.F., Fisher, E.C. and Parker, K.G. 1959a. Population levels of *Pratylenchus penetrans* and *Xiphinema americanum* in relation to potassium fertilization on Montmorency sour cherries on Mazzard rootstock (Abstr.). *Phytopathology* **49**: 543.

Kirkpatrik, J.D., Mai, W.F., Fisher, E.C. and Parker, K.G. 1959b. Relation of nematode population in nutrition of sour cherries (Abstr.). *Phytopathology* **49**: 543.

Kirkun, J. and Orion, D. 1977. Studies of the interaction of *Verticillium dahliae* and *Pratylenchus thornei* on potato. *Phytoparasitica* **5**: 67.

Kirmani, M.R., Saxena, S.K. and Khan, A.M. 1978. Growth and development of root-knot on eggplant as influenced by some fungi. *Indian Journal of Nematology* **8**: 153-155.

Klein-Lankhorst, R., Reitveld, P., Machiels, B., Verlek, R., Weide, R., Gebhardt, C., Koorneef, M. and Zabel, P. 1991. RFLP markers linked to the root-knot nematode resistance in gene *Mi* in tomato. *Theoretical and Applied Genetics* **81**: 661-667.

Kolodge, C., Radewald, J.D. and Shibaya, F. 1987. Revised host range and studies on life cycle of *Longidorus africanus*. *Journal of Nematology* **19**: 77-81.

Koshy, P.K. 1986. The burrowing nematode, *Radopholus similis* (Cobb, 1893) Thorne, 1949. In *Plant Parasitic Nematodes of India-Problems and Progress.* (G. Swarup and D.R. Dasgupta, eds.), Indian Agri. Res. Inst., New Delhi, India, pp. 223-248.

Koshy, P.K.2002. Nematode pests of spices and condiments and their control. In *Nematode Control in Crops* (M.M. Alam & N. Sharma, eds.), pp. 187-200. International Book Distributing Company, Lucknow, India.

Koshy, P.K. and Geetha, S.M. 1992. Nematode pests of palms and cocoa. In *Nematode Pests of Crops* (D.S. Bhatti and R.K. Walia, eds.), pp. 214-227. CBS Publishers & Distributors, Delhi.

Koshy, P.K. and Sosamma, V.K. 1987. *Indian Journal of Nematology* **17**: 108-118.

Koshy, P.K., Sosamma, V.K. and Samuel, R. 1999. Interactive effect of vesicular arbascular mycorrhizae (VAM) with *Radopholus similis* on coconut seedlings. In *Nematology – Challenges and Opportunities in 21st Century,* (Usha K. Mehta, ed.), pp. 106-110. Sugarcane Breeding Institute, Coimbatore.

Koshy, P.K., Sosamma, V.K. and Sundararaju, P. 1979. *Plant Disease Reporter* **63**: 433-435.

Krishnamurthy, G.V.G. and Elias, N.A. 1967. Host range of *Meloidogyne incognita* causing root-knot on tobacco in Hunsur, Mysore State. *Indian Phytopathology* **20**: 374-377.

Krishnamurthy, G.V.G. and Elias, N.A. 1969. *Indian Journal of Agricultural Sciences* **39**: 263-265.

Krishna Prasad, P.R. 1978. *Studies on the Root-knot Nematodes Infecting Patchouli* (Pogostemon cablin *Bench*). M. Sc. (Agri.) thesis, Univ. of Agri. Sci., Bangalore.

Krishna Rao, A.B.V., Padhi, N.N. and Acharya, A. 1987. Effect of different nematicides, oilcakes and urea in the control of *Rotylenchulus* in okra. *Indian Journal of Nematology* **17**: 171-173.

Kuhn, J. 1857. Uber das Vorkommen von Ruben – Nematoden an den Wurzeln der Halmfruchte. *Landwrits. Jahrb.* **3**: 47-50.

Kumar, A.C. 1988. *Indian Coffee* **52**: 12-19.

Kumar, S. and Nanjan, K. 1983. Control of spiral nematodes in patchouli. *Fungicide and Nematicide Tests* **39**: 93.

Kumar, S. and Sivakumar, C.V. 1981. Disease complex involving *Rotylenchulus reniformis* and *Rhizoctonia solani* in okra. *Second Nematology Symposium*, Tamil Nadu Agricultural University, Coimbatore, p.12.

Kumari, R., Verma, K.K., Dhindsa, K.S. and Bhatti, D.S. 1986. *Indian Journal of Nematology* **16**:226-240.

Kumutha, K. 2001. Symbiotic Influence of AM Fungi and Rhizobacteria on Biochemical and Nutritional Changes in Mulberry (Morus alba). Ph. D. thesis, Tamil Nadu Agri. Univ., Coimbatore.

Kushwaha, J.S., Prasad, D. and Vimal, O.P. 1983. Effect of urea alone and in combination with saw dust on tomato growth and nematode development in alluvial soil. *Indian Journal of Entomology* **49**:479-481.

Kyo, M., Miyauchi, Y., Fujimoto, T. and Mayama, S. 1990. Production of nematicidal compounds by hairy root cultures of *Tagetes patula* L. *Plant Cell Reports* 9: 393-397.

Lakshmana Murthy, K.V. 1983. *Studies on Burrowing Nematode,* Radopholus similis *(Cobb, 1983) Thorne, 1949 on Banana in Karnataka.* M. Sc. (Agri.) thesis, Univ. of Agri. Sci., Bangalore.

Lakshmanan, P.L. and Sivakumar, C.V. 1981. Control of *Rotylenchulus reniformis* in Bhendi by integrated method. *Second Nematology Symposium*, Tamil Nadu Agricultural University, Coimbatore, p.54.

Lal, A., Yadav, B.S. and Nandwana, R.P. 1977. Effect of chopped leaves of various plants and sewage on the plant growth and reniform nematode, *Rotylenchulus reniformis. Indian Journal of Mycology and Plant Pathology* **7**: 68-69.

Lal, B., Rana, B.P. and Dahia, R.S. 2002. Efficacy of antagonistic fungi on root-knot nematode, Meloidogyne incognita in okra. Proc. of the Natl. Symp. on Biodiversity and Management of Nematodes in Cropping Systems for Sustainable Agriculture, Rajasthan Agri. Univ., Bikaner.

Lamberti, F. 1981. Combating nematode vectors of viruses. *Plant Disease* **65**: 113-117.

Lanjewar, R.D. and Shukla, V.N. 1985. Parasitism and interaction between *Pythium myriotylum* and *Meloidogyne incognita* in soft rot complex of ginger. *Indian Journal of Nematology* **15**: 170-173.

Laqman Khan, M. 2001. Dr. Y.S. Paramar University of Horticulture and Forestry, Solan, Himachal Pradesh. In *Indian Nematology-Progress and Prospectives*. (S.C. Dhawan *et al*., eds.), pp. 153-158. Division of Nematology, Indian Agri. Res. Inst., New Delhi.

Lear, B. 1959. Application of castor pomace and castor cropping to soil to reduce nematode populations. *Plant Disease Reporter* **43**: 459-460.

Leela, N.K., Khan, R.M., Parvatha Reddy, P. and Nidiry, E.S.J. 1992. Nematicidal activity of essential oil of *Pelargonium graveolens* against the root-knot nematode, *Meloidogyne incognita. Nematologia Mediterranea* **20**: 57-58.

Leibscher, G. 1892. Beobachtungen Uber das Auftreten eines Nematoden an Erbsen. *Jour. Landwrits.* **40**: 357-368.

Lele, V.C., Durgapal, J.C., Agarwal, D.K. and Sethi, C.L. 1978. Crown and root gall of grape (*Vitis vinefera* L.) in Andhra Pradesh. *Current Science* **47**: 280.

Libman, G., Leach, J.G. and Adams, R.E. 1964. Role of certain plant parasitic nematodes in infection of tomatoes by *Pseudomonas solanacearum. Phytopathology* **54**: 151-153.

Linford, M.B. and Yap, F. 1940. Some host plants of the reniform nematode in Hawaii. *Proceedings of the Helminthological Society of Washington* **7**: 42-44.

Linford, M.B., Yap, F. and Olivera, J.M. 1938. Reduction of soil populations of the root-knot nematode during decomposition of organic matter. *Soil Science* **45**: 127-140.

Lingaraju, S. and Goswami, B.K. 1993. Influence of mycorrhizal fungus *Glomus fasciculatum* on the host parasite relationship of *Rotylenchulus reniformis* on cowpea. *Indian Journal of Nematology* **23**: 137-145.

Littrel, R.H. and Johnson, A.W. 1969. Pathogenicity of *Pythium aphanidermatum* to chrysanthemum in combined inoculations with *Belonolaimus longicaudatus* or *Meloidogyne incognita. Phytopathology* **59**: 115-116.

Loganathan, M., Swarnakumari, N., Sivakumar, M., Prakasam, V., Mohan, L., Ramraj, B. and Samiyappan, R. 2001. Biological suppression of fungal nematode complex diseases of major cruciferous vegetables. *South Indian Horticulture* **49**: Special issue.

Loos, C.A. 1953. *Meloidogyne brevicauda* n. sp. a cause of root-knot of mature tea in Ceylon. *Journal of Science (B)* **23**: 119-124.

Loos, C.A. 1959. Symptom expression of *Fusarium* wilt disease of the Gros Michel banana in the presence of *Radopholus similis* (Cobb, 1893) Thorne, 1949 and *Meloidogyne incognita acrita* Chitwood, 1949. *Proceedings of the Helminthological Society of Washington* **26**: 103 –111.

Loos, C.A. 1961. Eradication of the burrowing nematode, *Radopholus similis*, from bananas. *Plant Disease Reporter* **45**: 457-461.

Lordello, L.G.E. 1968. *Nematodes das Plantas Cultivadas*. Livaria Nobel SA, Sao Paulo, Brazil.

Lucas, G.B., Sasser, J.N. and Kelman, A. 1955. The relationship of root-knot nematodes to Granville wilt resistance in tobacco. *Phytopathology* **45**: 537-540.

Mac Guidwin, A.E., Bird, G.W. and Safir, G.R. 1985. Influence of *Glomus fasciculatum* on *Meloidogyne hapla* infecting *Allium cepa*. *Journal of Nematology* **17**: 389-295.

Maheswari, T.U. and Mani, A. 1988. Combined efficacy of Pasteuria penetrans and Paecilomyces lilacinus on the biocontrol of Meloidogyne javanica on tomato. International Nematology Network Newsletter **5**(3): 10-11.

Maheswari, T.U., Mani, A. and Rao, P.K. 1987. Combined efficacy of the bacterial spore parasite *Pasteuria penetrans* (Thorne, 1940) and nematicides in the control of *Meloidogyne javanica* on tomato. *Journal of Biological Control* **1**: 53-57.

Mahmood, K., Qamar, S., Naqvi, A. and Alam, M.M. 1974. The influence of bottle gourd mosaic virus and brinjal mosaic virus on the development of root-knot caused by *Meloidogyne incognita* (Kofoid and White, 1919) Chitwood, 1949. *Indian Phytopathology* **27**: 610-611.

Mani, A. 1986. Occurrence of *Meloidogyne javanica* on citrus in Andhra Pradesh (India). *International Nematology Network Newsletter* **3**(2): 9-10.

Mani, A. 1988a. Studies on the bacterial parasite *Pasteuria penetrans:* I. Spore viability after storage. II. Culture on citrus nematode *Tylenchulus semipenetrans*. *International Nematology Network Newsletter* **5** (2): 24-25.

Mani, A. 1988b. Effect of interculture of marigold and mustard with acid lime on citrus nematode, *Tylenchulus semipenetrans. International Nematology Network Newsletter* **5** (4): 14-15.

Mani, A., Murty, I.R., Rao, P.K. and Prasad, V.D. 1989. Growth of *Paecilomyces lilacinus* on natural substrates and its efficacy against citrus nematode, *Tylenchulus semipenetrans*. *Journal of Biological Control* **3**: 59-61.

Mani, A. and Murty, V.D. 1986. Citrus nematode and root-knot nematodes : Distribution, crop loss, symptoms and management. In *"Plant Parasitic Nematodes - Problems and Progress"* (G. Swarup and D.R. Dasgupta, eds.), pp. 249-260. Indian Agricultural Research Institute, New Delhi.

Mani, A. and Murty, V.D. 1989. *National Symposium on Insect Pests and Disease Management on Citrus*. Jawaharlal Nehru Krishi Viswa Vidyalaya, Chindwara.

Mani, A. and Reddy, G.S. 1986. Studies on the varietal reaction of some species of *Citrus* and *Poncirus* to *Tylenchulus semipenetrans* Cobb. *Indian Journal of Nematology* **16**: 267-268.

Mani, M.P. 1996. *Effect of* Pasteuria penetrans *(Thorne) Sayre & Starr and* Pseudomonas fluorescens *(Migula) Against* Meloidogyne incognita *(Kofoid & White) Chitwood in Grapevine* (*Vitis vinefera* Linn.). M. Sc. (Agri.) thesis, Tamil Nadu Agri. Univ., Coimbatore, 46 pp.

Mani, M.P., Rajeswari, S. and Sivakumar, C.V. 1998. Management of the potato cyst nematodes, *Globodera* spp. through plant rhizosphere bacterium *Pseudomonas fluorescens* Migula. *Journal of Biological Control* **12**: 131-134.

Mani, M.P., Sivakumar, C.V. and Ramakrishnan, S. 1999. Status of *Pasteuria penetrans* in root-knot nematode infested vineyards of Tamil Nadu. *Indian Journal of Nematology* **29**: 104.

Mankau, R. 1975. *Bacillus penetrans* n. comb. causing a virulent disease of plant parasitic nematodes. *Journal of Invertebrate Pathology* **26**: 333-339.

Mankau, R. 1980. Biocontrol: Fungi as nematode control agents. *Journal of Nematology* **12**: 244-252.

Manzoor, S., Sinha, A.K. and Bora, B.C. 2002. Management of citrus nematode, *Tylenchulus semipenetrans* on Khasi mandarin by *Paecilomyces lilacinus. Indian Journal of Nematology* **32**: 153-155.

Martin, M.J., Riedel, R.M. and Rowe, R.C. 1982. *Verticillium dahliae* and *Pratylenchus penetrans*: Interactions in the early dying complex of potato in Ohio. *Phytopathology* **72**: 640-644.

Martin, W.J. 1970. Elimination of root-knot and reniform nematodes and scurf infection from rootlets of sweet potato plants by hot water treatment. *Plant Disease Reporter* **54**: 1056-1058.

Masood, A. and Husain, S.I. 1976. Phenolic and ortho-dihydroxy phenolic changes and their role in the resistance and susceptibility of three tomato varieties to *Meloidogyne incognita. Indian Journal of Nematology* **6**: 86-93.

McBeth, C.W. and Taylor, A.L. 1944. Immune and resistant cover crops valuable in root-knot infested peach orchard. *Proceedings of the American Society of Horticultural Science* **45**: 158-166.

McGuire, J.M. and Good, M.J. 1970. The effect of benomyl on *Xiphinema americanum* and tobacco ring spot virus infection. *Phytopathology* **60**: 1150-1151.

McKeen, C.D. and Mountain, W.B. 1960. Synergism between *Pratylenchus penetrans* (Cobb) and *Verticillium albo-utrum* R. and B. in eggplant wilt. *Canadian Journal of Botany* **38**: 789-794.

Mc Kenry, M.V. 1980. Nature, mode of action and biological activity of nematicides. In *Hand Book of Pest Management in Agriculture,* Section D, Vol. 3 (D. Pimetel, ed.), pp. 59-73. CRC Press, Cleveland.

McKinley, R.R.T. and Tolboys, P.W. 1979. Effects of *Pratylenchus penetrans* on development of strawberry wilt caused by *Verticillium dahliae. Annals of Applied Biology* **92**: 347-357.

Meagher, J.W. and Jenkins, P.T. 1970. Interaction of *Meloidogyne hapla* and *Verticillium dahliae* and chemical control of wilt in strawberry. *Australian Journal of Experimental Agriculture and Animal Husbandry* **10**: 493-496.

Medhane, N.S., Jagdale, G.B., Pawar, A.B. and Darekar, K.S. 1985. Effect of *Tagetes erecta* on root-knot nematodes infecting betel vine. *International Nematology Network Newsletter* **2**(3): 11-12.

Meijnke, C.A.R. and Oostenbrink, M. 1958. *Tagetes* ter bestriding van aaltjesaantastingen. *Meded Dir Tuinb* **21**: 283-290.

Messeguer, R., Grand, M., de Vicente, M.C., Young, N.D. and Tanksley, S.T. 1991. High resolution RFLP map around root-knot nematode resistance gene (*Mi*) in tomato. *Theoretical and Applied Genetics* **82**: 529-536.

Midha, R.L. and Trivedi, P.C. 1991. Estimation of losses caused by *Meloidogyne incognita* on coriander, cumin and fennel. *Current Nematology* **2**: 159-162.

Midha, R.L., Trivedi, P.C. and Yadav, B.D. 2001. *Paecilomyces* spp. as biocontrol agents against *Meloidogyne incognita* on coriander. *Natl. Cong. on Centenary of Nematol. in India – Appraisal & Future Plans*, Indian Agri. Res. Inst., New Delhi, pp.146-147.

Midha, S.K. 1985. Efficacy of *Paecilomyces lilacinus* in controlling root-knot infestation on cowpea and mung bean. *Indian Journal of Nematology* **15**: 267.

Miller, P.M. 1975. Effect of the tobacco cyst nematode, *Heterodera tobaccum*, on severity of *Verticillium* and *Fusarium* wilts of tomato. *Phytopathology* **65**: 81-82.

Mishra and Shukla, B.N. 1997. Interaction between Glomus fasciculatum and Meloidogyne incognita on tomato. Journal of Mycology and Plant Pathology **27**: 199-202.

Mishra, S.D. and Prasad, S.K. 1977. Effect of neem cake water extract on root-knot nematode infecting tomato. *Indian Journal of Entomology* **39**: 228-231.

Mobin, M. and Khan, A.M. 1969. Effect of organic amendments on the population of rhizosphere fungi and nematodes around the roots of some fruit trees (Abstr.). *Proceedings of All India Nematology Symposium*, Indian Agri. Res. Inst., New Delhi, p. 68.

Mohandas, C. 2001. Nematode diseases of tuber crops and their management. *Natl. Cong. on Centenary of Nematol. in India – Appraisal & Future Plans*, Indian Agri. Res. Inst., New Delhi, pp. 35-36.

Mohandas, C. and Ramakrishnan, S. 2002. Grow sweet potato and manage root-knot nematode problems of vegetables (Abstr.). *International Conference on Vegetables*, Bangalore, p. 256.

Mohandas, C. and Ramana, K.V. 1991. Pathogenicity of *Meloidogyne incognita* and *Radopholus similis* on black pepper (*Piper nigrum* L.). *Journal of Plantation Crops* **19**: 41-43.

Mohandas, C., Sreeja, P., Nageswari, S. and Sheela, M.S. 1998. Identification of resistance in African yam against root-knot nematode. *Proc. of Natl. Symp. on Rational Approaches in Nematode Management for Sustainable Agriculture*, Nematol. Soc. of India, New Delhi, pp. 63-64.

Mohanty, K.C. and Mahapatra, S.N. 2001. Integrated management of root-knot nematode in banana. *Natl. Cong. on Centenary of Nematol. in India – Appraisal & Future Plans*, Indian Agri. Res. Inst., New Delhi, p.166.

Mohanty, K.C., Sahoo, N.K. and Ray, S. 1972. Occurrence of *Radopholus similis* (Cobb, 1893) Thorne, 1949 on banana and pepper in wide areas of Orissa, India. *Afro-Asian Nematology Network* **1**(1): 25-26.

Morgan, D.O. 1925. Investigations on eelworm in potatoes in South Lincolnshire. *Journal of Helminthology* **3**: 185-192.

Morgan-Jones, G. and Rodriguez-Kabana, R. 1985. Phytonematode Pathology: Fungal modes of action – A perspective. *Nematropica* **15**: 107-114.

Morgan-Jones, G., Goody, G. and Rodriguez-Kabana, R. 1981. *Verticillium chlamydosporium* fungal parasite of *Meloidogyne arenaria* females. *Nematropica* **11**: 115-119.

Morgan-Jones, G., Rodriguez-Kabana, R. and Tovar, J.G. 1984. Fungi associated with cysts of *Heterodera glycines* cysts in the Cacuca Valley, Columbia. *Nematropica* **14**: 173-177.

Morgan-Jones, G., White, J.F. and Rodriguez-Kabana, R. 1983. Phytonematode Pathology – Ultrastructural studies 1 – Parasitism of *Meloidogyne arenaria* eggs by *Verticillium chlamydosporium*. *Nematropica* **13**: 245-260.

Mortensten, J.A. 1985. Proceedings of Florida State Horticulture Society **98**: 166-169.

Mukhopadhyaya, M.C. and Dalal, M.R. 1971. Effect of two nematicides on *Tylenchulus semipenetrans* and on sweet lime yield. *Indian Journal of Nematology* **1**: 95-97.

Mukhopadhyaya, M.C. and Suryanarayana, D. 1969. Citrus decline in Haryana : Role of *Tylenchulus semipenetrans* and its control. *Indian Phytopathology* **22**: 495-497.

Muller, J. 1973. Zum Einfluss von Pratylenchus penetrans auf die Verticillium- Welke von Impatiens balsamina. Zeitschrift fur Pflanzenkrankheiten und Pflanzenschutz **80**: 295-311.

Munnecke, D.E., Chandler, P.A. and Starr, M.P. 1963. Hairy root (*Agrobacterium rhizogenes*) of field roses. *Phytopathology* **53**: 788-799.

Murant, A.F. and Taylor, C.E. 1965. Treatment of soil with chemicals to prevent transmission of tomato black ring and raspberry ring spot virus by *Longidorus elongatus*. *Annals of Applied Biology* **55**: 227-237.

Mustafa, U. and Khan, M.R. 2004. Management of disease complex caused by *Fusarium oxysporum* f. sp. *gladioli* and *Meloidogyne incognita* on gladiolus. *National Symposium on Paradigms in Nematological Research for Biodynamic Farming*, Univ. of Agri. Sci., Bangalore, p. 91-92.

Muthukrishnan, T.S., Rajendran, G. and Chandrasekaran, J. 1975. Pathogenicity of *Helicotylenchus dihystera* to chilli (*Capsicum annum*) and its host range. *Indian Journal of Nematology* **5**: 180-184.

Muthukrishnan, T.S., Vadivelu, S. and Rajendran, G. 1977. *Longidorus africanus* and its pathogenic effect on *Crossandra undulaefolia*. *Indian Journal of Nematology* **7**: 69-72.

Nagesh, M., Hussaini, S.S. and Gopinatha, K.V. 2001. Management of root-knot nematode, *Meloidogyne incognita* (Kofoid & White, 1919) Chitw. in chrysanthemum using formulations of *Paecilomyces lilacinus* in combination with neem cake. In *IPM in Horticultural Crops: Emerging Trends in the New*

Millenium (A. Verghese and P. Parvatha Reddy, eds.), pp. 149-150. Assocn. for Advancement of Pest Mangmt. in Hortil. Ecosystems, Bangalore.

Nagesh, M., Hussaini, S.S., Ramanujam, B. and Chidanandaswamy, B.S. 2006. Management of *Meloidogyne incognita* and *Fusarium oxysporum* f. sp. *lycopersici* wilt complex using antagonistic fungi in tomato. *Nematologia Mediterranea* **34**: 63-68.

Nagesh, M., Hussaini, S.S., Singh, S.P. and Biswas, S.R. 2003. Management of root-knot nematode, *Meloidogyne incognita* (Kofoid & White) Chitwood in chrysanthemum using *Paecilomyces lilacinus* (Thom.) Samson in combination with neem cake. *Journal of Biological Control* **17**: 125-131.

Nagesh, M. and Janakiram, T. 2004. Root-knot nematode problem in polyhouse roses and its management using dazomet, neem cake and *Pochonia chlamydosporia* (*Verticillium chlamydosporium*). *Journal of Ornamental Horticulture* **7**: 147-152.

Nagesh, M., Janakiram, T., Rao, T.M. and Parvatha Reddy, P. 1995. Evaluation of China aster varieties/pure lines for resistance to *Meloidogyne incognita* (Kofoid & White, 1919) Chitwood, 1949. *Pest Management in Horticultural Ecosystems* **1**: 101-104.

Nagesh, M., Meenakshi, S., Murthy, N. and Parvatha Reddy, P. 1995. Screening of some tuberose cultivars/hybrids for resistance against *Meloidogyne incognita* race-1. *Indian Perfumer* **39**: 138-140.

Nagesh, M. and Parvatha Reddy, P. 1995. Comparative efficacy of *Paecilomyces lilacinus* (Thom.) Samson and *Verticillium lecanii* (A. Zimmerman) Viegas in combination with botanicals against *Meloidogyne incognita* (Kofoid and White, 1919) Chitwood, 1949 infecting *Crossandra undulaefolia* L. *Journal of Biological Control* **9**: 109-112.

Nagesh, M. and Parvatha Reddy, P. 1996a. Management of *Meloidogyne incognita* on carnation and gerbera in commercial polyhouses. *Second Internatioanl Crop Science Congress*, New Delhi, p.349.

Nagesh, M. and Parvatha Reddy, P. 1996b. Eco-friendly management of *Meloidogyne incognita* on *Crossandra undulaefolia* using vesicular arbuscular mycorrhiza, *Glomus mosseae* and oil cakes. *Mycorrhiza News* **9**(1): 12-14.

Nagesh, M. and Parvatha Reddy, P. 1998. A modified method for mass multiplication of *Pasteuria penetrans* (Thorne) Sayre & Starr. *Journal of Biological Control* **12**: 63-66.

Nagesh, M. and Parvatha Reddy, P. 2000. Crop loss estimation in carnation and gerebera due to the root-knot nematode, *Meloidogyne incognita* (Kofoid & White) Chitwood. *Pest Management in Horticultural Ecosystems* **6**: 158-159.

Nagesh, M. and Parvatha Reddy, P. 2005. Management of carnation and gerbera to control the root-knot nematode, *Meloidogyne incognita*, in commercial polyhouses. *Nematologia Mediterranea* **33**: 157-162.

Nagesh, M., Parvatha Reddy, P. and Rama, N. 2001. Pathogenicity of selected antagonistic soil fungi on *Meloidogyne incognita* (Kofoid & White) eggs and egg masses under *in vitro* and *in vivo* conditions. *Journal of Biological Control* **15**: 63-68.

Nagesh, M., Parvatha Reddy, P. and Ramachandran, N. 1998. Integrated management of *Meloidogyne incognita* and *Fusarium oxysporum* f sp. *gladioli* in gladiolus using antagonistic fungi and neem cake. In *Nematology - Challenges & Opportunities in 21st Century* (Usha K. Mehta, ed.), pp. 263-266. Sugarcane Breeding Inst., Coimbatore.

Nagesh, M., Parvatha Reddy, P. and Rao, M.S. 1995. Management of Meloidogyne incognita on Crossandra undulaefolia using Paecilomyces lilacinus and Pasteuria penetrans formulations. National Symposium on Perspectives in Eco-friendly Approaches to Plant Protection. Indian Agri. Res. Inst., New Delhi, pp. 47-48.

Nagesh, M., Parvatha Reddy, P. and Rao, M.S. 1997. Integrated management of *Meloidogyne incognita* on tuberose using *Paecilomyces lilacinus* in combination with plant extracts. *Nematologia Mediterranea* **25**: 3-7.

Nagesh, M., Parvatha Reddy, P. and Rao, M.S. 1998. Comparative efficacy of VAM fungi in combination with neem cake against *Meloidogyne incognita* on *Crossandra undulaefolia*. *Mycorrhiza News* **11**: 11-13.

Nagesh, M., Parvatha Reddy, P., Vijaya Kumar, M. V. and Nagaraju, B. M. 1998. Management of *Meloidogyne incognita* on *Polianthes tuberosa* using parasitic fungus, *Paecilomyces lilacinus* in combination with oil cakes. In "*Advances in IPM for Horticultural Crops*" (P. Parvatha Reddy, N. K. Krishna Kumar and A. Verghese, eds.), pp. 339-344. Association for Advancement of Pest Management in Hortil. Ecosystems, Bangalore.

Nagesh, M., Parvatha Reddy, P., Vijaya Kumar, M. V. and Nagaraju, B. M. 1999. Studies on correlation between *Glomus fasciculatum* spore density, root colonization and *Meloidogyne incognita* infection on *Lycopersicon esculentum*. *Journal of Plant Disease and Protection* **106**: 82-87.

Naik, D. 2004. Biotechnological Approaches for the Management of Wilt Disease Complex in Capsicum (*Capsicum annuum* L.) and Egg plant (*Solanum melongena*) with Special Emphasis on Biological Control. Ph. D. thesis, Kuvempu Univ., Shimoga.

Nair, K.K.R. 1979. Studies on the chemical control of banana nematodes. *Agricultural Research Journal of Kerala* **17**: 232-235.

Nair, M.R.G.K., Das, N..M. and Menon, M.R. 1966. On the occurrence of the burrowing nematode *Radopholus similis* (Cobb, 1893) Thorne, 1949 on banana in Kerala. *Indian Journal of Entomology* **28**: 553-554.

Nakat, R.V., Acharya, A., Jonathan, E.I., Hazarika, K., Jha, S., Singh, U.S., Wasnikar, A.R., Subba Rao, D.V. and Sitaramaiah, K. 1995. Evaluation of *Paecilomyces lilacinus* for the control of root-knot nematodes *Meloidogyne incognita* on betel vine. *National Symp. on Nematode Problems of India – An Appraisal of the Nematode Mangmt. with Eco-friendly Approaches and Biocomponents*, Indian Agri. Res. Inst., New Delhi, p. 44.

Nanjappa, C.K., Panda, M. and Seshadri, A.R. 1978. *Third International Congress on Plant Pathology*, Germany.

Napiere, C.M. and Quinio, A.J. 1980. Influence of root-knot nematode on bacterial wilt severity in tomato. *Annals of Tropical Research* **2**: 29-39.

Nath, A., Sharma, N.K., Bhardwaj, S. and Thapa, C.N. 1982. Nematicidal property of garlic. *Nematologica* **28**: 253-255.

Nath, R., Khan, M.N., Wanshi, R.S.K. and Dwivedi, R.P. 1984. Influence of root-knot nematode, *Meloidogyne javanica* on pre and post-emergence damping off of tomato. *Indian Journal of Nematology* **14**: 135-140.

Nath, R.P., Sinha, B.K. and Haider, M.G. 1976. Studies on the nematodes of vegetables in Bihar-II. Combined effect of *Meloidogyne incognita* and *Ozonium texanum* var. *parasiticum* on germination of eggplant. *Indian Journal of Nematology* **6**: 177-179.

Nayak, G.M. 1986. *Studies on Papaya in Relation to Certain Plant Diseases.* M. Sc. Thesis, University of Agricultural Sciences, Bangalore, 106 pp.

Nayar, K., Mathew, J., Kuriyan, J., Gnanamanickam, S.S. and Mahadevan, A. (eds.). 1988. Role and association of root-knot nematode (*Meloidogyne incognita*) in the bacterial wilt of brinjal incited by *Pseudomonas solanacearum*. In *Advances in Research on Pant Pathogenic Bacteria*. Based on the proceedings of the *National Symposium on Phytobacteriology*, University of Madras, Madras, India, pp. 45-48.

Ndubizu, T.O.C. 1977. Effect of earthworms, nematodes, cultivation and host plants on *Verticillium* wilt of peach and cherry. *Annals of Applied Biology* **86**: 153-161.

Negron, J.A. and Acosta, N. 1989. *Fusarium oxysporum* f. sp. *coffeae* - *Meloidogyne incognita* complex in 'Bourbon' coffee. *Nematropica* **19**: 161-168.

Nemec, S. and Meridith, F.I. 1981. Amino acid content of leaves of mycorrhizal and non-mycorrhizal citrus rootstocks. *Annals of Botany* **47**: 351-358.

Netscher, C. 1983. Control of *Meloidogyne incognita* in vegetable production by crop rotation in Ivory Coast. *Acta Horticulture* **152**: 219-225.

Newhall, A.G. 1958. The incidence of Panama disease of banana in the presence of the root-knot and the burrowing nematodes (*Meloidogyne incognita* and *Radopholus similis*). *Plant Disease Reporter* **42**: 853-856.

Nidiry, E.S.J., Khan, R.M. and Parvatha Reddy, P. 1992. *In vitro* nematicidal activity of *Gloriosa superba* seed extract against *Meloidogyne incognita*. *Nematologia Mediterranea* **21**: 128-128.

Nigh, E.L. 1966. Incidence of crown gall infection in peach as affected by the Javanese root-knot nematode. *Phytopathology* **56**: 150.

Niknam, G.R. and Dhawan, S.C. 2001a. Induction of systemic resistance by *Bacillus subtilis* isolate Bs1 against *Rotylenchulus reniformis* in tomato. *Natl. Cong. on Centenary of Nematol. in India – Appraisal & Future Plans*, Indian Agri. Res. Inst., New Delhi, pp.143-144.

Niknam, G.R. and Dhawan, S.C. 2001b. Effect of seed bacterization, soil drench and bare root-dip application methods of *Pseudomonas fluorescens* isolate Pf1 on the suppression of *Rotylenchulus reniformis* infecting tomato. *Natl. Cong. on Centenary of Nematol. in India – Appraisal & Future Plans*, Indian Agri. Res. Inst., New Delhi, p. 144.

Nirmal Johnson, S.B. and Devarajan, K. 2004. Effect of *Pseudomonas fluorescens* on banana nematodes. *National Symposium on Paradigms in Nematological Research for Biodynamic Farming*, Univ. of Agri. Sci., Bangalore, p. 81-82.

Nirula, K.K. and Bassi, K.K. 1965. Thermotherapy for root-knot nematode, *Meloidogyne incognita* in potato tubers. *Indian Potato Journal* **7**: 9-11.

Nisha, M.S. and Sheela, M.S. 2004. Rhizome treatment – A low cost technology for the management of nematodes in Kacholam, an important medicinal plant. *National Symposium on Paradigms in Nematological Research for Biodynamic Farming*, Univ. of Agri. Sci., Bangalore, p. 62-63.

Nisha, M.S and Sheela, M.S. 2006. Bio-management of *Meloidogyne incognita* on Coleus, *Solenostemon rotundifolius* by integrating solarization, *Paecilomyces lilacinus, Bacillus macerans* and neem cake. *Indian Journal of Nematology* **36**: 136-138.

Nordmeyer, D. and Dickson, D.W. 1981. Effect of oximecarbamates and organophosphates and *one avermectin on the oxygen uptake of three Meloidogyne species. Journal of Nematology* **13**: 452-453.

O'Bannon, J.H. and Esser, R.P. 1975. Evaluation of citrus hybrids and relatives as hosts of *Pratylenchus coffeae* with comments on other hosts. *Nematologia Mediterranea* **3**: 113-122.

O'Bannon, J.H., Inserra, R.N., Nemec, S. and vovlas, N. 1979. The influence of *Glomus mosseae* on *Tylenchulus semipenetrans* infected and un infected *Citrus limon* seedlings. *Journal of Nematology* **11**: 247-250.

O'Bannon, J.H. and Nemec, S. 1979. the response of *Citrus limon* seedlings to a symbiont, *Glomus etunicatum* and a pathogen, *Radopholus similis. Journal of Nematology* **11**: 270-275.

Ohkawa, K. and Saigusa, T. 1981. Resistance of rose rootstocks to *Meloidogyne hapla, Pratylenchus penetrans* and *P. vulnus. Horticultural Science* **16**: 559-560.

Olivera, A.A.R. and Zambolin, L. 1987. Interaction between the endomycorrhizal fungus *Glomus etunicatum* and the gall nematode *Meloidogyne javanica* on beans with split roots. *Fitopatologia Brasileria* **12**: 222-225.

Oostenbrink, M. 1972. Evaluation and integration of nematode control methods. In *Economic Nematology* (J.M. Webster, ed.), pp. 497-514. academic Press, New York.

Oostendorp, M. and Sikora, R.A. 1989. Seed treatment with antagonistic rhizobacteria for the suppression of *Heterodera schachtii* early root infection of sugar beet. *Review of Nematology* **12**: 77-83.

Orion, D. and Krikum, J. 1976. Response of *Verticillium* resistant and *Verticillium* susceptible tomato varieties to inoculation with the nematode *Meloidogyne javanica* and with *Verticillium dahliae. Phytoparasitica* **4**: 41-44.

Osche, J.J. and Brewton, W.S. 1954. Preliminary report on *Crotalaria* versus nematodes. *Proceedings of Florida State Horticultural Society* **67**: 218-219.

Overman, A.J. and Jones, J.P. 1970. Effect of stunt and root-knot nematodes on *Verticillium* wilt of tomato. *Phytopathology* **60**: 1306.

Oyekan, P.O. and Mitchell, J.E. 1971. Effect of *Pratylenchus penetrans* on the resistance of pea variety to *Fusarium* wilt. *Plant Disease Reporter* **55**: 1032-1035.

Padhi, N.N. and Misra, R.P. 1987. Control of *Rotylenchulus reniformis* on French bean (*Phaseolus vulgaris* L.). *Indian Journal of Nematology* **17**: 130-131.

Palanisamy, S. and Sivakumar, C.V. 1981. Assessment of avoidable yield loss in cowpea, black gram, maize and finger millet. *Second Nematology Symposium*, Tamil Nadu Agricultural University, Coimbatore, p. 58.

Pandey, R. 1994a. Bionomics of phytonematodes in relation to medicinal and aromatic plants. *Indian Journal of Plant Pathology* **12**: 16-23.

Pandey, R. 1994b. Comparative performance of oil seed cakes and pesticides in management of root-knot nematode disease of davana. *Nematologia Mediterranea* **22**: 17-19.

Pandey, R. 1997. Ecofriendly management of Meloidogyne incognita on Hyoscyamus niger. Indian Journal of Nematology **27**: 175-178.

Pandey, R. 2002. Application of botanicals for management of root-knot nematode disease of *Ammi majus* L. *Indian Journal of Nematology* **32**: 198-200.

Pandey, R. 2003. Mint nematology – Current status and future needs. In *Advances in Nematology* (P.C. Trivedi, ed.), pp. 155-166.

Pandey, R. 2005a. Field application of bio-organics in the management of *Meloidogyne incognita* in *Mentha arvensis*. *Nematologia Mediterranea* **33**: 51-54.

Pandey, R. 2005b. Management of *Meloidogyne incognita* in *Artemisia pallens* with bioagents. *Phytoparasitica* **33**: 304-308.

Pandey, R., Gupta, M.L., Singh, H.N. and Kumar, S. 1999. The influence of vesicular arbuscular mycorrhizal fungi alone and IPM combination with *Meloidogyne incognita* on *Hyoscyamnus niger*. *Bioresource Technology* **69**: 275-278.

Pandey, R., Haseeb, A. and Hussain, A. 1992. Distribution, pathogenicity and management of *Meloidogyne incognita* on *Mentha arvensis* cv. MAS-1. *Afro Asian Journal of Nematology* **2**: 27-34.

Pandey, R. and Kalra, A. 2005a. Chemical activators: A novel and sustainable management approach for *Meloidogyne incognita* (Kofoid and White) Chitwood in *Chamomile recutita* L. *Archives of Phytopathology and Plant Protection* **38**: 107-111.

Pandey, R. and Kalra, A. 2005b. Root-knot disease of ashwagandha *Withania somnifera* and its eco-friendly cost effective management. *Journal of Mycology and Plant Pathology* **33**: 240-245.

Pandey, R., Kalra, A.,Gupta, M.L and Sharma, P. 2003. Phytonematodes: Major pest of MAPs. In *Proceedings of First National Interactive Meet on Medicinal and Aromatic Plants* (A.K. Mathur, S. Dwivedi, D.D. Patra, G.D. Bagchi, N.S. Sangwan, A. Sharma and S.P.S. Khanuja, eds.), pp. 188-197. CIMAP, Lucknow, India.

Pandey, R., Kalra, A., Singh, H.N., Gupta, M.L., Singh, H.B. and Kumar, S. 2000. Suppressive response of bioagent, organic material and pesticide on population development of *Meloidogyne incognita* on *Matricartia chamomilla*. *Journal of Spices and Aromatic Plant Sciences* **22**: 649-651.

Pandey, R. and Patra, N.K. 2001. Screening mint (*Mentha* spp.) accessions against root-knot nematode infection. *Journal of Spices and Aromatic Crops* **10**: 55-56.

Pandey, R., Singh, H.B., Gupta, M.L., Singh, H.N., Sharma, S. and Kumar, S. 1998. Efficacy of VAM fungi, green manure, organic materials and pesticides in the management of root-knot nematode in menthol mint. *Proc. of National Symp. on Rational Approaches in Nematode Mangmt. for Sustainable Agri* (S.C. Dhawan and K.K. Kaushal, eds.), pp. 47-50. Gujarat Agri. Univ., Anand.

Pandey, R., Singh, H.N. and Kumar, S. 1999. Pathogenicity of *Meloidogyne incognita* (Kofoid & White) Chitwood on *Matricartia chamomilla* L. *Journal of Spices and Aromatic Crops* **8**: 201-203.

Pandey, R.C. and Dwivedi, B.K. 2001. Study on the effect of biocontrol agents against root-knot disease of brinjal. *Current Nematology* **12**: 73-74.

Pandey, R.C. and Trivedi, P.C. 1992. Biological control of Meloidogyne incognita by Paecilomyces lilacinus in Capsicum annum. Indian Phytopathology **45**: 134-135.

Pani, A.K. and Das, S.N. 1972. Studies on etiological complexes in plant disease. I. Association of root-knot nematodes in bacterial wilt of tomato. *Journal of Research Orissa University of Agriculture & Technology* **2**: 54-59.

Paruthi, L.J., Jain, R.K. and Gupta, D.C. 1987. Effect of different periods of degradation of subabool leaves alone and in combination with nematicides in okra. *Indian Journal of Nematology* **17**: 30-32.

Parvatha Reddy, P. 1981. Pathogenicity of *Meloidogyne incognita* to passion fruit. *Indian Journal of Nematology* **11**: 80-81.

Parvatha Reddy, P. 1983. Reaction of selected citrus rootstocks to the citrus nematode. *Proceedings of International Citrus Symposium*, Bangalore, pp. 379-381.

Parvatha Reddy, P. 1984. Efficacy of seed treatment with three nematicides for the control of *Meloidogyne incognita* infecting cowpea, French bean and peas. *Indian Journal of Nematology* **14**: 39-40.

Parvatha Reddy, P. 1985a. Chemical control of root-knot nematodes infecting peas. *Indian Journal of Nematology* **15**: 120-121.

Parvatha Reddy, P., Iyer, C.P.A. and Subramanyam, M.D. 1988. Evaluation of papaya cultivars and hybrids against *Meloidogyne incognita. Indian Journal of Nematology* **18**: 381-382.

Parvatha Reddy, P. 1985b. Estimation of crop loss in peas due to *Meloidogyne incognita. Indian Journal of Nematology* **15**: 226.

Parvatha Reddy, P. and Agarwal, P.K. 1987. Resistance in citrus rootstocks to the citrus nematode, *Tylenchulus semipenetrans. Indian Journal of Horticulture* **44**: 111-114.

Parvatha Reddy, P., Govindu, H.C. and Setty, K.G.H. 1975a. Studies on the effect of amino acids on the root-knot nematode, *Meloidogyne incognita* infecting tomato. *Indian Journal of Nematology* **5**: 36-41.

Parvatha Reddy, P., Govindu, H.C. and Setty, K.G.H. 1975b. Studies on the action of DL-methionine on *Meloidogyne incognita* infecting tomato. *Indian Journal of Nematology* **5**: 42-48.

Parvatha Reddy, P. and Khan, R. M. 1988. Evaluation of *Paecilomyces lilacinus* for the biological control of *Rotylenchulus reniformis* infecting tomato as compared with carbofuran. *Nematologia Mediterranea* **16**: 113-116.

Parvatha Reddy, P. and Khan, R.M. 1989. Evaluation of biocontrol agent *Paecilomyces lilacinus* and carbofuran for the management of *Rotylenchulus renfiromis* infecting brinjal. *Pakistan Journal of Nematology* **7**: 55-59.

Parvatha Reddy, P. and Khan, R.M. 1991. Integrated management of root-knot nematodes infecting okra. *Current Nematology* **2**:115-116.

Parvatha Reddy, P., Khan, R.M. and Agarwal, P.K. 1987. Selection of citrus rootstocks and hybrids resistant to the citrus nematode, *Tylenchulus semipenetrans*. *Pakistan Journal of Nematology* **5**(2): 69-72.

Parvatha Reddy, P., Khan, R.M. and Agarwal, P.K. 1989. Screening of banana germplasm against the burrowing nematode, *Radopholus similis*. *Indian Journal of Horticulture* **46**: 276- 278.

Parvatha Reddy, P., Khan, R.M. and Rao, M.S. 1991a. Integrated management of the citrus nematode, *Tylenchulus semipenetrans* using oil cakes and *Paecilomyces lilacinus*. *Afro-Asian Journal of Nematology* **1**: 221-222.

Parvatha Reddy, P., Khan, R.M. and Rao, M.S. 1991b. Evaluation of chemicals for the management of the burrowing nematode, *Radopholus similis* infecting banana. *Current Nematology* **2**: 1-2.

Parvatha Reddy, P., Khan, R.M. and Rao, M.S. 1993. Integrated management of the citrus nematode, *Tylenchulus semipenetrans* using neem cake and the parasitic fungus, *Paecilomyces lilacinus*. *Proceedings of World Neem Conference*, Bangalore, pp. 647-650.

Parvatha Reddy, P. and Nagesh, M. 1995. Nematode problems of ornamental plants in India. In *Nematode Pest Management – An Appraisal of Eco-friendly Approaches* (G. Swarup, D.R. Dasgupta and J.S. Gill, eds.), pp. 244-251. Nematol. Soc. of India, New Delhi.

Parvatha Reddy, P. and Nagesh, M. 2000. Integrated management of the citrus nematode using bacterial (*Pasteuria penetrans)* and fungal (*Paecilomyces lilacinus)* biocontrol agents. *Proceedings of International Symposium on Citriculture*, Nagpur, pp. 825-829.

Parvatha Reddy, P. and Nagesh, M. 2002. *Avermectins – Isolation, Fermentation, Preliminary Charecterization and Screening for Nematicidal Activity*. Tech. Bull.17, Indian Inst. of Hortil. Research, Bangalore, 28 pp.

Parvatha Reddy, P., Nagesh, M. and Devappa, V. 1997. Effect of integration of *Pasteuria penetrans, Paecilomyces lilacinus* and neem cake for the management of root-knot nematodes infecting tomato. *Pest Management in Horticultural. Ecosystems* **3**: 100-104.

Parvatha Reddy, P., Nagesh, M. and Devappa, V. 1998. Management of *Meloidogyne incognita* on tomato by integration of *Trichoderma harzianum, Glomus fasciculatum* and neem cake. In "*Advances in IPM for Horticultural Crops*" (P. Parvatha Reddy, N. K. Krishna Kumar & A. Verghese, eds.), pp. 349-352. Association for Advancement of Pest Management in Hortil. Ecosystems, Bangalore.

Parvatha Reddy, P., Nagesh, M., Devappa, V. and Vijaya Kumar, M. V. 1998. Management of *Meloidogyne incognita* on tomato by integrating endomycorrhiza, *Glomus mosseae* with oil cakes under nursery and field conditions. *Journal of Plant Disease and Protection* **105**: 53-57.

Parvatha Reddy, P., Nagesh, M., Rao, M.S. and Devappa, V. 1997. Integrated management of the burrowing nematode *Radopholus similis* using endomycorrhiza, *Glomus mosseae* and oil cakes. *Pest Management in Horticultural Ecosystems* **3**: 25-30.

Parvatha Reddy, P., Nagesh, M., Rao, M.S. and Rama, N. 2000. Management of *Tylenchulus semipenetrans* by integration of *Pseudomonas fluorescens* with oil cakes. *Proc. International Symposium on Citriculture*, Nagpur, pp. 830-833.

Parvatha Reddy, P., Rangaswamy, S.D. and Nagesh, M. 2000. Integrated management of *Meloidogyne incognita* (Kofoid & White) Chitwood on tomato in nursery and main field. *Pest Mangment in Horticultural Ecosystems* **6**: 47-49.

Parvatha Reddy, P., Rao, M.S., Mohandas, S. and Nagesh, M. 1995. Integrated management of the citrus nematode, *Tylenchulus semipenetrans* Cobb using VA mycorrhiza, *Glomus fasciculatum* (Thaxt.) Gerd & Trappe and oil cakes. *Pest Management in Horticultural Ecosystems* **1**: 37-41.

Parvatha Reddy, P., Rao, M.S. and Nagesh, M. 1996a. Management of the citrus nematode, *Tylenchulus semipenetrans* by integration of *Trichoderma harzianum* with oil cakes. *Nematologia Mediterranea* **24**: 265-267.

Parvatha Reddy, P., Rao, M.S. and Nagesh, M. 1996b. Management of the citrus nematode on acid lime by integration of parasitic fungi and oil cakes. *Pest Management in Horticultural Ecosystems* **2**: 15-18.

Parvatha Reddy, P., Rao, M.S. and Nagesh, M. 1996c. Integrated management of the citrus nematode, *Tylenchulus semipenetrans* using pesticides and parasitic fungus, *Paecilomyces lilacinus*. *Pest Management in Horticultural Ecosystems* **2**: 61-63.

Parvatha Reddy, P., Rao, M.S. and Nagesh, M. 1996d. Crop loss estimation in banana due to the burrowing nematode, *Radopholus similis*. *Pest Management in Horticultural Ecosystems* **2**: 85-86.

Parvatha Reddy, P., Rao, M.S. and Nagesh, M. 1998. Effect of bare-root dip treatment of tomato seedlings in plant leaf extracts mixed with *Paecilomyces lilacinus* spores for the management of *Meloidogyne incognita*. In "*Advances in IPM for Horticultural Crops*" (P. Parvatha Reddy, N. K. Krishna Kumar and A. Verghese, eds.), pp. 334-338. Association for Advancement of Pest Management in Hortil. Ecosystems, Bangalore.

Parvatha Reddy, P., Rao, M.S. and Nagesh, M. 1999. Eco-friendly management of *Meloidogyne incognita* on tomato by integration of *Verticillium chlamydosporium* with neem and calotropis leaves. *Journal of Plant Disease and Protection* **106**: 530-533.

Parvatha Reddy, P., Rao, M.S. and Nagesh, M. 2002. Integrated management of burrowing nematode (*Radopholus similis*) using endomycorrhiza (*Glomus mosseae*) and oil cakes. In *Banana – Improvement, Production and Utilization* (H.P. Singh and K.L. Chadha, eds.), pp. 344-348. AIPUB, Trichy.

Parvatha Reddy, P. and Singh, D.B. 1978. Evaluating the reaction of some species and varieties of *Citrus* and *Poncirus* to the citrus nematode. *Indian Journal of Nematology* **8**: 82-84.

Parvatha Reddy, P. and Singh, D.B. 1979. Control of root-knot nematode, *Meloidogyne incognita* on tomato by chemical bare-root dips. *Indian Journal of Nematology* **9**: 40-43.

Parvatha Reddy, P. and Singh, D.B. 1980. Plant parasitic nematodes associated with banana in South India. *Proceedings of National Seminar on Banana Production Technology*, Tamil Nadu Agricultural University, Coimbatore, pp. 161-163.

Parvatha Reddy, P. and Singh, D.B. 1981. Assessment of avoidable yield loss in okra, brinjal, French bean and cowpea due to root-knot nematodes. *Third International Symposium on Plant Pathology*, New Delhi, pp.93-94.

Parvatha Reddy, P. and Singh, D.B. 1984. Reaction of selected cultivars/rootstocks of grapevine to *Meloidogyne incognita. Indian Journal of Nematology* **14**: 73-74.

Parvatha Reddy, P., Singh, D.B. and Ram Kishun. 1979. Effect of root-knot nematodes on the susceptibility of Pusa Purple Cluster brinjal to bacterial wilt. *Current Science* **48**: 915-916.

Parvatha Reddy, P., Singh, D.B. and Rao, V.R. 1979. The root-knot nematode on Gladiolus from India. *Current Science* **48**: 82.

Parvatha Reddy, P., Singh, D.B. and Sharma, S.R. 1979. Interaction of *Meloidogyne incognita* and *Rhizoctonia solani* in a root rot disease complex of French bean. *Indian Phytopathology* **32**: 651-652.

Pathak, K.N. and Keshari, N. 2004. Interaction of *Meloidogyne incognita* with *Fusarium oxysporum* f. *conglutinans* on cauliflower. *Indian Journal of Nematology* **34**: 85-87.

Patel, A.D., Ghelani, Y.H. and Patel, D.J. 2004. Assessment of avoidable yield losses due to reniform nematode, *Rotylenchulus reniformis* in cowpea. *National Symposium on Paradigms in Nematological Research for Biodynamic Farming,* Univ. of Agri. Sci., Bangalore, p. 33.

Patel, B.A., Patel, R.G., Patel, H.R., Vyas, R.V. and Patel, B.N. 2001. Management of root-knot nematodes in turmeric. *Natl. Cong. on Centenary of Nematol. in India – Appraisal & Future Plans*, Indian Agri. Res. Inst., New Delhi, p.165.

Patel, G.J., Shah, H.M. and Patel, D.J. 1979. Reaction of some tomato cultivars to root-knot nematode. *Indian Journal of Nematology* **9**: 60-61.

Patel, H.V., Patel, D.J. and Patel, B.A. 1989. Reaction of papaya varieties to *Rotylenchulus reniformis. International Nematology Network Newsletter* **6** (3): 24.

Patel, R.M. and Desai, M.V. 1964. A possible biological control of root-knot. *Plant Disease Reporter* **48**: 167-178.

Patel, S.K. Patel, H.V. and Patel, A.D. 2006. Integrated management of root-knot nematode in tomato nursery through botanical plant materials. *Indian Journal of Nematology* **36**: 307-308.

Patnaik, P.R., Mohanty, K.C., Mahapatra, S.N. and Sahoo, S. 2004. Evaluation of chilli varieties against root-knot nematode, *Meloidogyne incognita. National Symposium on Paradigms in Nematological Research for Biodynamic Farming,* Univ. of Agri. Sci., Bangalore, p. 100.

Perveen, S., Ehteshamul-Haque, S. and Ghaffar, A. 1998. Efficacy of *Pseudomonas aeruginosa* and *Paecilomyces lilacinus* in the control of root rot-root knot disease complex on some vegetables. *Nematologia Mediterranea* **26**: 209-212.

Perwez, M.S., Rehman, M.F. and Haider, S.R. 1988. Effect of *Tagetes erecta* on *Meloidogyne javanica* infecting lettuce. *International Nematology Network Newsletter* **5**(3): 18-19.

Pinochet, J., Calvet, C., Camprubi, A. and Fernandez, C. 1995. Growth and nutritional response of Nemared peach rootstock infected with *Pratylenchus vulnus* and the mycorrhizal fungus, *Glomus mosseae*.

Pinochet, J. and Rowe, P.R. 1979. Progress in breeding for resistance to *Radopholus similis* in banana. *Nematropica* **9**: 76-78.

Pireda, O., Bonierbale, M.W., Plaisted, R.L., Brodie, B.B. and Tanksley, S.D. 1993. Identification of RFLP markers linked to the H1 gene conferring resistance to the potato cyst nematode (*Globodera rostochiensis*). *Genome* **36**: 152-156.

Pitcher, R.S. and Crosse, J.E. 1958. Studies on the relationship of eelworms and bacteria to certain plant diseases II. Further analysis of the strawberry cauliflower disease complex. *Nematologica* 3: 244-256.

Poornima, K., Kumar, N. and Balamohan, T.N. 2004. Biological control of lesion nematode in banana. *National Symposium on Paradigms in Nematological Research for Biodynamic Farming,* Univ. of Agri. Sci., Bangalore, p. 99.

Poornima, K. and Vadivelu, S. 1998. Pathogenicity of *Meloidogyne incognita* to turmeric (*Curcuma longa* L.). In *Nematology-Challenges and Opportunities in 21st Century* (Usha K. Mehta, ed.), pp. 29-33. Sugarcane Breeding Institute, Coimbatore.

Poucher, C., Ford, H.W., Suit, R.F. and Du Charme, E.P. 1967. *Burrowing Nematode in Citrus*. Fla. Dept. Agri., Divn. of Plant Ind. Bull. No. 7, 63 pp.

Pradhan, P., Das, S.N. and Dora, D.K. 1989. Effect of some organic oils on the infectivity of root-knot nematode (*Meloidogyne incognita*) affecting tomato. *Indian Journal of Nematology* **19**: 162-165.

Prakash Babu, N. 2000. *Role of VAM in the Management of* Radopholus similis *on Banana*. M. Sc. (Agri.) thesis, University of Agricultural Sciences, Bangalore, 89 pp.

Prakash Babu, N., Kantharaju, V., Krishnappa, K. and Reddy, B.M.R. 2002. Biodiversity in indigenous isolates of vesicular arbuscular mycorrhiza, *Glomus fasciculatum* against *Radopholus similis* on banana (Abstr.). *Symposium on Biodiversity and Management of Nematodes in Cropping Systems for Sustainable Agriculture,* Jaipur, p. 108.

Prakash Babu, N., Reddy, B.M.R. and Krishnappa, K. and 2001. Role of VAM in the management of *Radopholus similis* on banana (Abstr.). *National Conference on Centenary of Nematology in India-Appraisal and Future Plans*, Indian Agri. Res. Inst., New Delhi, p. 149.

Pramila, T. and Goswami, B.K. 1992. Effect of VAM on the disease due to *Macrophomina phaseolina* and *Meloidogyne incognita* in cowpea. *Annals of Agricultural Research* **13**: 253-256.

Prasad, D., Nagia, D.K., Kumar, S. and Saini, M.L. 1994. Integrated approach for management of plant parasitic nematodes in brinjal, *Solanum melongena* L. *Plant Protection Bulletin* **46**: 31-33.

Prasad, K.S.K. 1989. Nematological problems and progress of research on potato in India. *Fourth Group Meeting on Nematological Problems of Plantation Crops,* University of Agricultural Sciences, Bangalore.

Prasad, K.S.K. and Krishnappa, K. 1981. Effect of bare-root dip treatment with pesticides on *Rotylenchulus reniformis* affecting brinjal. *Indian Journal of Nematology* **11**:154-158.

Prasad, K.S.K., Raj, D. and Sharma, R.K. 1983. Effect of planting time on *Meloidogyne incognita* infesting potato at Shimla Hills. *Third Nematology Symposium*, Himachal Pradesh Agricultural University, Solan, pp. 5-6.

Prasad, K.S.K., Siddaramaiah, A.L. and Hegde, R.K. 1980. Notes on the association of *Corticium rolfsii* and *Meloidogyne incognita* in the wilt complex of *Solanum khasianum*. *Indian Journal of Agricultural Sciences* **50**: 454.

Prasad, M. B. N. V., Parvatha Reddy, P., Rao, M. S. and Rekha, A. 1998. Evaluation of some intergeneric citrus rootstock hybrids for resistance to the citrus nematode *Tylenchulus semipenetrans*. In "*Advances in IPM for Horticultural Crops*" (P. Parvatha Reddy, N. K. Krishna Kumar and A. Verghese, eds.), pp. 328-331. Association for Advancement of Pest Management in Hortil. Ecosystems, Bangalore.

Prasad, M.B.N.V., Sawant, S. D., Parvatha Reddy, P., Palaniappan, R., Srinivasa Rao, N. K. and Rekha, A. 1999. Citrus rootstock hybrids resistant to *Phytophthora* and citrus nematodes and also tolerant to salinity and draught. *Proceedings of National Symposium on Citriculture*, Nagpur, pp. 59-62.

Prasad, P.R.K. and Reddy, D.D.R. 1984. Pathogenicity and analysis of crop losses on patchouli due to *Meloidogyne incognita*. *Indian Journal of Nematology* **14**: 36-38.

Prasad, S.K. and Dasgupta, D.R. 1964. Plant parasitic nematodes associated with roses in Delhi. *Indian Journal of Entomology* **26**: 12-122.

Putter, I., Mac Connell, J.G., Preiser, F.A., Ristich, A.A. and Dybas, R.A. 1981. Avermectins: Novel insecticides, acaricides and nematicides from a soil microorganism. *Experientia* **37**: 963-964.

Raaijmakers, J.M. and Weller, D.M. 1998. Natural plant protection by 2,4-diacetylphloroglucinol-producing *Pseudomonas* spp. in take-all decline soils. *Molecular Plant Microbe Interactions* **11**: 144-152.

Raaijmakers, J.M., Weller, D.M., and Thomashow, L.S. 1997. Frequency of antibiotic producing *Pseudomonas* spp. in natural environments. *Applied and Environmental Microbiology* **63**: 881-887.

Racke, J. and Sikora, R.A. 1992. Isolation, formulation and antagonistic activity of rhizosphere bacteria toward the potato cyst nematode, *Globodera pallida*. *Soil Biology and Biochemistry* **24**: 521-526.

Raj, B.T. and Nirula, K.K. 1969. Trials on the eradication of root-knot nematodes from potato fields with DD. *All India Nematol. Symp.*, Indian Agri. Res. Inst., New Delhi, India.

Raj, D. and Gill, J.S. 1983. Reaction of some potato varieties to root-knot nematode *Meloidogyne incognita*. *Third Nematology Symposium*, Himachal Pradesh Agricultural University, Solan, pp. 27-28.

Rajagopalan, P. and Naganathan, T.G. 1977a. Studies on nematode parasites of grapevine. *Tamil Nadu Agricultural University Annual Report* **6**: 129.

Rajagopalan, P. and Naganathan, T.G. 1977b. Studies on nematode parasites of banana. *Tamil Nadu Agricultural University Annual Report* **6**: 131.

Rajendran, B. and Rajendran, G. 1979. Report of Meloidogyne incognita in Jasminum flexile. South Indian Horticulture **27**: 70.

Rajendran, G. and Naganathan, T.G. 1981. Evaluation of systemic chemicals for the control of *Rotylenchulus reniformis* on papaya. *Indian Journal of Nematology* **11**: 130.

Rajendran, G., Naganathan, T.G. and Sivagami, V. 1979. Studies on banana nematodes. *Indian Journal of Nematology* **9**: 54.

Rajendran, G., Vadivelu, S. and Muthukrishnan, T.S. 1976. Pathogenicity of *Meloidogyne incognita* on crossandra. *Indian Journal of Nematology* **6**: 115-116.

Ram, B. and Baheti, B.L. 2004. Management of reniform nematode, *Rotylenchulus reniformis* on cowpea through seed and soil treatment with plant products. *Indian Journal of Nematology* **34**: 193-195.

Ram, S. and Gupta, D.C. 2001. Studies on the effect of intercropping of onion on the growth of tomato and population dynamics of root-knot nematode, *Meloidogyne javanica*. *Natl. Cong. on Centenary of Nematol. in India – Appraisal & Future Plans*, Indian Agri. Res. Inst., New Delhi, p. 94.

Ramakrishnan, C. and Balasubramanian, P. 1981. Control of *Meloidogyne incognita* in chilli *(Capsicum annum* L.) nursery with nematicides. *Second Nematology Symposium*, Tamil Nadu Agricultural University, Coimbatore, p. 54.

Ramana, K.V., Eapen, S.J. and Sarma, Y.R. 1998. Effect of *Meloidogyne incognita, Pythium aphanidermatum* and *Ralstonia solanacearum* alone and in combinations in ginger. In *Nematology-Challenges and Opportunities in 21st Century* (Usha K. Mehta, ed.), pp. 87-93. Sugarcane Breeding Institute, Coimbatore.

Ramana, K.V. and Mohandas, C. 1988. Occurrence of potato cyst nematode, *Globodera pallida* (Stone, 1973) in Kerala. *Indian Journal of Nematology* **18**: 141.

Ramappa, H.K. 1988. *Studies on Nematode Parasites of French Bean and Their Control*. M. Sc. (Agri.) thesis, Univ. of Agri. Sci., Bangalore.

Rameshchand and Gill, J.S. 2002. Evaluation of various application methods of *Pasteuria penetrans* against *Meloidogyne incognita* in tomato. *Indian Journal of Nematology* **32**: 23-26.

Rangswamy, S.D., Parvatha Reddy, P. and Nagesh, M. 2000. Evaluation of biocontrol agents *(Pasteuria penetrans* and *Trichoderma viride)* and botanicals for the management of root-knot nematodes, *Meloidogyne incognita* infecting tomato. *Pest Management in Horticultural Ecosystems* **6**: 135-138.

Rangaswamy, S.D., Parvatha Reddy, P., Nagesh, M. and Nanje Gowda, D.N. 1999. Integrated management of *Meloidogyne incognita* using *Pasteuria penetrans, Trichoderma viride* and oil cakes in tomato. *Pest Management in Horticultural Ecosystems* **5**: 122-126.

Rangaswamy, S.D., Parvatha Reddy, P. and Nanje Gowda, D.N. 1999. Management of root-knot nematode, *Meloidogyne incognita* in tomato by intercropping with marigold and mustard. *Pest Management in Horticultural Ecosystems* **5**: 118-121.

Rangaswamy, S.D., Parvatha Reddy, P., Nanje Gowda, D.N. and Narayana Swamy, BC. 1999. Management of *Meloidogyne incognita* in tomato by growing trap/antagonistic crops in rotation. *Current Nematology* **6**: 9-12.

Rankin, D.J., Hewitt, W.B. and Schmitt, R.V. 1971. Controlling fan leaf virus-dagger nematode disease complex in vineyards by soil fumigation. *California Agriculture* **25**: 11-14.

Rao, A.B.V.K., Padhi, N.N. and Acharya, A. 1987. Effect of different nematicides, oilcakes and urea in the control of *Rotylenchulus reniformis* on okra. *Indian Journal of Nematology* **17**: 171-173.

Rao, M.S., Gowen, S.R., Pembroke, B. and Parvatha Reddy, P. 1997. Relationship of *Pasteuria penetrans* spore encumbrance on juveniles of *Meloidogyne incognita* and their infection in adults. *Nematologia Mediterranea* **25**: 129-131.

Rao, M.S., Kerry, B.R., Gowen, S.R., Bourne, J.M. and Parvatha Reddy, P. 1997. Management of *Meloidogyne incognita* in tomato nurseries by integration of *Glomus deserticola* with *Verticillium chlamydosporium*. *Journal of Plant Disease and Protection* **104**: 419-422.

Rao, M.S., Mohandas, S. and Parvatha Reddy, P. 1993. Management of reniform nematode on eggplant by integrating botanical (castor cake), mycorrhiza (*Glomus fasciculatum*) and bioagent (*Paecilomyces lilacinus*). *Golden Jubilee Symposium on Horticultural Research - Changing Scenario,* Bangalore, pp. 259-260.

Rao, M.S., Mohandas, S., Parvatha Reddy, P. and Khan, R.M. 1993. Synergistic effect of endomycorrhizal fungus (*Glomus fasciculatum*) and biocontrol fungus (*Paecilomyces lilacinus*) in the management of root-knot nematode infecting tomato. In "*Soil Organisms and Sustainability*" (D. Rajagopal, R. D. Kale and K. Bano, eds.), pp. 119-123. Indian Society of Soil Biology and Ecology, Univ. of Agri. Sci., Bangalore.

Rao, M.S. and Naik, D. 2003. Effect of Trichoderma harzianum and Paecilomyces lilacinus on Meloidogyne incognita on papaya (*Carica papaya* L.) nursery seedlings. Pest Management in Horticultural Ecosystems **9**: 155-160.

Rao, M.S., Naik, D., Shylaja, M. and Parvatha Reddy, P. 2002. Prospects for the management of the root-knot nematode and bacterial wilt disease complex in capsicum by integration of fungal and bacterial biocontrol agents. *Proceedings of International Conference on Vegetables,* Bangalore, pp. 347-351.

Rao, M.S., Naik, D., Shylaja, M. and Reddy, P.P. 2003. Management of *Meloidogyne incognita* on eggplant by integrating endomycorrhiza, *Glomus fasciculatum* with bio-agent, *Verticillium chlamydosporium* under field conditions. *Indian Journal of Nematology* **33**: 29-32.

Rao, M.S. and Parvatha Reddy, P. 1992a. Studies on the comparative efficacy of certain plant leaves and carbofuran in the management of *Meloidogyne incognita* on tomato. *Current Nematology* **3**: 5-6.

Rao, M.S. and Parvatha Reddy, P. 1992b. Prospects of management of root-knot nematode on tomato through the integration of biocontrol agents and botanicals. *Seminar on Current Trends in Diagnosis and Management of Plant Diseases,* Bangalore, p.10.

Rao, M.S. and Parvatha Reddy, P. 1992c. Innovative approaches of utilization of biocontrol agents-*Paecilomyces lilacinus* and *Verticillium chlamydosporium* against root-knot nematode on tomato (Abstr.). *First Afro-Asian Nematology Symposium,* Aligarh Muslim University, Aligarh, pp. 25-26.

Rao, M.S. and Parvatha Reddy, P. 1993a. Effective use of 'Welgro' and 'RD-9 Repelin' for the management of root-knot nematode on tomato. In *"Neem and Environment"* (R. P. Singh, M. S. Chari, A. K. Raheja and W. Kraus, eds.), **2**: 651-656. Oxford & IBH Publishing Co. Pvt. Ltd., New Delhi.

Rao, M. S. and Parvatha Reddy, P. 1993b. Interactive effect of *Verticillium chlamydosporium* and castor cake on the control of root-knot nematode on eggplant. In *"Soil Organisms and Sustainability"* (D. Rajagopal, R.D. Kale and K. Bano, eds.), pp. 114-118. Indian Society of Soil Biology and Ecology, Univ. of Agri. Sci., Bangalore.

Rao, M.S. and Parvatha Reddy, P. 1994. A method for conveying *Paecilomyces lilacinus* to soil for the management of root-knot nematodes on eggplant. *Nematologia Mediterranea* **22**: 265-267.

Rao, M.S. and Parvatha Reddy, P. 2001. Control of *Meloidogyne incognita* on eggplant using *Glomus mosseae* integrated with *Paecilomyces lilacinus* and neem cake. *Nematologia Mediterranea* **29**: 153-157.

Rao, M.S., Parvatha Reddy, P. and Mohandas, S. 1994. Studies on integration of *Glomus fasciculatum* (endomycorrhiza) with castor cake (botanical) for the management of *Meloidogyne incognita* on tomato. *Proc. 2nd Afro-Asian Nematol. Symp.*, pp. 54-58.

Rao, M.S., Parvatha Reddy, P. and Mohandas, S. 1995. Studies on development of a biorational management strategy against root-knot nematode attacking tomato. *National Symposium on Nematode Problems of India – An Appraisal of the Nematode Management with Eco-friendly Approaches* (G. Swarup, D.R. Dasgupta and J.S. Gill, eds.), p. 14. Nematol. Soc. of India, New Delhi.

Rao, M.S., Parvatha Reddy, P. and Mohandas, S. 1996. Effect of integration of *Calotropis procera* leaf and *Glomus fasciculatum* on the management of *Meloidogyne incognita* infesting tomato. *Nematologia Mediterranea* **24**: 59-61.

Rao, M.S., Parvatha Reddy, P. and Nagesh, M. 1994. Use of biotic agent, *Paecilomyces lilacinus* along with seed treatment of okra with neem cake suspension for the management of root-knot nematode under field conditions. *National Symposium on Emerging Trends in Pest Management*, Y.S. Paramar Univ. of Hort. and Forestry, Nauni, Himachal Pradesh.

Rao, M.S., Parvatha Reddy, P. and Nagesh, M. 1997a. Integrated management of *Meloidogyne incognita* on okra by castor cake suspension and *Paecilomyces lilacinus*. *Nematologia Mediterranea* **25**: 17-19.

Rao, M.S., Parvatha Reddy, P. and Nagesh, M. 1997b. Integration of *Paecilomyces lilacinus* with neem leaf suspension for the management of root-knot nematodes on eggplant. *Nematologia Mediterranea* **25**: 249-252.

Rao, M.S., Parvatha Reddy, P. and Nagesh, M. 1997c. Management of root-knot nematode, *Meloidogyne incognita* on tomato by integration of *Trichoderma harzianum* with neem cake. *Journal of Plant Disease and Protection* **104**: 423-425.

Rao, M.S., Parvatha Reddy, P. and Nagesh, M. 1998a. Integrated management of Meloidogyne incognita on tomato using Verticillium chlamydosporium and Pasteuria penetrans. Pest Management in Horticultural Ecosystems **4**: 32-35.

Rao, M.S., Parvatha Reddy, P. and Nagesh, M. 1998b. Integrated management of *Meloidogyne incognita* on tomato using oil cakes and a bioagent, *Verticillium chlamydosporium*. In "*Advances in IPM for Horticultural Crops*" (P. Parvatha Reddy, N.K. Krishna Kumar and A. Verghese, eds.), pp. 345-348. Association for Advancement of Pest Management in Hortil. Ecosystems, Bangalore.

Rao, M.S., Parvatha Reddy, P. and Nagesh, M. 1998c. Evaluation of plant based formulations of *Trichoderma harzianum* for the management of *Meloidogyne incognita* on eggplant. *Nematologia Mediterranea* **26**: 59-62.

Rao, M.S., Parvatha Reddy, P. and Nagesh, M. 1998d. Evaluation of *Paecilomyces lilacinus* cultured on neem cake extract for the management of root-knot nematodes on eggplant. *Pest Management in Horticultural Ecosystems* **4**:116-119.

Rao, M.S., Parvatha Reddy, P. and Nagesh, M. 1999. Bare root-dip treatment of tomato seedlings in calotropis or castor leaf extracts mixed with *Paecilomyces lilacinus* spores for the management of *Meloidogyne incognita*. *Nematologia Mediterranea* **27**: 323-326.

Rao, M.S., Parvatha Reddy, P. and Nagesh, M. 2000. Management of *Meloidogyne incognita* on tomato by integrating *Glomus mosseae* with *Pasteuria penetrans* under field conditions. *Pest Management in Horticultural Ecosystems* **6**: 130-134.

Rao, M.S., Parvatha Reddy, P., Somasekhar, N. and Nagesh, M. 1997. Management of root-knot nematodes, *Meloidogyne incognita* in tomato nursery by integration of endomycorrhiza, *Glomus fasciculatum* with castor cake. *Pest Management in Horticultural Ecosystems* **3**: 31-35.

Rao, M.S., Parvatha Reddy, P. and Sukhada, M. 1998. Biointensive management of *Meloidogyne incognita* on eggplant by integrating *Paeciloymces lilacinus* and *Glomus mosseae*. *Nematolologia Mediterranea* **26**: 213-216.

Rao, M.S., Parvatha Reddy, P., Sukhada, M., Nagesh, M. and Pankaj 1998. Management of root-knot nematodes on eggplant by integrating endomycorrhiza (*Glomus fasciculatum*) and castor (*Ricinus communis*) cake. *Nematologia Mediterranea* **26**: 217-219.

Rao, M.S., Reddy, P.P. and Mohandas, S. 1995. Effect of integration of endomycorrhiza (*Glomus mosseae*) and neem cake on the control of root-knot nematode on tomato. *Journal of Plant Disease and Protection* **102**: 526-529.

Rao, M.S. and Shylaja, M. 2004. Role of *Pseudomonas fluorescens* (Migula) in induction of systemic resistance (ISR) and managing *Rotylenchulus reniformis* (Linford and Oliveira) on carrot. *Pest Management in Horticultural Ecosystems* **10**: 87-93.

Rao, M.S., Tewari, R.P. and Pandey, M. 1992. Occurrence of mushroom nematodes in South India. *Mushroom Research* **1**: 135-136.

Rao, V.R. and Singh, D.B. 1978. Chemical control of root-knot nematodes in okra with the aid of nematicides. *Indian Journal of Entomology* **35**: 151-154.

Rao, V.R. and Tikoo, S.K. 1973. Resistance of tomato cultivars to the root-knot nematode, *Meloidogyne incognita* (Kofoid and White, 1919) Chitwood, 1949. *Current Science* **44**: 282-283.

Rashid, A. and Khan, A.M. 1975. Studies on the lesion nematode, *Pratylenchus coffeae* associated with chrysanthemum. *Indian Phytopathology* **28**: 22-28.

Rashid, A. and Khan, A.M. 1992. *Indian Phytopathology* **28**: 22-28.

Raveendran, V. and Nadakal, A.M. 1975. An additional list of plants infected by the root-knot nematode, *Meloidogyne incognita* (Kofoid & White, 1919) Chitwood, 1949 in Kerala. *Indian Journal of Nematology* **5**: 126-127.

Ravi, K., Nanje Gowda, D.N. and Parvatha Reddy, P. 2000. Integrated management of the borrowing nematode, *Radopholus similis* (Cobb, 1993) Thorne 1949 on banana. *Pest Management in Horticultural Ecosystems* **6**: 124-129.

Ravi, K., Nanje Gowda, D.N. and Parvatha Reddy, P. 2001. Integrated management of disease complex in banana involving borrowing nematode and Panama wilt – an ecofriendly approach. In *IPM in Horticultural Crops: Emerging Trends in the New Millenium* (A. Verghese and P. Parvatha Reddy, eds.), p. 154. Association for Advancement of Pest Management in Horticultural Ecosystems, Bangalore.

Ravichandra, N.G. and Krishnappa, K. 1985. Effect of various treatments, both individually and in integration, in controlling the burrowing nematode, *Radopholus similis*, infesting banana. *Indian Journal of Nematology* **15**: 62-65.

Ravi Dutt and Bhatti, D.S. 1986. Determination of effective doses and time of application of nematicides and castor leaves for controlling *Meloidogyne javanica* in tomato. *Indian Journal of Nematology* **16**: 8-11.

Ray, S. and Parija, A.K. 1987. Community analysis of nematodes in the rhizosphere of banana in Puri district, Orissa. *Indian Journal of Nematology* **17**: 159-161.

Reddy, D.D.R. 1985. Analysis of crop losses in tomato due to *Meloidogyne incognita*. *Indian Journal of Nematology* **15**: 55-59.

Reddy, D.D.R. and Seshadri, A.R. 1972. Studies on some systemic nematicides. II. Further studies on the action of thionazin and aldicarb on *Meloidogyne incognita* and *Rotylenchulus reniformis*. *Indian Journal of Nematology* **2**: 182-190.

Reddy, D.D.R. and Seshadri, A.R.1975. Elimination of root-knot nematode infestation from tomato seedlings by chemical bare-root dips or soil application. *Indian Journal of Nematology* **5**: 170-175.

Reka Arya and Saxena, S.K. 1998. *Trichothecium roseum* together with *Rhizoctonia solani* and *Meloidogyne incognita* on germination of seeds of tomato cv. Pusa Ruby. *Indian Journal of Nematology* **28**: 217-221.

Renjhen, P.L. 1973. *History of Plant Quarantine and its Functioning in India*. Govt. of India, Minist. of Agri., Directorate of Pl. Prot., Quarantine and Storage, Faridabad.

Ribeiro, C.A.G. and Ferraz, S. 1983. Studies on the interaction between *Meloidogyne javanica* and *Fusarium oxysporum* f. sp. *phaseoli* on bean. *Fitopatologia Brasileira* **8**: 439-446.

Ritzema Bos, J. 1891. Zwei Nematodenkrankeiten der Erdbeer Pflanze. *Zeitscher. Pflanzenkrank* **1**: 1-10.

Roberts, P.A., Magyarosy, A.C., Mathews, W. and May, D.W. 1988. Effect of metham sodium applied by drip irrigation on root-knot nematodes, *Pythium ultimum* and *Fusarium* sp. in soil and carrot and tomato roots. *Plant Disease* **72**: 213-217.

Rom, R.C., Dozier, W.A, Knowles, J.W., Carlton, C.C., Arrington, E.H., Wehunt, E.J., Yadava, U.L., Doud, S.L., Ritchie, D.F., Clayton, C.N., Zehr, E.L., Gamrbell, C.E., Britton, J.A. and Lockwood, D.W. 1985. *Compact Fruit Tree* **18**: 85-91.

Rossner, J. and Zebitz, C.P.W. 1987. Effect of neem products on nematodes and growth of tomato (*Lycopersicon esculentum*) plants. In: *Natural Pesticides from the Neem Tree* (Azadirachta indica *A. Juss.*) *and Other Tropical Plants.* (H. Schmutterer and K.R.S. Ascher, eds.), pp.611-621. Eschborn, Germany.

Routaray, B.N. and Das, S.N. 1988. Population management of root-knot nematode *Meloidogyne incognita* on papaya. *International Nematology Network Newsletter* **5**(3): 25-27.

Routaray, B.N. and Sahoo, H. 1985. Integrated control of root-knot nematode, *Meloidogyne incognita* with neem cake and granular nematicides on tomato. *Indian Journal of Nematology* **15**: 261.

Rowe, R.C., Davis, J.R., Powelson, M.L. and Rouse, D.I. 1987. Potato early dying: causal agents and management strategies. *Plant Disease* **71**: 482-489.

Roy, A.K. 1984. Effect of nitrogenous soil amendments on the parasitic activity of *Catenaria anguillulae* on root-knot nematode. *Indian Journal of Nematology* **14**: 190-193.

Saeed, M., Ahmad, M. and Khan, H.A. 1972. A complex disease of tomato and papaya caused by nematode fungi association in Pakistan. *Pakistan Journal of Science and Industrial Research* **15**: 312-313.

Sahoo, S., Mohanty, K.C., Sahoo, N.K., Mohapatra, A.K.B. and Mishra, H.N. 2004a. Effect of different cropping sequences on root-knot nematode (*Meloidogyne incognita*) and wilt incidence in tomato. *National Symposium on Paradigms in Nematological Research for Biodynamic Farming,* Univ. of Agri. Sci., Bangalore, p. 37.

Sahoo, S., Mohanty, K.C., Sahoo, N.K., Mohapatra, A.K.B. and Mishra, H.N. 2004b. Comparative efficacy of different eco-friendly approaches for the management of root-knot nematode in okra. *National Symposium on Paradigms in Nematological Research for Biodynamic Farming,* Univ. of Agri. Sci., Bangalore, p. 79.

Saifullah. 1996. Effect of nematophagous fungi on growth of potato and *Globodera rostochiensis* and their environmental tolerance. *Afro-Asian Journal of Nematology* **6**: 128-132.

Saikia, B. 1992. Pathogenicity, Crop Loss Assessment and Biological Control of Meloidogyne incognita (Kofoid and White, 1919) Chitwood, 1949 on Betel Vine (Piper betle L.). M. Sc. (Agri.) thesis, Assam Agri. Univ., Jorhat.

Samiyappan, R. 2003. Biological control of fungal and nematode complex diseases by plant growth promoting rhizobacteria (PGPR). *Winter School on Biological Control of Plant Parasitic Nematodes,* Tamil Nadu Agri. Univ., Coimbatore, pp. 69-77.

Samuel, M. and Mathew, J. 1983. *Indian Phytopathology* **36**: 398-399.

Sangwan, N.K., Verma, K.K., Verma, B.S., Malik, M.S. and Dhindsa, K.S. 1985. Nematicidal activity of essential oils of *Symbopogon* grasses. *Nematologica* **13**: 93-99.

Santhosh Kumar, T. and Sheela, M.S. 2004. Status of nematode pests of Chethikoduveli, *Plumbago rosea* L. *National Symposium on Paradigms in*

Nematological Research for Biodynamic Farming, Univ. of Agri. Sci., Bangalore, p. 28-29.

Sanwal, K.C. and Mathur, V.K. 1975. *Indian Journal of Nematology* **5**: 251-254.

Sarwar, M., Prameswaran, T.N., Shanmugam, C. and Narayan, M.R. 1982. Appraisal of Patchouli in India. Root-knot nematode problem and control. *Indian Perfumer* **26** (2-4): 117-121.

Saxena, S.K., Alam, M.M. and Khan, A.M. 1974. Chemotherapeutic control of nematodes with Vydate oxamyl, VC-13 and Dazomet on chilli plants. *Indian Journal of Nematology* **4**: 235-238.

Schacht, H. 1859. Ueber einige Feinde und Krankheiten der Zuckerrube. *Zeitscher. Ver. Rubenzucker-Ind. Zoolver* **9**: 390.

Schieber, E. and Sosa, O.N. 1960. Nematodes on coffee in Gautemala. *Plant Disease Reporter* **44**: 722-723.

Schindler, A.F., Stewart, R.N. and Semeniuk, P. 1961. A synergistic *Fusarium* – nematode interaction on carnations. *Phytopathology* **51**: 143-146.

Schlang, J. and Sikora, R.A. 1978. Investigations on the interaction between Meloidogyne arenaria, Verticillium dahliae and Phytophthora cryptoge on Gerbera jomesonii. Zeits-chrift fur Pflanzenkrenkheiten und Pflanzenschutz **85**: 203-209.

Schmidli-Sacherer, P., Keel, C. and Defago, G. 1997. The global regulator GacA of *Pseudomonas fluorescens* CHA0 is required for suppression of root diseases in dicotyledons but not in Gramineae. *Plant Pathology* **46**: 80-90.

Scholte, K. 1990. Causes of differences in growth pattern, yield and quality of potatoes (*Solanum tuberosum* L.) in short rotations on sandy soil as affected by crop rotation, cultivar and application of granular nematicides. *Potato Research* **33**: 181-190.

Scholte, K. and s'Jacob, 1989. Synergistic interactions between *Rhizoctonia solani* Kuhn, *Verticillium dahliae* Kleb., *Meloidogyne* spp. and *Pratylenchus neglectus* (Rensch) Chitwood & Oteifa, in potato. *Potato Research* **32**: 387-395.

Schuster, M.L. 1959. Relation of root-knot nematodes and irrigation water to the incidence and dissemination of bacterial wilt of bean. *Plant Disease Reporter* **43**: 27-32.

Seenivasan, N., Parameswaran, S., Sridar, P., Gopalakrishnan, C. and Gnanamurthy, P. 2001. Application of bioagents and neem cake as soil application for the management of root-knot nematode in turmeric. *Natl. Cong. on Centenary of Nematol. in India – Appraisal & Future Plans*, Indian Agri. Res. Inst., New Delhi, pp. 164.

Sehgal, M., Chawla, M.L., Singh, S. and Balakrishnan, K. 1990. Studies on interactive effect of moisture and temperature on *Tylenchulus semipenetrans*. *Current Nematology* **1**: 35-36.

Sen, K. and Dasgupta, M.K. 1977. Additional hosts of the root-knot nematode, *Meloidogyne* spp. from India. *Indian Journal of Nematology* **7**: 74.

Sen, K. and Dasgupta, M.K. 1981. Effect of rice hull ash on root-knot nematodes in tomato. *Tropical Pest Management* **27**: 518-519.

Senthamarai, M., Poornima, K. and Subramanian, S. 2006a. Nematode-fungal disease complex involving *Meloidogyne incognita* and *Macrophomina phaseolina* on *Coleus forskohlii* Briq. *Indian Journal of Nematology* **36**: 181-184.

Senthamarai, M., Poornima, K. and Subramanian, S. 2006b. Bio-management of root-knot nematode, *Meloidogyne incognita* on *Coleus forskohlii* Briq. *Indian Journal of Nematology* **36**: 206-208.

Senthamarai, M., Poornima, K. and Subramanian, S. 2006c. Assessment of avoidable yield loss on *Coleus forskohlii* due to *Meloidogyne incognita. Indian Journal of Nematology* **36**: 296-297.

Senthil Kumar, T. and Rajendran, G. 2004. Biocontrol agents for the management of disease complex involving root-knot nematode, *Meloidogyne incognita* and *Fusarium moniliforme* on grapevine (*Vitis vinefera*). *Indian Journal of Nematology* **34**: 49-51.

Sethi, C.L and Gaur, H.S. 1986. Nematode management: An overview. In *Plant Parasitic Nematodes of India – Problems and Progress.* Nematol. Soc. of India, New Delhi, pp. 424-445.

Sethi, C.L., Nath, R.P., Mathur, V.K. and Ahuja, S. 1972. Interceptions of plant parasitic nematodes from imported seed/plant materials. *Indian Journal of Nematology* **2**: 89-93.

Shahzad, S. and Gaffer, A. 1989. Use of *Paecilomyces lilacinus* in the control of root rot and root-knot complex in okra and mungbean. *Pakistan Journal of Nematology* **7**: 47-53.

Shanthi, A. 2003. *Management of* Radopholus similis, Helicotylenchus multicinctus *and* Pratylenchus coffeae *in Banana.* Ph. D. thesis, Tamil Nadu Agri. Univ., Coimbatore.

Shanthi, A., Rajendran, G. and Sivakumar, M. 2004. Influence of biofertilizers on lesion nematodes in banana. *National Symposium on Paradigms in Nematological Research for Biodynamic Farming,* Univ. of Agri. Sci., Bangalore, p. 97.

Shanthi, A. and Rajeswari, S. 1998. Effect of chopped leaves and nematicides on the interaction of *Glomus fasciculatum* with *Meloidogyne incognita* on cowpea. *Indian Journal of Nematology* **28**: 114-117.

Shanthi, A., Rajeswari, S. and Sivakumar, C.V. 1998. Soil application of *Pseudomonas fluorescens* (Migula) for the control of root-knot nematode (*Meloidogyne incognita*) on grapevine (*Vitis vinefera* Linn.). In *Nematology – Challenges and Opportunities in 21st Century,* (Usha K. Mehta, ed.), pp. 203-206. Sugarcane Breeding Institute, Coimbatore.

Shanthi, A. and Sivakumar, C.V. 1995. Biocontrol potential of *Pseudomonas fluorescens* (Migula) against root-knot nematode, *Meloidogyne incognita* (Kofoid and White, 1919) Chitwood, 1949 on tomato. *Journal of Biological Control* **9**: 113-115.

Shanthi, A., Sundarababu, R. and Sivakumar, C.V. 1999. Field evaluation of rhizobacterium, *Pseudomonas fluorescens* for the management of the citrus nematode, *Tylenchulus semipenetrans.* In *Proc. of the Natl. Symp. on Rational*

Approaches in Nema Mangmt. for Sustainable Agri., Indian Agri. Res. Inst., New Delhi, pp. 38-42.

Sharma, A. and Trivedi, P.C. 1989. Influence of inoculum levels of fungus *Paecilomyces lilacinus* (Thom.) Samson on the biocontrol of root-knot nematode, *Meloidogyne incognita* (Chitwood). *International Nematology Network Newsletter* **6**(2): 27-29.

Sharma, A. and Trivedi, P.C. 1994. Interaction of root-knot nematode with VAM on *Abelmoschus esculentus* and their effect on fruit yield. In *Vistas in Seed Biology* (T. Singh and P.C. Trivedi, eds.), Vol. II, pp. 61-81. Printwell, Jaipur.

Sharma, G.C. and Bhatia, M. 2004. Impact of sowing time on the nematode populations of French bean and assessment of avoidable yield losses to the crop due to their infestation. *Indian Journal of Nematology* **34**: 90-92.

Sharma, G.L. 1989. Estimated losses due to root-knot nematodes, *Meloidogyne incognita* and *M. javanica* in pea crop. *International Nematology Network Newsletter* **6**(1): 28-29.

Sharma, G.L., Trivedi, P.C., Sharma, M.K. and Tiagi, B. 1998. Fungus *Paeciloymces lilacinus* as a biocontrol agent of root-knot nematode, *Meloidogyne incognita*. *The Indian Zoologist* **12**: 259-262.

Sharma, H.K. and Kamra, A. 2001. Impeccability to circumvent the root-knot *Meloidogyne incognita* by *Paeciloymces lilacinus* on okra. *Natl. Cong. on Centenary of Nematol. in India – Appraisal & Future Plans*, Indian Agri. Res. Inst., New Delhi, p. 168.

Sharma, H.K., Pankaj and Mishra, S.D. 2005. Polyethylene mulching in the management of plant parasitic nematodes. *Indian Journal of Nematology* **35**: 82-85.

Sharma, J.K., Singh, I. and Chhabra, H.K. 1980. Observations on the influence of *Meloidogyne incognita* and *Rhizoctonia bataticola* on okra. *Indian Journal of Nematology* **10**: 148-151.

Sharma, N.K. and Gill, J.S. 1979. Interaction between *Meloidogyne incognita* and *Rhizoctonia solani* on potato. *Indian Phytopathology* **32**: 293-298.

Sharma, N.K. and Sethi, C.L. 1975. Leghaemoglobin content of cowpea nodules as influenced by *Meloidogyne incognita* and *Heterodera cajani*. *Indian Journal of Nematology* **5**: 113-114.

Sharma, N.K. and Sethi, C.L. 1976a. Reaction of certain cowpea varieties to *Meloidogyne incognita* and *Heterodera cajani*. *Indian Journal of Nematology* **6**: 99-102.

Sharma, N.K. and Sethi, C.L. 1976b. Interrelationship between *Meloidogyne incognita*, *Heterodera cajani* and *Rhizobium* sp. on cowpea (*Vigna sinensis* (L.) Savi). *Indian Journal of Nematology* **6**: 117-123.

Sharma, S.B., Sharma, H.K. and Pankaj. 2002. Nematode problems in India. In *Crop Pest and Disease Management: Challenges for the Millennium* (D. Prasad and S.N. Puri, eds.), pp.267-275. Jyoti Publishers, New Delhi, India.

Sharma, S.K., Singh, I. and Sakhuja, P.K. 1980. *Indian Journal of Nematology* **10**: 53-58.

Sheela, M.S., Jiji, T. and Nisha, M.S. 2002. Evaluation of different control strategies for the management of nematodes associated with vegetables (brinjal) (Abstr.). *International conference on Vegetables,* Bangalore, p. 268.

Sheela, M.S. and Nisha, M.S. 2004. Impact of biological control agents for the management of root-knot nematode in brinjal. *National Symposium on Paradigms in Nematological Research for Biodynamic Farming,* Univ. of Agri. Sci., Bangalore, p. 63-64.

Sheela, M.S., Nisha, M.S. and Mohadas, C. 2002. Impact of trap cropping and crop rotation for the management of root-knot nematode in vegetables. *Proceedings of International Conference on Vegetables,* Bangalore, pp. 344 -346.

Sheela, M.S., Nisha, M.S. and Mohandas, C. 2004. Eco-friendly management of nematodes associated with Chinese potato (Coleus), *Solanostemon rotundifolius* (Poir) Morton. *National Symposium on Paradigms in Nematological Research for Biodynamic Farming,* Univ. of Agri. Sci., Bangalore, p. 90.

Sheela, M.S. and Rajani, T.S. 1998. Status of phytonematodes as a part of medicinal plants in Kerala. In *Nematology – Challenges and Opportunities in 21st Century* (Usha K. Mehta, ed.), pp.2-3. Sugarcane Breeding Inst., Coimbatore.

Shetty, K.D. and Reddy, D.D.R. 1984. Pathogenicity of *Meloidogyne incognita* on *Solanum khasianum. Indian Journal of Nematology* **14**: 185-186.

Shetty, K.D. and Reddy, D.D.R. 1985a. Control of root-knot nematodes on *Solanum khasianum* by chemical bare-root dips. *Indian Journal of Nematology* **15**: 227-229.

Shetty, K.D. and Reddy, D.D.R. 1985b. Resistance in *Solanum* species to root-knot nematode, *Meloidogyne incognita. Indian Journal of Nematology* **15**: 230.

Shivaprasad, B., Reddy, B.M.R. and Krishnappa, K. 2001. Effect of trap cropping in management of root-knot nematode in brinjal. *Natl. Cong. on Centenary of Nematol. in India – Appraisal & Future Plans,* Indian Agri. Res. Inst., New Delhi, pp. 94-95.

Shukla, P.K. and Haseeb, A. 1996. Effectiveness of some nematicides and oil cakes in the management of *Pratylenchus thornei* on *Mentha citrata, M. piperita* and *M. spicata. BITE* **57**: 307-309.

Shukla, V.N. and Swarup, G. 1970. Interrelationships of the root-knot nematode, *Meloidogyne incognita* (Kofoid and White, 1919) Chitwood, 1949 and *Sclerotium rolfsii* Sacc. on the emergence of tomato seedlings. *Bulletin of Indian Phytopathological Society* **6**: 52-54.

Shylaja, M., Rao, M.S. and Rahiman, A.B. 2004. Studies on the Development of Bio-intensive Disease Management Strategies in Tuberose (Polyanthes tuberosa L.) and Carnations (Dianthus caryophyllus L.) using Biological Control Agents. Ph. D. thesis, Kuvempu Univ., Shimoga.

Shylaja, M. 2004. Studies on the development of bio-intensive management strategies of nematode induced wilt disease complex in Tuberose (*Polyanthes tuberosa*) and Carnations (*Dianthus caryophyllus*) using biological control agents. *National Symposium on Paradigms in Nematological Research for Biodynamic Farming,* Univ. of Agri. Sci., Bangalore, p. 136-137.

Siddiqi, M.A. and Alam, M.M. 1987. Efficacy of seed dressing with extracts of neem and Persian lilac against *Meloidogyne incognita* and *Rotylenchulus reniformis*. *Nematologia Mediterranea* **15**: 399-403.

Siddiqi, M.A. and Alam, M.M. 1988a. Control of root-knot and reniform nematodes by bare-root dip in leaf extracts of margosa and Persian lilac. *Zeitschrift fur Pflanzenkrankheiten und Pflanzenschutz* **95**: 138-142.

Siddiqi, M.A. and Alam, M.M. 1988b. Effect of seed treatment with azadirachtin on root-knot development and growth of some vegetables. *Annals of Applied Biology* **112** : 20-21.

Siddiqi, M.A. and Alam, M.M. 1989a. Seed treatment with azadirachtin for the control of the stunt nematode attacking cabbage and cauliflower. *Annals of Applied Biology* **113**: 4-5.

Siddiqi, M.A. and Alam, M.M. 1989b. Control of stunt nematode by bare-root dip in leaf extracts of margosa and Persian lilac. *Pakistan Journal of Nematology* **7**: 33-38.

Siddiqi, M.A. and Alam, M.M. 1990a. Control of root-knot, reniform and stunt nematodes with nimbin seed dressing. *Nematologia Mediterranea* **18**: 19-22.

Siddiqi, M.A. and Alam, M.M. 1990b. Saw dust as soil amendment for control of nematodes infesting some vegetables. *Biological Wastes* **33**: 123-129.

Siddiqi, M.A. and Saxena, S.K. 1987a. Effect of interculture of margosa and Persian lilac with tomato and eggplant on root-knot and reniform nematodes. *International Nematology Network Newsletter* **4**(2): 5-8.

Siddiqi, M.A. and Saxena, S.K. 1987b. Studies on the control of the stunt nematode, *Tylenchorhynchus brassicae* by interculture of margosa and Persian lilac. *International Nematology Network Newsletter* **4**(4): 27-29.

Siddiqi, M.R. 1961. Occurrence of the citrus nematode, *Tylenchulus semipenetrans* Cobb. 1913 and the reniform nematode, *Rotylenchulus reniformis* in India. *Proceedings of 48th Indian Science Congress* **3**: 504.

Siddiqi, M.R. 1964. Studies on nematode root-rot of citrus in Uttar Pradesh, India. *Proceedings of Zoological Society, Calcutta* **17**: 67-75.

Siddiqui, I.A., Ehteshamul-Haque, S. and Ghaffar, A. 1999. Root dip treatment with *Pseudomonas aeruginosa* and *Trichoderma* spp. in the control of root rot-root knot disease complex in chilli (*Capsicum annum* L.). *Pakistan Journal of Nematology* **17**: 67-75.

Siddiqui, Z.A., Khan, A.M. and Saxena, S.K. 1973. Studies on *Tylenchorhynchus brassicae* II - Effect of temperature and moisture on multiplication of nematode. *Indian Journal of Entomology* **26**: 139-147.

Siddiqui, Z.A., Khan, A.M. and Saxena, S.K. 1976. Nematode population and yield of certain vegetables as influenced by oil-cake amendments. *Indian Journal of Nematology* **6**: 179-182.

Siddiqui, Z.A. and Mahmood, I. 1999. Role of bacteria in the management of plant parasitic nematodes: A review. *Bioresource Technology* **69**: 167-179.

Sikora, R.A. 1978. Effect of the endotrophic mycorrhizal fungus, *Glomus mosseae* on the host parasite relationship of *Meloidogyne incognita* in tomato. *Journal of Plant Disease and Protection* **85**: 197-202.

Sikora, R.A. 1981. Interaction between plant parasitic nematodes, plant roots and vesicular arbuscular mycorrhizae. *Proc. of Symp. on Biological and Chemical Interactions in the Rhizosphere*, Stockholm, pp. 115-136.

Sikora, R.A. and Schlosser, E. 1973. Nematodes and fungi associated with root systems of banana in a state of decline in Lebanon. *Plant Disease Reporter* **57**: 615-618.

Sikora, R.A. and Schlonbeck, F. 1975. Mycorrhiza (*Endogone mosseae*) on the population dynamics of the root-knot nematode (*Meloidogyne incognita* and *M. hapla*). *Proceedings of the International Congress on Plant Protection* **5**: 80-86.

Singh, B., Bhatti, D.S. and Singh, K. 1974. Resistance to root-knot nematodes, *Meloidogyne* spp. in vegetable crops. *Pest Articles and News Summaries* **20**: 58-67.

Singh, B. and Dhawan, S.C. 1998. Compatibility of *Pasteuria penetrans* with carbofuran in controlling *Heterodera cajani* on cowpea. *Indian Journal of Nematology* **28**: 110-113.

Singh, D. 1991. *Studies on the Effect of* Tagetes *Species on* Meloidogyne javanica *Infectivity in Brinjal (Solanum melongena)*. Ph. D. thesis, Haryana Agri. Univ., Hisar, 83 pp.

Singh, D.B. and Parvatha Reddy, P. 1981a. Influence of *Meloidogyne incognita* infestation on *Rhizobium* nodule formation in French bean. *Nematologia Mediterranea* **9**: 1-5.

Singh, D.B. and Parvatha Reddy, P. 1981b. Note on chemical control of root-knot nematodes infesting French bean. *Indian Journal of Agricultural Science* **51**: 534-535.

Singh, D.B. and Parvatha Reddy, P. 1982a. Cultivars of cowpea resistant to root-knot nematode, *Meloidogyne incognita* in India. *Tropical Pest Management* **28**: 185-186.

Singh, D.B. and Parvatha Reddy, P. 1985. Nature of resistance to *Meloidogyne incognita* in cowpea, *Vigna unguiculata*. *Nematologia Mediterranea* **13**: 127-132.

Singh, D.B., Parvatha Reddy, P. and Rao, V.R. 1981. Reaction of some varieties and selections of French bean to *Meloidogyne incognita*. *Indian Journal of Nematology* **11**: 81-83.

Singh, D.B., Parvatha Reddy, P. and Sharma, S.R. 1981. Effect of the root-knot nematode, *Meloidogyne incognita* on *Fusarium* wilt of French beans. *Indian Journal of Nematology* **11**: 84-85.

Singh, D.B., Rao, V.R. and Parvatha Reddy, P. 1978. Evaluation of nematicides for the control of root-knot nematodes on brinjal. *Indian Journal of Nematology* **8**: 64-66.

Singh, D.B., Reddy, P.P. and Shyamasundar Joshi. 1984. Histological, histopathological and histochemical investigations on root-knot nematode resistant and susceptible lines of cowpea. *Nematologia Mediterranea* **12**: 213-219.

Singh, M., Samra, R. and Sharma, G.L. 1998. Effect of various soil types on the development of *Pasteuria penetrans* on *Meloidogyne incognita* in brinjal crop. *Indian Journal of Nematology* **28**: 35-40.

Singh, R.P., Khan, A.M. and Saxena, S.K. 1979. Effect of oilcakes on the movement of nematicides in soil. *Nematologia Mediterranea* **7:** 209-215.

Singh, R.S. 1965. Control of root-knot of tomato with organic soil amendments. *FAO Plant Protection Bulletin* **13**:35-37.

Singh, R.S., Singh, b. and Beniwal, S.P.S. 1967. Observations on the effect of saw dust on the incidence of root-knot and yield of okra and tomatoes in nematode infested soil. *Plant Disease Reporter* **50**: 108-124.

Singh, R.S. and Sitaramaiah, K. 1966. Incidence of root-knot of okra and tomato in oil-cake amended soil. *Plant Disease Reporter* **50**: 668-672.

Singh, R.S. and Sitaramaiah, K. 1967. Effect of decomposing green leaves, sawdust and urea on the incidence of root-knot of okra and tomato. *Indian Phytopathology* **20:** 349-355.

Singh, R.S. and Sitaramaiah, K. 1970. Control of plant parasitic nematodes with organic soil amendments. *PANS* **16**: 287-297.

Singh, R.S. and Sitaramaiah, K. 1971a. Control of root-knot through organic and inorganic amendments of soil: Effect of oilcakes and sawdust. *Indian Journal of Mycology and Plant Pathology* **1**: 20-29.

Singh, R.S. and Sitaramaiah, K. 1971b. Control of root-knot through organic and inorganic amendments of soil: Effect of saw dust and inorganic nitrogen. *Indian Journal of Nematology* **1**: 80-84.

Singh, R.S. and Sitaramaiah, K. 1973. *Control of Plant Parasitic Nematodes with Organic Amendments of Soil.* Final technical report on effect of organic amendments, green manuring and inorganic fertilizers on root-knot of vegetable crops, G.B. Pant Univ. of Agri. & Technol., Bull. No. 6, 289 pp.

Singh, R.V., Midha, S.K. and Kumar, V. 2003. Major nematode pests of economically important crops in India and their management. In *Advances in Nematology* (P.C. Trivedi, ed.), pp. 233-256.

Singh, R.V. and Gill, J.S. 1998. Important nematode problems on vegetables and their management. In *Pathological Problems of Economic Crop Plants and Their Management* (S.M. Paul Khurana, ed.), pp. 557-570. Scientific Publishers (India), Jodhpur, India.

Singh, R.V., Sharma, H.K. and Gill, J.S. 1998. Management of *Meloidogyne incognita* in okra with neem-based products alone and with carbofuran. *Proc. of Natl. Symp. on Rational Approaches in Nematode Management for Sustainable Agriculture,* Nematol. Soc. of India, New Delhi, pp.58-62.

Singh, S.P., Ahmad, M., Khan, A.M. and Saxena, S.K. 1980. Effect of seed treatment with certain oilcakes or nematicide on the growth of tomato and on rhizosphere population of nematodes and fungi. *Nematologia Mediterranea* **8**: 193-199.

Singh, S.P., Pant, V., Khan, A.M. and Saxena, S.K. 1984. Efficacy of certain nematicides and oil cakes for the control of stunt nematode, *Tylenchorhynchus brassicae* and parasitic fungi in the rhizosphere of cauliflower. *Indian Phytopathology* **37**: 647-651.

Sirohi, A. and Mohan, S. 2002. Nematode parasites of Solanaceous vegetables. In *Crop Pest and Disease Management: Challenges for the Millennium* (D. Prasad and S.N. Puri, eds.), pp.332-345. Jyoti Publishers, New Delhi, India.

Sitaramaiah, K. and Naidu, H. 2003. Plant parasitic nematodes – management opportunities and challenges. In *Nematode Management in Plants* (P.C. Trivedi, ed.), pp. 57-80. Scientific Publishers (India), Jodhpur.

Sitaramaiah, K. and Sikora, R.A. 1981. Influence of the endomycorrhizal fungus *Glomus mosseae* on *Rotylenchulus reniformis* penetration and development on bush bean, cucumber and muskmelon. *Meded. Rajksfac. Land Bouwwet. Gent.* **46**: 495-502.

Sitaramaiah, K. and Sikora, R.A. 1982. Effect of the mycorrhizal fungus, *Glomus fasciculatum* on the host-parasite relationship of *Rotylenchulus reniformis* in tomato. *Nematologica* **28**: 412-419.

Sitaramaiah, K. and Sinha, S.K. 1984. Histological aspects of Pseudomonas solanacearum and root-knot nematode wilt complex in brinjal. *Indian Journal of Nematology* **14**: 175-178.

Sivakumar, C.V. 1998. Bacterial antagonists for suppression of plant parasitic nematodes. In *'Biological Suppression of Plant Diseases, Phytoparasitic Nematodes and Weeds'*, Project Directorate of Biological Control, Bangalore, pp. 128-137.

Sivakumar, C.V. and Meerazainuddin, M. 1974. Influence of N, P and K on the reniform nematode, *Rotylenchulus reniformis* and its effect on the yield of okra. *Indian Journal of Nematology* **4**: 243-244.

Sivakumar, C.V., Palanisamy, S. and Balasubramanain, M. 1977. Control of *Rotylenchulus reniformis* in tomato nursery. *Indian Journal of Nematology* **7**: 78-80.

Sivakumar, C.V., Palanisamy, S. and Naganathan, T.G. 1976. Persistence of nematicidal activity in seed treatment of okra with carbofuran and aldicarb sulfone. *Indian Journal of Nematology* **6**: 106-108.

Sivakumar, M. 1998. Non chemical management of *Meloidogyne hapla* Chitwood on carrot (*Dacus carotae*). In *Nematology – Challenges and Opportunities in 21st Century* (Usha K. Mehta, ed.), pp. 209-212. Sugarcane Breeding Institute, Coimbatore.

Sivakumar, M. and Marimuthu, T. 1987. Preliminary studies on the effect of soil solarization on phytonematodes of betel vine. *Indian Journal of Nematology* **17**: 58-59.

Sivakumar, M. and Vadivelu, S. 1999. Management of *Meloidogyne incognita* on grapevine using biocontrol agents, botanicals and biofertilizers. *Pest Management in Horticultural Ecosystems* **5**: 127-131.

Sivaprasad, P. and Sheela, M.S. 1998. Distribution of phytonematodes and VAM fungi in the rhizosphere of pepper and its impact on nematode management. In *Nematology – Challenges and Opportunities in 21st Century* (Usha K. Mehta, ed.), pp. 102-105. Sugarcane Breeding Institute, Coimbatore.

Smith, G.S. 1987. Interactions of nematodes with mycorrhizal fungi. In *Vistas on Nematology* (J.A. Veech and D.W. Dickson, eds.), pp. 292-300. Society of Nematologists, Hyattsville, Maryland, USA.

Smith, P.G. 1944. Embryo culture of a tomato species hybrid. *Proceedings of the American Society of Horticultural Science* **44**: 413-416.

Smits, B.G. and Noguera, R. 1982. Effect of *Meloidogyne incognita* on the pathogenicity of different isolates of *Fusarium oxysporum* on brinjal (*Solanum melongena* L.). *Agronomia Tropical* **32**: 284-290.

Sobita Devi, L. and Dutta, U. 2002. Effect of *Pseudomonas fluorescens* on root-knot (*Meloidogyne incognita*) on okra plant. *Indian Journal of Nematology* **32**: 215.

Somasekhar, N. and Gill, J.S. 1991. Efficacy of *Pasteuria penetrans* alone and in combination with carbofuran in controlling *Meloidogyne incognita*. *Indian Journal of Nematology* **21**: 61-65.

Somasekhar, N., Mehta, K.U. and Hari, K. 1999. Solubilization of insoluble phosphate by nematode antagonistic fungi. In *Nematology – Challenges and Opportunities in 21st Century* (Usha K. Mehta, ed.), pp. 213-215. Sugarcane Breeding Institute, Coimbatore.

Somasekhar, P. 1959. Nematodes associated with Arabica coffee in India. *FAO Plant Protection Bulletin* **7**(6): 78-79.

Sosamma, V.K., Geetha, V.K. and Koshy, P.K. 1994. Effect of the fungus *Paecilomyces lilacinus* on the burrowing nematode, *Radopholus similis* infesting betel vine. *Indian Journal of Nematology* **24**:50-54.

Sosamma, V.K. and Koshy, P.K. 1978. *Indian Phytopathology* **31**: 381-382.

Sosamma, V.K. and Koshy, P.K. 1995. Effect of Pasteuria penetrans and Paecilomyces lilacinus on population build up of root-knot nematode, Meloidogyne incognita on black pepper. National Symp. on Nematode Problems of India – An Appraisal of the Nematode Mangmt. with Eco-friendly Approaches and Biocomponents. Indian Agri. Res. Inst., New Delhi, p. 47.

Sosamma, V.K., Koshy, P.K. and Bhaskar Rao, E.V.V. 1980. Susceptibility of coconut cultivars and hybrids to *Radopholus similis* in the field. *Indian Journal of Nematology* **10**: 250-252.

Sosamma, V.K., Koshy, P.K. and Samuel, R. 1998. Effect of vesicular arbuscular mycorrhizae, *Glomus mosseae* on nematodes infecting banana under coconut based high density multispecies cropping system. In *Nematology – Challenges and Opportunities in 21st Century* (Usha K. Mehta, ed.), pp. 94-97. Sugarcane Breeding Institute, Coimbatore.

Sosamma, V.K., Sundararaju, P. and Koshy, P.K. 1979. Effect of *Radopholus similis* on turmeric. *Indian Journal of Nematology* **9**: 27-31.

Spiegel, Y., Cohn, E., Galper, S., Sharon, E. and Chet, I. 1991. Evaluation of a newly isolated bacterium, *Pseudomonas chitinolytica* sp nov. for controlling the root-knot nematode *Meloidogyne javanica*. *Biocontrol Science and Technology* **1**: 115-125.

Sreeja, P., Nageswari, S. and Mohandas, C. 1998. Reaction of cassava cultivars to root-knot nematode, *Meloidogyne incognita* (Kofoid & White, 1919) Chitwood, 1949. In *Nematology-Challenges and Opportunities in 21st Century* (Usha K. Mehta, ed.), pp.137-179. Sugarcane Breeding Institute, Coimbatore.

Srinivasan, A. 1974. *Studies on Root Lesion Nematode,* Pratylenchus delattrei. M. Sc. (Agri.) thesis, Tamil Nadu Agricultural University, Coimbatore.

Srinivasan, A. and Muthurksihnan, T.S. 1975. A note on nematode fungal complex in crossandra (*Crossandra undulaefolia*) in Coimbatore. *Current Science* **44**: 743-744.

Srinivasan, A. and Muthukrishnan, T.S. 1976. Control of root lesion nematode *Pratylenchus delattrei* Luc 1958 on crossandra. *Indian Journal of Nematology* **6**: 111-114.

Srivastava, A.K. and Singh, R.B. 1990. Effect of organic amendment on interaction of *Macrophomina phaseolina* and *Meloidogyne incognita* on French bean (*Phaseolus vulgaris*). *New Agriculturist* **1**: 99-100.

Srivastava, A.S., Pandey, R.C. and Ram, S. 1971. Application of organic amendments for the control of root-knot nematode, *Meloidogyne javanica* (Treub). *Labdev Journal of Science and Technology* **9B**: 203-205.

Srivastava, S.S. and Swarup, G. 1986. Biomanagement of root-knot nematode (*Meloidogyne incognita*) through the predatory fungus, *Arthrobotrys conoides*. *National Conference on Plant Parasitic Nematodes of India-Problems and Progress*, Indian Agri. Res. Inst., New Delhi, p. 85.

Starr, J.L. and Mai, W.F. 1976. Efficacy of pyroxychlor in controlling *Pythium polymorphon* – *Meloidogyne hapla* root disease complex of celery. *Plant Disease Reporter* **63**: 390-392.

Stephen, Z.A., El-Behadi, A.H., Al-Zaharoon, H.H., Antoon, B.G. and Georges, M.S. 1996. Control of root-knot and wilt disease complex on tomato plants. *Dirasat Series B. Pure and Applied Sciences* **23**: 13-16.

Stewart, R.N and Schindler, A.F. 1956. The effect of some ectoparasitic and endoparasitic nematodes in the expression of bacterial wilt in carnations. *Phytopathology* **46**: 419-422.

Stirling, G.R. and White, A.M. 1982. Distribution of a parasite of root-knot nematodes in South Australian vineyards. *Plant Disease* **66**: 52-53.

Stratford, R. and Shields, R. 1995. WO 9530017; 09.11.95, Unilever, Rotterdam, The Netherlands.

Strobel, N.E., Hussey, R.S. and Roncadori, R.S. 1982. Interaction of vesicular arbuscular fungi, *Meloidogyne incognita* and soil fertility on peach. *Phytopathology* **72**: 690-694.

Subramanian, S. and Selvaraj, P. 1988. Effect of *Tagetes patula* L. leaf extract on *Radopholus similis* (Cobb, 1893) Thorne, 1949. *Indian Journal of Nematology* **18**: 337-338.

Subramanyam, S., Rajendran, G. and Vadivelu, S. 1998. Estimation of loss in tomato due to *Meloidogyne incognita* and *Rotylenchulus reniformis*. *Indian Journal of Nematology* **20**: 239-240.

Sudha, S. and Sundararaju, P. 1998. Effect of neem oil cake and nematicide for the control of burrowing nematode, *Radopholus similis* in the areca nut based cropping system. In *Nematology - Challenges & Opportunities in 21st Century* (Usha K. Mehta, ed.), pp. 251-257. Sugarcane Breeding Inst., Coimbatore.

Suit, R.F. and DuCharme, E.P. 1953. The burrowing nematode and other parasitic nematodes in relation to spreading decline of citrus. *Plant Disease Reporter* **37**: 379-383.

Suma, M.R., Krishnappa, K., Reddy, B.M.R. and Kantharaju, V. 2002. VAM in the management of the burrowing nematode on banana (Abstr.). *Southern Zone Annual Meeting of the Indian Phytopath. Soc.*, Bangalore, pp. 19-21.

Sumathi, M., Parthiban, S. and Samuel, S.D. 2006. Effect of graded doses of nitrogen and VAM on herbage, oil yield and nematode population on patchouli (*Pogostemon patchouli* Pellet). *Indian Journal of Nematology* **36**: 209-213.

Sumner, D.R. and Johnson, A.W. 1973. Effect of root-knot nematode on *Fusarium* wilt of watermelon. *Phytopathology* **63**: 361-364.

Sundarababu, R. 1992. Nematodes in jasmine. *Tamil Nadu Agricultural University Newsletter* **21**(11): 2.

Sundarababu, R. and Kumar, V. 2003. Management of root lesion nematode, *Pratylenchus coffeae* in six commercial cultivars of banana through organic and inorganic amendments. *Infomusa* **12**: 21-25.

Sundarababu, R., Mani, M.P. and Sivakumar, C.V. 1999. A new biocontrol agent for plant parasitic nematodes. *Indian Horticulture* **12**: 25.

Sundarababu, R., Sakaranarayanan, C. and Santhi, A. 1993. Interaction between vesicular arbuscular mycorrhiza and *Meloidogyne javanica* on tomato as influenced by time of inoculation. *Indian Journal of Nematology* **23**: 125-127.

Sundarababu, R. and Suguna, N. 1998. Effective optimum dose of *Glomus mosseae* for the management of *Meloidogyne incognita* on tomato, chillies and bhendi. *International Journal of Tropical Plant Diseases* **16**: 127-131.

Sundarababu, R. and Vadivelu, S. 1988a. Studies on the nematodes associated with rose. *Indian Journal of Nematology* **18**: 139.

Sundarababu, R. and Vadivelu, S. 1988b. Pathogenicity of *Meloidogyne* species to tuberose (*Polyanthes tuberosa* L.). *Indian Journal of Nematology* **18**: 146-148.

Sundaram, H.D. and Arangarasan, V. 1996. Role of vesicular arbuscular mycorrhizal fungi on the growth, nematode (*Meloidogyne incognita*) disease incidence and fruit yield of tomato CO 3 grown in loamy soils of Arcot District (Tamil Nadu). *Mycorrhizae: Biofertilizer for the Future*, pp. 100-103.

Sundaram, R. and Tangaraju, T. 2001. Biointensive management of *Meloidogyne incognita* and *Pythium aphanidermatum* in tomato nurseries. *Natl. Cong. on Centenary of Nematol. in India – Appraisal & Future Plans*, Indian Agri. Res. Inst., New Delhi, pp. 169-170.

Sundaram, R. and Velayutham, B. 1988. Relative efficacy of some insecticides and neem cake in the control of *Rotylenchulus reniformis* and *Helicotylenchus dihystera* affecting garden bean. *Indian Journal of Nematology* **18**: 329-331.

Sundararaju, P. and Cannayane, I. 2002. Integrated nematode management on banana in India. In *Global Conference on Banana and Plantain*, Bangalore, pp. 73-76.

Sundararaju, P., Cannayane, I. and Sathiamoorthy, S. 2004. *Nematode Management in Banana and Plantains*. Tech. Bull. No. 10, National Res. Centre for Banana, Thiruchirapalli, Kerala, India, 14 pp.

Sundararaju, P., Cannayane, I. and Thangavelu, R. 2002. Biomanagement of *Pratylenchus coffeae* and *Meloidogyne incognita* on banana by using endophytic fungi. In *Global Conference on Banana and Plantain*, Bangalore, p. 191

Sundararaju, P. and Koshy, P.K. 1982. *Indian Journal of Nematology* **12**: 404-405.

Sundararaju, P., Mustaffa, M.M., Kumar, V., Cannayane, I. and Tanuja Priya, B. 2002. Effect of organic farming on plant parasitic nematodes infesting banana cv. Karpuravalli. *Current Nematology* **13**: 39-43.

Sundararaju, P., Sosamma, V.K. and Koshy, P.K. 1979. Pathogenicity of *Radopholus similis* on ginger. *Indian Journal of Nematology* **9**: 91-94.

Suresh, C.K. and Bagyaraj, D.J. 1984. Interaction between a vesicular arbuscular mycorrhiza and a Root-knot Nematode and its effect on growth and chemical composition of tomato. *Nematologia Mediterranea* **12**: 31-39.

Susan, M.L., Daniel, R.P., David, C.J., Lyn, C.K., Robert, L.D. and Weili, M. 2002a. *Application of* Burkholderia cepacea *and* Trichoderma virens *Alone and in Combination Against* Meloidogyne incognita. USDA, ARS Publication.

Susan, M.L., Samie, M.I., David, C.J. and Daniel, R.P. 2002b. *Evaluation of* Trichoderma virens *and* Burkholderia cepacea *for Antagonistic Activity Against Root-knot Nematode,* Meloidogyne incognita. USDA, ARS Publication.

Swarup, G., Dasgupta, D.R. and Koshy, P.K. 1989. *Plant Diseases.* Anmol Publications, New Delhi, 395 pp.

Tanda, A.S. and Atwal, A.S. 1988. *Nematologica* **34**: 484-492.

Taylor, A.L. and Sasser, J.N. 1978. *Biology, Identification and Control of Root-knot Nematodes (Meloidogyne* spp.). North Carolina State University Graphics, Raleigh, 111 pp.

Taylor, C.E. and Gordon, S.C. 1970. A comparison of four nematicides for the control of *Longidorus elongatus* and *Xiphinema diversicaudatum* and the virus they transmit. *Horticultural Research* **10**: 133-134.

Tewari, S.P. and Dave, G.S. 1985. Burrowing nematode, *Radopholus similis* (Cobb) Thorne, 1949 on banana in Madhya Pradesh. *Current Science* **54**: 1286.

Thakar, N.A. and Patel, C.C. 1984. *Indian Journal of Nematology* **14**: 204.

Thakar, N.A. and Patel, C.C. 1985. Control of root-knot nematode, *Meloidogyne javanica* on tomato by chemical bare-root dips. *Indian Journal of Nematology* **15**: 110-111.

Thapa, C.D., Munjal, R.L. and Seth, P.K. 1987. The mushroom nematodes and their control. *Indian Journal of Mushrooms* **3**: 1-11.

Thangaraju, D. 1983. Distribution of potato cyst nematodes in Kodaikanal Hills, Madurai district, Tamil Nadu. *Indian Journal of Nematology* **13**: 222-223.

Thirumala Rao, V. 1956. Stray notes on some crop outbreaks of South India. *Indian Journal of Entomology* **18**: 123-126.

Thomason, I.J., Erwin, D.C. and Garber, M.J. 1959. The relationship of the root-knot nematode, *Meloidogyne javanica,* to *Fusarium* wilt of cowpea. *Phytopathology* **49**: 602-606.

Tiyagi, S.A., Anver, S., Bano, M. and Siddiqi, M.A. 1986. Feasibility of growing zinnia as a 'mix-crop' along with tomato for control of root-knot and reniform nematodes. *International Nematology Network Newsletter* **3**(3): 6-7.

Trifitt, M.J. 1930. On the bionomics of *Heterodera schachtii* on potatoes with special reference to the influence of mustard on the escape of the larvae from the cysts. *Journal of Helminthology* **8**: 19-48.

Trivedi, P.C. 1992. Evaluation of a fungus, *Paecilomyces lilacinus* for the biological control of root-knot nematode, *Meloidogyne incognita* on *Solanum melongena*. In *Proc. of the 3rd Intnl. Conf. on Plant Prot. in the Tropics* (F.A.C. Ooi and G.S. Lim, eds.), Vol. 6, pp. 29-33. Genting Highlands, Malaysia.

Trivedi, P.C. and Tiagi, A. 1984. Effect of some plants on *Meloidogyne incognita* population in chilli soil. *Indian Journal of Botanical Science* **63**: 416-419.

Tsay, T.T., Wu, S.T. and Lin, Y.Y. 2004. Evaluation of Asteraceae plants for control of *Meloidogyne incognita*. *Journal of Nematology* **36**: 36-41.

Umesh, K.C., Krishnappa, K. and Bagyaraj, D.J. 1988. Interaction of burrowing nematode, *Radopholus similis* and VA mycorrhiza, *Glomus fasciculatum* in banana. *Indian Journal of Nematology* **18**: 6-11.

Upadhyay, K.D. and Dwivedi, K. 1987. Analysis of crop losses in pea and gram due to *Meloidogyne incognita*. *International Nematology Network Newsletter* **4**(4): 6-7.

Vadivelu, R.S. and Muthukrishnan, T.S. 1979. Control of root lesion nematode, *Pratylenchus delattrei* on crossandra. *Indian Journal of Nematology* **9**: 86.

Vadivelu, R.S. and Rajendran, G. 1986. Damage potential of the root-knot nematode, *Meloidogyne arenaria* to onion. *International Nematology Network Newsletter* **4**(2): 5-8.

Vadivelu, R.S., Rajendran, G., Naganathan, T.G. and Jayaraj, S. 1987. Studies on the nematode pests of banana and their management - A review. *Third Group Discussion on the Nematological Problems of Plantation Crops*, Sugarcane Breeding Institute, Coimbatore, pp.32-33.

Varma, M.K., Sharma, H.C. and Pathak, V.N. 1978. Evaluation of systemic granular nematicides against *Tylenchorhynchus* spp. parasitic on cabbage. *Indian Journal of Nematology* **8**: 59-60.

Venkata Ram, C.S. 1963. The Root-knot Eelworm of Mature Tea Meloidogyne brevicauda, its Importance in South India and an Evaluation of Certain Chemical Treatments for its Control. Tea Scientific Report for 1962-63, United Planters Association of South India, Appendix pp. 31-36.

Venkata Rao, A., Narasimhan, V., Rajagopalan, P., Vidyasekaran, P. and Muthukrishnan, T.S. 1973. Efficacy of nematicides for the control of betel vine wilt. *Indian Journal of Agricultural Sciences* **43**: 1072-1075.

Venkitesan, T.S. 1976. Studies on the Burrowing Nematode, *Radopholus similis* (Cobb, 1893) Thorne, 1949 on Pepper *(Piper nigrum* L.) and its Role in Slow Wilt Disease. Ph. D. thesis, Univ. of Agri. Sci., Bangalore.

Venkitesan, T.S. and Charles, J.S.K. 1983. Sucker dip treatment for banana nematodes control. *Third Nematology Symposium,* Himachal Pradesh Agricultural University, Solan, p. 45.

Verma, A.C. 2001. N.D. University of Agriculture and Technology, Kumarganj, Faizabad, Uttar Pradesh. In *Indian Nematology-Progress and Perspectives* (S.C. Dhawan *et al*., eds.), pp. 121-125. Division of Nematology, Indian Agri. Res. Inst., New Delhi.

Verma, A.C., Singh, H.K. and Khan, N. 2005. Management of root-knot nematode, *Meloidogyne incognita* through antagonistic approaches in pointed gourd. *Indian Journal of Nematology* **35**: 78-79.

Verma, K.K. 2001. Integrated approach for controlling root-knot nematode, *Meloidogyne incognita* on okra. *Natl. Cong. on Centenary of Nematol. in India – Appraisal & Future Plans*, Indian Agri. Res. Inst., New Delhi, p.167.

Verma, K.K., Gupta, D.C. and Paruthi, I.J. 1999. Preliminary trial on the efficacy of Pseudomonas fluorescens as seed treatment against Meloidogyne incognita in tomato. In Proc. of the Natl. Symp. on Rational Approaches in Nema Mangmt. for Sustainable Agri., Indian Agri. Res. Inst., New Delhi, pp. 79-81.

Verma, K.K., Verma, B.S., Sangwan, N.K. and Dhindsa, K.S. 1989. Toxicity of some indigenous plant extracts to root-knot, seed gall and citrus nematode. *Pesticides* **23**(11): 25-27.

Verma, R.R. 1986. Efficacy of organic amendments against *Meloidogyne incognita* infesting tomato. *Indian Journal of Nematology* **16**: 105-106.

Verma, R.R. 1987. *Indian Journal of Nematology* **17**: 123-124.

Vetrivelkalai, P. and Subramanian, S. 2006. Plant parasitic nematode populations as affected by onion based coping sequences. *Indian Journal of Nematology* **36**: 10-14.

Victoria, K.J., Sanchez, P.A. and Barriga, O.R. 1970. Eradication de palms de cocotero afectadas por el anillo rojo (*Rhadinaphelenchus cocophilus* (Cobb, 1919) Goodey, 1960, Nematoda: Aphelenchoididae)mediante la utilization de sustancias quimicas. *Revista Inst. Colomb. Agropec.* **5**: 185-197.

Vidya, K. and Reddy, B.M.R. 1998. Integrated management of *Radopholus similis* infecting banana using plant product, nematicide and biocontrol agents. *Mysore Journal of Agricultural Sciences* **32**: 186-190.

Vijayalakshmi, K. and Goswami, B.K. 1987. Effect of root-dip treatment of tomato seedlings in aqueous extracts of some oilseed cakes on root-knot nematode infestation. *Annals of Agricultural Research* **8**: 168-171.

Walia, R.K., Bansal, R.K. and Bhatti, D.S. 1991. Effect of *Paecilomyces lilacinus* application time and method in controlling *Meloidogyne javanica* on okra. *Nematologia Mediterranea* **19**: 247-149.

Walia, R.K., Bansal, R.K. and Bhatti, D.S. 1992. Efficacy of bacterium (*Pasteuria penetrans*) and fungus (*Paecilomyces lilacinus*) alone and in combination against root-knot nematode (*Meloidogyne javanica*) infecting brinjal. *Proceedings of the First Afro-Asian Nematology Symposium*, Aligarh Muslim Univ., Aligarh, p. 38.

Walia, R.K. and Dalal, M.R. 1994. Efficacy of bacterial parasite *Pasteuria penetrans* application as nursery soil treatment in controlling root-knot nematode *Meloidogyne javanica* on tomato. *Pest Management and Economic Zoology* **2**: 19-21.

Weischer, B. 1967. *Report to the Government of India on Plant Parasitic Nematodes.* PL 24/52 Report No. TA 2332, FAO, Rome, 15 pp.

Winslow, R.D. and Willis, R.J. 1968. Nematode diseases of potatoes. In *Economic Nematology* (J.M. Webster, ed.), pp. 17-48. Academic Press, New York.

Wood, F.C. and Goodey, J.B. 1957. Effects of gamma-ray irradiation on nematodes infesting cultivated mushroom beds. *Nature* **180**: 760-761.

Xiuuan, Y., Yuxian, H., Furn, C. and Zhengliang. 2000. Isolation and selection of egg-mass of fungi *Meloidogyne* species in Fujian Province. *Fujian Journal of Agriculture Science* **15**: 12-15.

Yamashita, T. and Akiya, T. 1968. Studies on amino acids generated from rape seed cake in the soil. *Soil Science and Plant Nutrition* **14**: 225-230.

Zaki, F.A. 1998. Biological control of *Meloidogyne javanica* in tomato by *Paecilomyces lilacinus* and castor leaves. *Indian Journal of Nematology* **28**: 132-139.

Zaki, F.A. and Bhatti, D.S. 1989. Effect of castor (*Ricinus communis*) leaves in combination with different fertilizer doses on *Meloidogyne javanica* infecting tomato. *Indian Journal of Nematology* **19**: 171-176.

Zaki, F.A. and Bhatti, D.S. 1991a. Effect of castor *Ricinus communis* and the biocontrol fungus, *Paecilomyces lilacinus* on *Meloidogyne javanica. Nematologica* **36**: 114-122.

Zaki, M.J. and Maqbool, M.A. 1992. *Pakistan Journal of Nematology* **10**: 75-79.

Zuckerman, B.M., Dicklow, M.B. and Acosta, N. 1993. A strain of *Bacillus thuringiensis* for the control of plant parasitic nematodes. *Biocontrol Science and Technology* **3**: 41-46.

Zutra, D. and Orion, D. 1982. Crown gall bacteria (*Agrobacterium radiobacter* var. *tumefaciens*) on cotton in Israel. *Plant Disease* **66**: 1200-1201.

SUBJECT INDEX